# Agricultural Development: A Comprehensive Approach

# Agricultural Development: A Comprehensive Approach

Editor: Dexter Watson

STATES
ACADEMIC PRESS
www.statesacademicpress.com

**States Academic Press,**
109 South 5th Street,
Brooklyn, NY 11249, USA

Visit us on the World Wide Web at:
www.statesacademicpress.com



ISBN: 978-1-63989-031-6 (Hardback)



**Cataloging-in-Publication Data**

Agricultural development : a comprehensive approach / edited by Dexter Watson.
p. cm.
Includes bibliographical references and index.
ISBN 978-1-63989-031-6
1. Agriculture. 2. Agricultural development projects. 3. Agricultural innovations. I. Watson, Dexter.
S439 .A37 2022
630--dc23

# Table of Contents

# Preface

This book aims to highlight the current researches and provides a platform to further the scope of innovations in this area. This book is a product of the combined efforts of many researchers and scientists, after going through thorough studies and analysis from different parts of the world. The objective of this book is to provide the readers with the latest information of the field.

It is estimated that by the end of the 21st century human population would reach an estimated 10 to 11 billion. To compensate for the ensuing food and energy shortage, it is imperative that measures should be taken to initiate a global phenomenon for agricultural development. Modern agriculture is heavily dependent on intensive agricultural practices. Relying exclusively on a small number of productive crops can potentially lead to significant loss of biodiversity. Therefore, measures such as efficient pest control, research initiatives, support to crop producers, engagement with the latest production techniques and preservation of diversity need to be adopted. Developing crop varieties that are more resilient to abiotic stresses, enhancing livestock health and productivity by improving animal genetics and veterinary care, and improved crop and natural resource management are strategies that can provide global food and energy security. The book studies, analyzes and upholds the pillars of agricultural development and its utmost significance in modern times. The various studies that are constantly contributing towards advancing technologies and evolution of agriculture are examined in detail. This book will prove to be immensely beneficial to agrologists, agroeconomists, horticulturists, agriculturists, conservationists, naturalists and students in this field.

I would like to express my sincere thanks to the authors for their dedicated efforts in the completion of this book. I acknowledge the efforts of the publisher for providing constant support. Lastly, I would like to thank my family for their support in all academic endeavors.

**Editor**

# 1

# The role of institutions as actors influencing Uganda's cassava sector

Cedric Mutyaba[a,*], Moses H. Lubinga[b], Richard O. Ogwal[a], Steven Tumwesigye[c]

[a]*National Agricultural Research Laboratories (NARL), Post-Harvest Research Programme, Kampala, Uganda*
[b]*Agricultural Research Council, Agricultural Economics and Capacity Development Division, Pretoria 0001, South Africa*
[c]*National Agricultural Research Laboratories (NARL), Food Bioscience Research Center, Kampala, Uganda*

**Abstract**

We aim at mapping out a detailed framework that reveals the proportionate flow of cassava and its products along the value chain (VC). Furthermore, we aim at establishing the role of institutions and the linkages between institutions and other VC actors that influence the cassava VC in Uganda. We use both primary and secondary data obtained from four regions in Uganda. Results show that farmers, processors, transporters, traders, consumers and institutions are the major actors. There are four categories of institutions, *viz*, government, non-government, community based organisations and international agencies. Roles performed by institutions include: development and enforcement of policies, Research and Development (R&D), capacity building, and creation of market access linkages for cassava and its products. Findings reveal that there is no clear nexus and no coordination among farmers/producers, processors, traders, transporters and consumers. However, institutions are well coordinated and play various roles along the VC to influence the dynamics of actors. Policy-wise it is important to establish strong private-public partnerships to bridge the impaired linkages between the actors (farmers/producers, processors, traders, transporters, and consumers) and institutions. Strong partnerships are envisaged to reduce the associated transaction costs amongst the actors.

*Keywords:* farmers, processors, traders, linkages, value chain, institutions, Uganda

## 1 Introduction

In Africa cassava (*Manihot esculenta* Crantz) is an important source of food and income. According to Adebayo *et al.* (2010), cassava's importance is attributable to its high resilience and adaptability under the many different ecological conditions exhibited in many countries on the continent. Gaffney *et al.* (2012); Kehinde (2006), and Sayre *et al.* (2011) argue further that its ability to be kept for a long period (approximately 2–3 years) in the ground prior to harvesting makes it a reliable food security crop. Besides cassava being consumed as food, it is used as raw material in the production of an assortment of processed products. Balagopalan (2002), Nzigamasabo & Zhou (2006), and Ogwal *et al.* (2012) mention products such as confectionery, animal feeds, and industrial products like paper and starch.

Uganda is Africa's sixth largest producer of cassava (Kilimo Trust, 2012). Farmers have produced between 4.7 and 5.2 million tons of cassava annually over the past five years, and have harvested around 400,000 hectares (FAO, 2014). In Uganda, cassava output is ranked second after bananas and is known to be a crop that

* Corresponding author
National Agricultural Research Laboratories (NARL),
Post-Harvest Research Programme, P.O. Box 7065 Kampala, Uganda
Email: cjmutyaba@gmail.com

generates better income and profit than maize (USAID, 2010). Gaffney *et al.* (2012) noted that income from cassava accounted for 6 % of the total income generated among cassava-producing homesteads in Uganda in the early 1990s while New Vision (2013) postulated that cassava has the potential of saving the country an estimated $300m if adopted as a commercial crop.

Given cassava's importance and its potential in alleviating poverty in many African countries, a number of studies assessing various aspects of the commodity have been done. For instance, Adebayo & Sangosina (2005) and Nzigamasabo & Zhou (2006) evaluated cassava-processing innovations and the utilisation of cassava. Other scholars, such as Lemchi *et al.* (2011), assessed the determinants of marketability of cassava products while Adebayo *et al.* (2010) examined how Nigerian smallholder farmers could be sustainably included in high quality cassava flour (HQCF) VCs. Karugia *et al.* (2009) evaluated the VC of cassava flour and other related products.

In Uganda, the existing literature by Collinson *et al.* (2000, 2002, 2003), Kimathi *et al.* (no date), Kilimo Trust (2012), and Kleih *et al.* (2012) indicates that some analysis of the cassava VC has been carried out. These studies reveal that commonly identified challenges encountered along the VC include pests and diseases during the production phase, poor post-harvest handling technologies as well as poor physical infrastructure. Literature also reveals that farmers operate at small scale such that cassava production is below the desired output for industrial purposes and farmers are the most marginalised actors along the chain. Other actors identified include traders (organised at various levels), processors (millers and manufacturers), transporters and marketing agents. Furthermore, it was also noted that linkages between the private sector players and other institutions were very limited. Despite the existence of a number of institutions which influence the cassava VC's dynamics and actors' behaviour, none of these studies examine their role and the nexus between these institutions, and the various actors along the chain. In this paper, we refer to an institution as an organisation, establishment, foundation or society, dedicated to the promotion of a particular cause or program, especially one of a public, educational or charitable character. Furthermore, we have not found in literature any cassava VC study that disentangles the proportionate flow of products in Uganda.

Thus, there exists a knowledge gap regarding the role played by the various institutions and the linkages between institutions and the various VC actors, as well as a detailed understanding of the proportionate flow of cassava products at the different stages along the chain. It is envisaged that understanding this framework will be crucial in ascertaining how smooth or otherwise the VC operates. This knowledge gap is bridged by mapping a framework that reveals the detailed proportionate flow of cassava products from one actor to another. In addition, this paper aims at presenting the roles of institutions and their linkage with the other VC actors. Attainment of this objective will foster better informed policies, with the aim of ensuring full exploitation of the cassava sector so as to enhance economic development.

## 2 Methodology

### *2.1 Sampling and data collection*

Based on the country's geographical mapping used by UBOS & MAAIF (2010), this study covered all four regions of Uganda (Central, East, North and West). Primary and secondary data were used for this study. The main sources of secondary data included Civil Society Organisations (CSOs). According to the Organisation for Economic Cooperation and Development (OECD) (2006), CSOs refer to a number of associations that influence how society voluntarily organises itself and which represent a wide range of society's interests and ties. Information regarding institutions (public and private) was gathered from secondary data sources such as the internet and government documents. Government documents used include the National Agricultural Advisory Services (NAADS) implementation guidelines by MAAIF (2010). Primary data were collected from farmers, traders and processors using a semi-structured questionnaire and checklists. In order to validate information provided by the individual respondents, more information was sought from Key Informants (KI), that is, representatives from Non-Government Organisations (NGOs), local council leaders, and community councillors. Furthermore, Focus Group Discussions (FGDs) were conducted within each chosen community.

Five hundred and forty eight respondents (Table 1) under the farmers' category were randomly selected from any two purposefully chosen sub-counties from each district. Comprehensive lists of all households within each of the selected sub-county were accessed through the electoral commission registers. The choice of the sub-counties from each district was guided by a

**Table 1:** *Districts surveyed and the number of respondents interviewed by category*

| *Region* | *District* | *Sub-county* | *Sample category* | | | |
|---|---|---|---|---|---|---|
| | | | *Farmers* | *Traders* | *Processors* | *KIs* |
| Central | Luwero | Kikyusa<br>Kamira | 58 | 6 | 6 | 4 |
| | Rakai | Kifamba<br>Kagamba | 61 | 5 | 6 | 4 |
| North | Apac | Aduku<br>Apac | 61 | 10 | 6 | 5 |
| | Nebbi/ Zombo | Panyango<br>Paidha | 51 | 16 | 8 | 4 |
| West | Kasese | Kisinga<br>Nyakiyumbu | 41 | 10 | 10 | 2 |
| | Rubirizi | Kichwamba<br>Kyabakara | 59 | 12 | 8 | 2 |
| | Masindi | Pakanyi<br>Kigumba | 49 | 25 | 7 | 3 |
| East | Serere | Olio<br>Kyere | 48 | 4 | 3 | 4 |
| | Budaka | Naboa<br>Lyama | 60 | 10 | 3 | 2 |
| | Busia | Busiime<br>Masafu | 60 | 10 | 6 | 4 |
| Total | | | 548 | 108 | 63 | 34 |

KIs = Key Informants

simple criterion based on whether the production of cassava is significant by volume in the sub-county. The main cassava producing sub-counties from each district were selected in collaboration with the district production office. Except for the Central and Northern regions, three districts were chosen from each region, with key emphasis on the major cassava producing districts.

From Central Uganda, only two districts were selected given that during the second season of 2008 and the first season of 2009, the region exhibited less than 50 % distribution of pure stands of cassava plots as compared to the Western, Northern and Eastern regions which registered 57.2 %, 59.7 % and 70.2 % of pure cassava plot stands, respectively (UBOS & MAAIF, 2010). Despite the fact that the Northern region registered a high proportion of pure cassava stands (59.7 %), two districts were chosen because it exhibited the least percentage distribution of households that produced cassava in both the second season and first season of 2008 and 2009, respectively. According to UBOS & MAAIF (2010), 18 % (2nd season of 2008) and 14.3 % (1st season of 2009) of the households in the Northern region engaged in cassava production relative to an average of 25.4 %, 33.1 % and 25.4 % for the Central, Eastern and Western region respectively for the two seasons.

Traders and processors were also randomly selected from the purposefully chosen sub-counties and municipalities/town councils using comprehensive lists of all registered traders and processors obtained from the head offices of the traders' associations. By use of a checklist, two FGDs per district were held with members from the various categories of VC actors, leading to a total of 20 FGDs. Key informants were selected from either the respective district production offices, the NGOs/CBOs or the government institutions that were influential in the cassava VC. Table 1 shows that the study was based on 773 respondents (n=773), with farmers having the largest proportion (70.9 %).

### 2.2 *Data analysis*

Spreadsheets were used to clean and edit the data from any outliers and entry errors. STATA version 9 was then used for the detailed data analysis. Following Kilimo Trust (2012) and FAO (2005), this study used a VC approach based on the functional analysis to identify all actors that influence the VC, as well as the nexus between institutions and other VC actors. Within the functional analysis domain, the different actors and their roles in the VC were identified. That is, *i)* physical flows of cassava and its related products were identified, *ii)* the various actors as well as institutions and their roles at each stage were also identified, and *iii)* the linkages between the producers/farmers, processors, traders and institutions were mapped.

Notably, VC mapping entails generating visual illustration between actors in the VC as well as other market players. It enables to illustrate and understand the process(es) that a commodity goes through from the farm gate until it reaches the final consumer. Furthermore, a VC map is also useful for recognizing and categorizing key market actors and support organisations as well as revealing the various market channels a product takes before reaching the final consumer. Information generated from focus group discussions and interviews with key informants was analysed using descriptive statistics and used to validate the information about institutions obtained from the individual respondents. Secondary data especially in form of existing literature was reviewed to better understand the initiatives carried out in the cassava sector in Uganda.

## 3 Results

### 3.1 *Proportionate use of cassava and its products*

About 95 % of farmers' cassava output is made up of fresh cassava roots and the remaining 5 % by fresh leaves (Figure 1). Fresh cassava leaves are rarely commercialised, given that almost 100 % of the leaves are consumed at household level. About 70 % of cassava leaves are used for human consumption as vegetables by households while 30 % are used as animal feed supplement usually fed to pigs without any form of value addition. Out of the total output, a large proportion (40 %) of cassava roots is consumed in fresh form. The roots may be roasted, steamed or mashed. Of fresh cassava roots consumed, the proportion that is directly sold by farmers greatly differs between small scale, medium and large scale farmer categories. Small scale farmers directly commercialise less than 10 %, while medium and large scale farmers directly sell 20–40 % and more than 60 %, respectively. About 9 % of the fresh cassava is exported to regional markets, including southern Sudan, Tanzania, Kenya, D.R. Congo, Rwanda and Burundi. The remaining proportion (51 %) of fresh roots is processed into an assortment of products with cassava chips being the most important (31 %), followed by local gin (11 %), flour (5 %), starch (2 %) and gari (2 %).

It was noted that much of the cassava chips (46 %) are processed into flour. Figure 1 shows that cassava flour and chips are sent to the largest number of destinations – six in the case of flour and five in the case of chips. About 63 % of the flour is used for household human consumption, a small proportion of which is sold through supermarkets. Furthermore, 34 % of the flour is exported to neighbouring countries while the lowest proportion (0.1 %) is used in the paper manufacturing industry. Cassava flour is also used for brewing, baking and confectionary. With regard to chips, the largest proportion (51 %) is exported to the regional markets. More than 90 % of the local gin is for human consumption and the remaining ends up in schools, for laboratory use. All gari is consumed at household level, while starch is only used in schools for laboratory experiments and by the food industry.

### 3.2 *Key actors*

Analytical results (Figure 2) show that the cassava VC consists of five key actors: farmers/producers, traders, processors, consumers and institutions. Farmers were mainly responsible for cassava production, with 85 % being small scale farmers while the others were either medium (5 %) or large (10 %) scale farmers. Small scale farmers produce cassava on less than one hectare of cultivated area, medium scale farmers use between 2–10 hectares while large scale farmers grow cassava on more than 10 hectares. Processors comprise of individuals and business entities that transform fresh cassava into other products, mainly chips, chops, slices, grits, gari, flour, starch or snacks. Processors are basically categorised as primary, intermediary or tertiary. Primary processors, mainly women, use basic tools like pangas, chippers, and graters to convert small quantities of fresh cassava into chips, chops, slices and grits.

Intermediary processors, mainly men, are responsible for bulking intermediate cassava products (granules, chips, chops and slices), milling them into flour and then selling it to wholesale traders. Such processors often have motorised hammer mills and are organised in well-coordinated groups comprising a manageable number of people. Tertiary processors are characterised by turning intermediary products into products such as

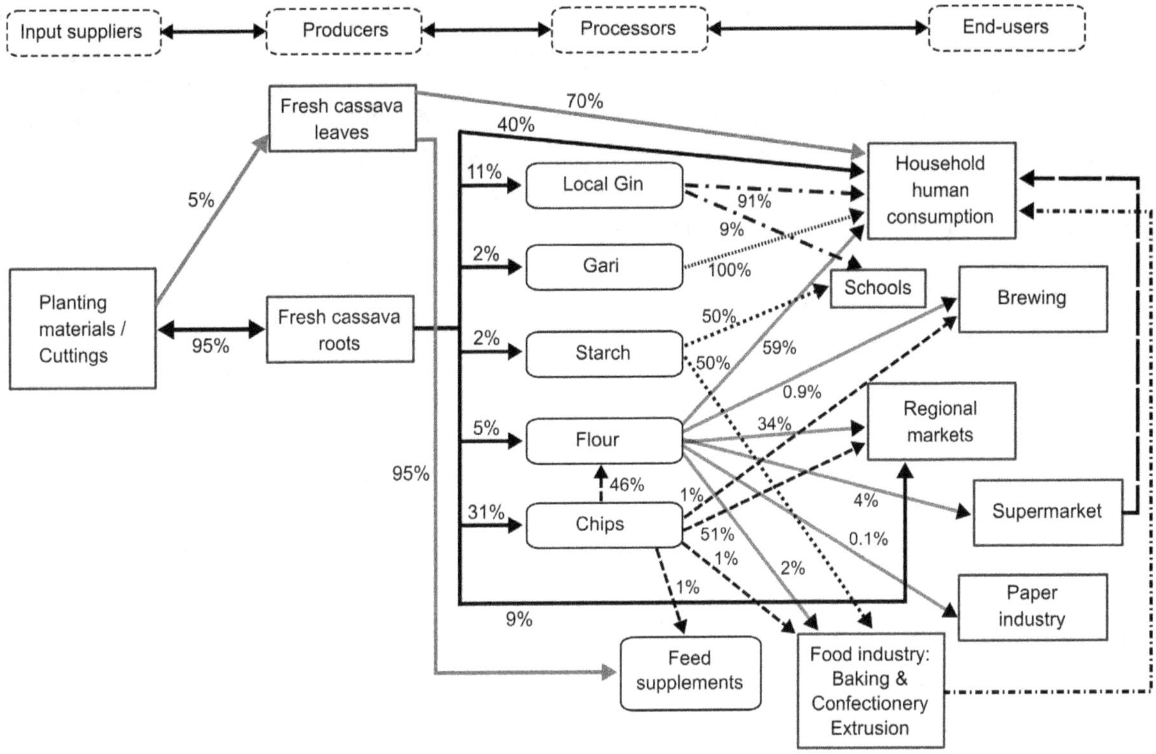


**Fig. 1:** *The use of cassava and its products in Uganda*

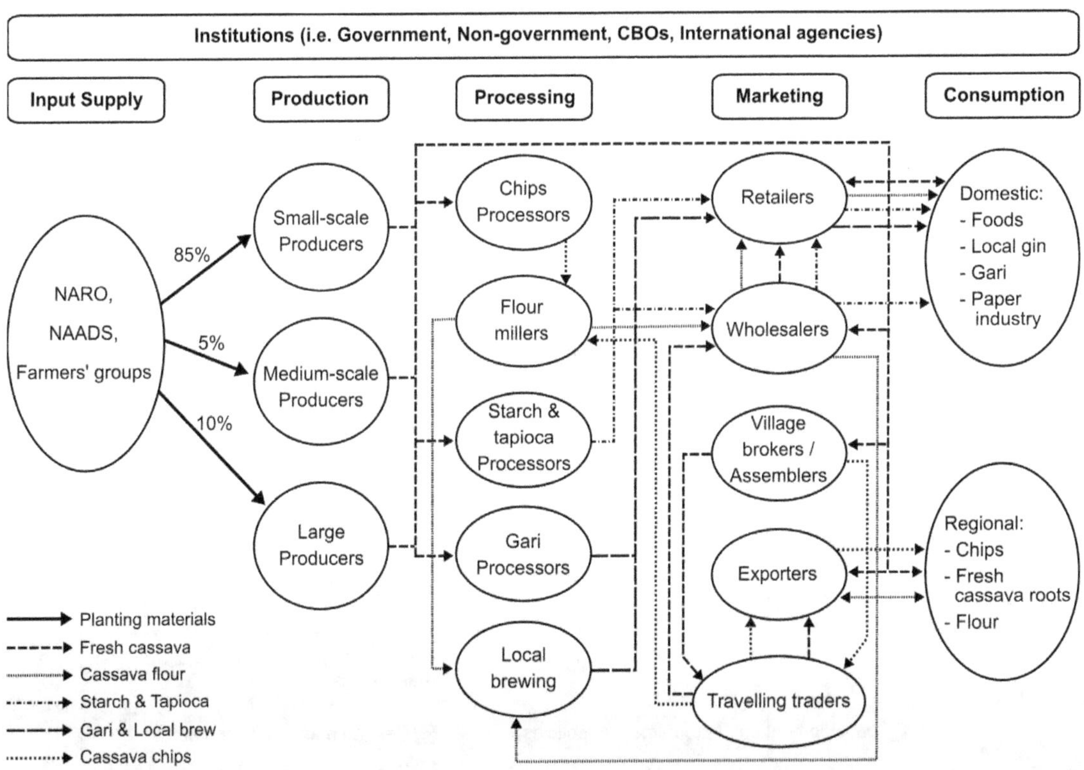


CBOs: Community Based Organisations; NAADS: National Agricultural Advisory Services;
NARO: National Agricultural Research Organisation

**Fig. 2:** *Interaction of actors along the cassava value chain in Uganda*

local gin, as well as other flour based products (like porridge, bread and pancakes).

Traders consist of retailers, village assemblers/brokers, transporters/travelling traders, wholesalers, and exporters (Figure 2). Retailers operate in local open markets and small shops. Village assemblers/brokers in collaboration with transporters/travelling traders obtain cassava products from villages and deliver it to markets. Travelling traders move from place to place with lorries or pick-up trucks procuring fresh cassava and cassava slices/grits/ chops/chips from producers. The traders then transport the fresh and the related intermediate products to markets where they either sell directly to millers (in the case of dry slices/grits/chops/chips) or to wholesalers in the case of fresh cassava roots. In some instances the travelling traders use agents to negotiate with producers, collect the produce as well as dry cassava products on their behalf.

Often wholesale traders also assume the role of processors given that they can afford to build and run stores/warehouses to act as bulking points of sale. They deal with large volumes of cassava chips and flour. Brokers are responsible for bulking cassava chips at designated stores/warehouses at the discretion of appointing wholesalers or travelling traders in the various parts of the country. They work as commission agents and are paid based on the amounts of chips collected. There are no contractual agreements between traders and these agents which sometimes lead to loss of capital by traders, especially when traders pay brokers upfront. Cassava end-users can be grouped into two categories, namely intermediate and final consumers. Intermediate consumers are those who purchase cassava products and then turn them into products of higher value, such as starch and beer, while final consumers purchase cassava products and prepare them in various forms for household consumption.

A number of institutions are also involved in the cassava VC. However, unlike the other VC actors that are directly involved in the cassava VC, institutions play a supporting/ facilitating role at various stages of the VC. Some institutions perform roles that influence more than one actor along the VC. Four institutions can be categorised: government bodies, non-government agencies, Community Based Organisations (CBOs) and international agencies. With respect to the roles played by institutions, the results (Figure 3) show that government institutions are mainly responsible for the development and enforcement of policies, while non-government agencies and CBOs mainly focus on capacity building. In this context, capacity building refers to a process of developing and strengthening the skills, instincts, abilities, processes and resources of the VC actors so as to enhance the development of the cassava sector.

International agencies are more concerned with Research and Development (R&D), followed by capacity building. A large contrast however exists between government institutions and the other categories of institutions (Figure 3) with respect to the development and enforcement of policies. Results also show that some respondents were not sure of the roles played by institutions across all the four categories. Furthermore, results reveal that the various categories of institutions are equally responsible for providing technical support to the value chain actors. With the exception of govern-

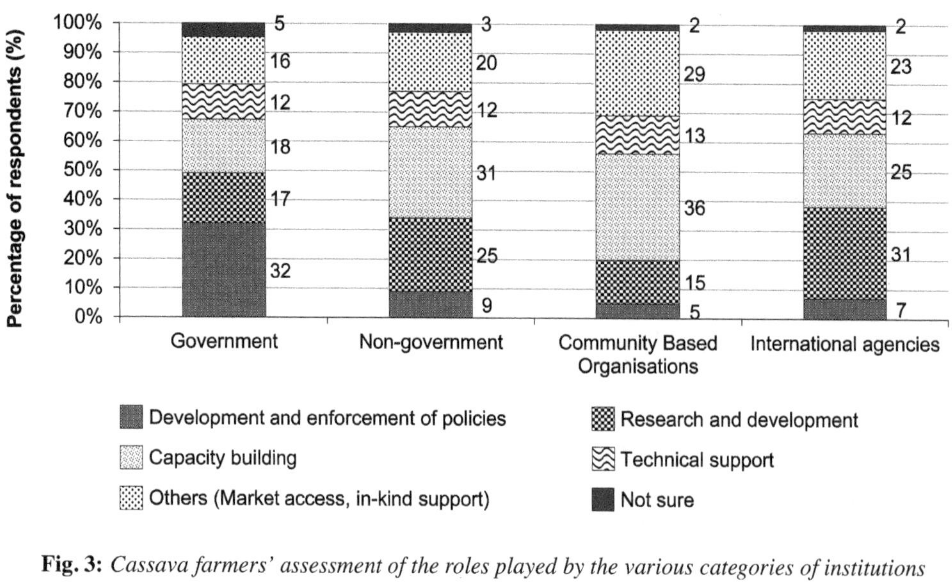


**Fig. 3:** *Cassava farmers' assessment of the roles played by the various categories of institutions*

ment institutions, other categories of institutions seem to equally play other roles like enhancing market access and providing in-kind support.

As a means of validating information availed by the individual farmers, data sought from Focus Group Discussions (FGDs) and Key Informants (KIs) was also analysed. Results of that analysis concur with the individual farmers' findings, except for the non-government agencies for which 9 % of the key informants acknowledged that they take part in the development and enforcement of policies yet findings based on FGDs suggest otherwise.

### 3.3 *Linkages amongst key actors in the cassava value chain*

Markets are a result of many different actors interacting to enable the exchange of goods and services. Assessment of the vertical and horizontal linkages amongst the actors reveals that formal vertical linkages based on contracts are almost non-existent yet they enhance the movement of products to end markets. Although such linkages come with a transaction cost, they enhance trust among the various actors along the VC and enhance effective planning as well as efficiency along the chain. Furthermore, formal horizontal linkages amongst these actors are either minimal or non-existent, thus leading to high transaction costs and low economies of scale. This finding relates to results by Kilimo Trust (2012).

Detailed analysis shows that over 95 % of the actors (producers/farmers, traders and processors) along the VC have no formal agreements with their customers, and about 99 % of the traders have no specific markets for their cassava products. Only about 5 % of the actors secure formal contracts with customers and these are mainly institutions. The middlemen dominate the category of traders as actors and are indispensable in bridging the gap between the farmers, processors as well the consumers. They buy cassava and its related products and resale it or make it available to consumers. However, they are noted to deter the development of strong horizontal and vertical linkages among the various actors due to their shrewd behaviour.

In contemporary market chains, institutions create policies which influence and shape other actors in taking particular actions which in turn influence or change VC. Institutions are linked to the other actors along the VC through the provision of a conducive environment. In this context, a conducive environment refers to the services, that is, the policies or standards, in-kind support, R&D and technical support which institutions render to the other actors. Unlike the linkages between other actors that directly influence the activities along the value chain, the linkages between the various institutions and the VC actors are formal, based on contractual agreements. This formal relationship may be associated with the fact that institutions tend to operate in an organised manner and have the capacity to mobilise farmers into clusters that can reliably supply good quality cassava and its related products in large quantities. For instance, Nakasongola- and Masindi- District Farmers' Associations (DFAs) established formal contractual agreements with firms like Triad links/Tullow Oils to supply cassava products. Formal linkages tend to lower transaction costs and increase trust amongst the other actors. Conversely, non-functional or poor institutions increase transaction costs amongst the VC actors.

## 4 Discussion

Based on the results, more than 50 % of fresh cassava roots are processed into other products given that it is highly perishable after being harvested (Naziri *et al.*, 2014). Thus, in order to minimise post-harvest losses, cassava has to either be consumed or processed within a few days. The high importance of cassava chips may be attributable to its potential to substitute maize bran in animal feed rations while if processed into HQCF, it can replace wheat flour in the confectionery industry. This concurs with earlier work by Graffham *et al.* (2000). The low proportion of fresh cassava devoted to starch and gari production may be associated with the high costs incurred in starch extraction and the relatively small niche market which for gari in Uganda, respectively. Graffham *et al.* (2000) urged that Uganda's market size for starch is too small for the investors to be able to recover investment costs. With reference to VC actors, the high proportion of small scale farmers relative to the other categories may be attributable to the high level of land fragmentation in Uganda.

In terms of roles played by institutions, the high proportion of respondents (32 %) arguing that government institutions are responsible for policies may be attributable to the fact that government institutions, particularly MAAIF and UNBS are at the forefront of developing and enforcing agriculture related policies in the country. This argument concurs with views of USAID & EAT (2013). The relatively low proportion of government institutions' participation in providing other services (16 %) may be due to the small proportion of funds allocated to the agriculture sector in the national budget. Rwakakamba (2013) argues that the agriculture sector has never received more than 5 % share of the national budget since the 2009/10 financial year. Agriculture is the lowest ranked sector in the economy and the allocation is below the Maputo Comprehensive Africa Agriculture Development Program (CAADP) target of at least 10 % (Rwakakamba, 2013).

The importance accorded to capacity building by both NGOs and CBOs may be attributed to the fact that donor agencies such as USAID, and International Development Research Centre (IDRC) tend to contract CBOs and non-government agencies that focus on helping smallholder farmers to overcome bottlenecks encountered along specific VCs. Arguably, CBOs and NGOs understand community dynamics at grass root level. Ulleberg (2010) opines that such institutions are innovative and adapt more quickly than government institutions.

The relatively high proportion of respondents acknowledging the participation of international agencies in R&D may be associated with the eminent work done by bodies like the International Institute of Tropical Agriculture (IITA) and Natural Research Institute (NRI). The high focus of international agencies on capacity building as well as R&D may be attributable to Uganda's need to establish sustainable systems among the various actors along the cassava sector. Thus, the various institutional categories seem to be working within the country's strategic plan of enhancing productivity and efficiency along various commodity VCs.

The peculiar results (Figure 3) that respondents are not sure of how institutions take part in the cassava value chain may be a result of the many people especially in rural areas who can not easily distingush between the various categories of institutions, mainly due to the overlapping nature of work being done in the cassava sector by institutions. Moreover, government and international donor agencies tend to use CBOs and locally established NGOs to engage the local populace. Generally, the close proportion of responses across the various institutional categories (i.e. government (35 %), NGOs (35 %), and international agencies (30–34 %)) with respect to the provision of R&D services may be associated with the fact that after the cassava mosaic virus outbreak, government bodies like MAAIF and international agencies like IITA strongly sensitized people about the disease.

With regards to the vertical and horizontal linkages amongst the actors, the marginal existence of formal vertical and horizontal linkages based on contracts may be attributable to the informal nature of trade in cassava and its products. The mal-functionality of such linkages thus influences the inefficiency and poor competitiveness of the cassava sector. The small proportion of actors with formal contracts may be explained by the fact that few institutions are able to consistently supply large consignments of good quality cassava products.

The roles played by the various institutional categories are exemplified and discussed in detail below.
Key government bodies identified by farmers include the Ministry of Agriculture, Animal Industry and Fisheries (MAAIF), National Agricultural Advisory Services (NAADS), the Uganda National Bureau of Standards (UNBS), National Agricultural Research Organization (NARO) and Local Governments. These bodies play various roles which may in some instances be cross cutting.

In Uganda for instance, MAAIF together with UNBS were at the forefront of setting and monitoring of cassava related policies and by the end of 2006, six product quality standards were drafted (Ntawuruhunga & Okidi, 2010). UNBS also builds the technical capacity of other actors along the VC, especially the processors through trainings. NARO engages in R&D through the multiplication of improved varieties like NASE 1, NASE 2, NASE 3, NASE 4, NASE 10, and NASE 12 that are resistant to Cassava Brown Streak Disease (CBSD) and Cassava Mosaic Virus (CMV). NAADS supplies farmers with improved cassava planting materials (cuttings) free of charge through the Village Farmer Forum (VFF).

Non-government agencies identified include: Uganda National Farmers' Federation (UNFFE), Sasakawa Global 2000 (SG2000), the Dutch Agricultural Development and Trading Company (DADTCO) and Agency for Accelerated Rural Development (AFARD). For instance, SG2000 organised farmers into clusters, forming a One Stop Center Associations (OSCAs), through which farmers get access to inputs, training, and processing equipment while UNFFE works closely with the departments of agriculture in local government and District Farmers Associations to train farmers in various agronomic practices and value addition. DADTCO introduced an autonomous mobile processing unit (AMPU) to overcome the constraint of high perishability of fresh cassava roots.

Community Based Organisations (CBOs) include District Farmers' Associations (DFAs), Pioneer Action for Sustainable Development and Serere Orange-fleshed Sweet potato Producers' Association (SOSPA). The CBOs process cassava into various products and sold locally in the neighbouring districts and markets. International agencies such as the International Institute of Tropical Agriculture (IITA), Natural Research Institute (NRI), African Agricultural Technology Foundation (AATF), Catholic Relief Services (CRS) and the

Private Sector Foundation (PSF) play different roles ranging across training, input supply, financial support and creation of market linkages. For example, IITA in collaboration with Ugachick Poultry Breeders Limited screened yellow genotypes that could resist cassava mosaic disease and cassava brown streak disease (Ntawuruhunga & Okidi, 2010).

In partnership with the National Crop Resources Research Institute (NaCRRI), AATF facilitates access to the machinery and other agro-inputs like fertilisers and herbicides in Apac and Nwoya districts. Currently, NRI in collaboration with national partners is implementing the "Cassava Adding Value for Africa (C:AVA)" project with the aim of developing the VC of High Quality Cassava Flour (HQCF). International institutions like USAID also provide financial support to facilitate activities within the cassava sector. Private sector agencies like the Private Sector Foundation Uganda (PSFU) are mainly interested in providing investment services, value addition, business development services and enhancing market linkages.

## 5 Conclusion and policy implications

Uganda's evolving cassava VC is largely a supply push rather than a demand driven chain. Among the farmers/producers, processors, traders, transporters and consumers, hardly any cooperation exists, resulting in mistrust, information asymmetry and individualism in the cassava trade. Institutions however, play a very critical role to ensure that the other VC actors (farmers, processors, traders, transporters, consumers) are well aligned so as to enhance both horizontal and vertical linkages. Government institutions are mainly at the forefront of the development and enforcement of policies as well as R&D, among other roles played. Other categories of institutions play cross cutting roles like capacity building, establishing market linkages, and R&D. Due to its numerous uses cassava presents a high potential for transforming rural livelihoods in Uganda through job creation and food security.

Policy-wise, it is prudent to design complementary interventions that ensure strong linkages and coordination amongst all actors. For instance, farmers should sell cassava to traders directly from the field. Well-coordinated VC actors will ensure smooth flow, cost-effective and timely delivery of raw materials to processors and products to end markets. Better logistic operations will improved trust among chain actors and will reduce post-harvest losses. Institutions in charge of trade policies, like UNBS, should work together with the various actors to develop and popularise cassava standards. These are very crucial within the sector given that high quality cassava products can easily be sold within the East African regional market and at a higher price. Stronger linkages and coordination will stimulate synergies, hence value and good performance along the chain. In the long run, this may lead to a demand-driven rather than a supply-driven VC, where producers and processors produce in anticipation of markets.

There is need to establish a culture of nurturing business ideas at the various levels along the chain into commercially viable and profitable ventures. This could be achieved through the expansion of the bio-business incubation and entrepreneurship centre. Similarly, the country could encourage and support establishment of Small and Medium Enterprises (SMEs) so as to improve the cassava VC. This approach will encourage youths in participating in cassava-related business. The SME approach may be implemented through imparting skills to the youth and helping them to establish cassava-based businesses such as the production of cassava chips, starch, and bio-ethanol. It is also important that Uganda establishes a website through which the various actors can easily access useful information (such as market prices and tonnage produced) in order to reduce the information gap among the actors.

It is also strongly recommended that farmers are put in the conditions to profitably produce more cassava so as to ensure consistent supply of fresh root to the other VC actors. Increased cassava production may be achieved through clustering of small scale producers into small groups, as this is bound to increase the productivity because groups find it easier to mobilise the required manual labour and to access agro-inputs. This implies that small scale farmers will be in position to sufficiently supply materials for industrial purposes. All that said, increased cassava production will inherently require a reliable supply of high-yielding planting, disease-resistant varieties, mechanisation of the farming practices and adequate technical expertise at the various stages along the chain. Thus, it is important to establish strong private-public partnerships in order to tackle all the above mentioned aspects and enable the development of the cassava sector in Uganda.

**Acknowledgement**

We acknowledge the financial support obtained from East Africa Agricultural Productivity Project (EAAPP)/World Bank, contributions of the Value-addition research team, and the respondents who provided the desired information.

## References

Adebayo, K., Abayomi, L., Abass, A., Dziedzoave, N. T., Forsythe, L., Hillocks, R. J., Gensi, R., Gibson, R. W., Graffham, A. J., Ilona, P., Kleih, U. K., Lamboll, R. I., Mahende, G., Martin, A. M., Onumah, G. E., Orr, A. W., Posthumus, H., Sanni, L. O., Sandifolo, V. & Westby, A. (2010). Sustainable Inclusion of Smallholders in the Emerging High Quality Cassava Flour Value Chains in Africa: Challenges for Agricultural Extension Services. *Journal of Agricultural Extension*, 14 (1).

Adebayo, K. & Sangosina, M. A. (2005). Processors' perception of the effectiveness of some cassava processing innovations in Ogun state, Nigeria. Paper prepared for presentation at the Farm Management Association of Nigria Conference, Asaba, Nigeria, October 18–20, 2005.

Balagopalan, C. (2002). Cassava utilization in food, feed and Industry. *In:* Hillocks, R. J., Tresh, J. M. & Belloti, A. C. (eds.), *Cassava, biology, production and utilization*. Cali, Colombia.

Collinson, C. D., Kleih, U., Burnett, D. G., Ferris, R. S. B., Jagwe, J. N. & Muganga, A. K. (2002). Transaction cost analysis. Final report. Kampala: NRI and IITA.

Collinson, C. D., Wanda, K., Muganga, A. & Ferris, R. S. B. (2000). *Cassava marketing in Uganda: Constraints and opportunities for growth and development*. IITA and NRI, Kampala, Uganda.

Collinson, C. D., Wanda, K., Muganga, A. & Ferris, R. S. B. (2003). A market opportunities survey for value-added utilization of cassava-based products in Uganda. Part II: Supply chain analysis constraints and opportunities for growth and development. ASARECA/IITA Monograph 4 IITA, Ibadan, Nigeria. URL `http://www.iita.org/c/document_library/get_file?uuid=d5650bca-e68b-4c85-9bb9-bb6e9e360f65&groupId=25357` (last accessed: 08.12.2015).

FAO (2005). Commodity chain analysis: Constructing the commodity chain functional analysis and flow charts. Food and Agriculture Organization of the United Nations (FAO). URL `http://www.fao.org/tc/easypol` (last accessed: 10.07.2014).

FAO (2014). Food and Agricultural Organization, online statistical database. URL `http://www.fao.org/statistics/en/` (last accessed: 10.05.2014).

Gaffney, A., Kpaka, C., Slakie, E. & Anderson, C. L. (2012). Cassava Integrated Value Chain. EPAR Brief No. 223. Prepared for the Agricultural Policy Team of the Bill & Melinda Gates Foundation. University of Washington. URL `http://evans.uw.edu/sites/default/files/public/EPAR_UW_Request_223_Cassava_Integrated_Value_Chain_Public_Version_03.05.13_af.pdf` (last accessed: 04.03.2016).

Graffham, A., Kleih, U., Jagwe, J., Nabawanuka, J., Wanda, K., Kalunda, P. & Ntibarikure, G. (2000). *Industrial markets for Starch-based products: An assessment of the Industrial potential for cassava in Uganda*. Natural Resources Institute (NRI), FOODNET / International Institute of Tropical Agriculture (IITA) and Kawanda Agricultural Research Institute (KARI).

Karugia, J., Waithaka, M., Freeman, A., Prabhu, R., Shiferaw, B., Gbegbelegbe, S., Massawe, S., Kyotalimye, M., Wanjiku, J. & Macharia, E. (2009). Responding to the food price crisis in Eastern and Southern Africa: Policy options for national and regional action. ReSAKSS Working Paper No. 27. ReSAKSS-Eastern and Central Africa. URL `https://cgspace.cgiar.org/handle/10568/184` (last accessed: 04.03.2016).

Kehinde, A. T. (2006). Utilization Potentials of Cassava in Nigeria: The Domestic and Industrial Products. *Food Reviews International*, 22, 29–42.

Kilimo Trust (2012). Development of inclusive markets in agriculture and trade (DIMAT): The nature and markets of cassava value chains in Uganda. URL `http://www.undp.org/content/dam/uganda/docs/UNDPUg_%20PovRed_Value%20Chain%20Analysis%20Report%20Cassava%202013%20Report.pdf` (last accessed: 09.12.2015).

Kimathi, M., Ngcli, P. & Wanjiru, J. (no date). Value chain analysis for cassava flour and related products: A case study of Uganda and Kenya. Study by Farm Concern International presented to ASARECA. Study by Farm Concern International presented to ASARECA. Study by Farm Concern International presented to ASARECA. URL `http://s3.amazonaws.com/zanran_storage/asareca.org/ContentPages/106589412.pdf` (last accessed: 06.09.2014).

Kleih, U., Phillips, D., Jagwe, J. & Kirya, M. (2012). Cassava Market and Value Chain Analysis Uganda

Case Study. Natural Resources Institute (NRI) and Africa Innovations Institute (AII). URL http://cava.nri.org/images/documents/publications/UgandaCassavaMarketStudy-FinalJuly2012_anonymised-version2.pdf (last accessed: 04.03.2016).

Lemchi, J., Ifeanyi-Obi, C. C. & Olatunji, S. O. (2011). Socio-economic factors affecting the marketing of Garri in Port Harcourt city of Rivers state. *Journal of Agriculture and Social Research*, 11 (2), 9–14.

MAAIF (2010). NAADS implementation guidelines: Farmer selection and support for progression from subsistence to commercialization. National Agricultural Advisory Services (NAADS), Kampala, Uganda. URL http://www.altusfoods.com/wp-content/uploads/2015/05/NAADS_phaseii_implementation_guidelines.pdf (last accessed: 04.03.2016).

Naziri, D., Quaye, W., Siwoku, B., Wanlapatit, S., Phu, T. V. & Bennett, B. (2014). The diversity of postharvest losses in cassava value chains in selected developing countries. *Journal of Agriculture and Rural Development in the Tropics and Subtropics*, 115 (2), 111 – 123.

New Vision (2013). Cassava commercialisation to save Uganda $300m. New Vision, Vision Group, Uganda. URL http://www.newvision.co.ug/news/648297-cassava-commercialisation-to-save-uganda-300m.html

Ntawuruhunga, P. & Okidi, J. (2010). Eastern Africa Root Crops Research Network (EARRNET) legacy. International Institute of Tropical Agriculture (IITA), Nigeria. URL http://www.iita.org/c/document_library/get_file?uuid=593f240d-7fc9-4061-bbd4-f37de7aaa8eb&groupId=25357 (last accessed: 12.05.2015).

Nzigamasabo, A. & Zhou, H. M. (2006). Traditional Cassava Foods in Burundi: A Review. *Food Reviews International*, 22, 1–27.

Ogwal, O. R., Kashub, S., Okello, A. S. & Mutyaba, C. (2012). Cassava value chain analysis in Uganda. A report prepared for the Cassava Regional Centre of Excellence (CRCoE). Unpublished report, National Agricultural Research Organization, Uganda.

Organisation for Economic Cooperation and Development (OECD) (2006). DAC Guidelines and Reference Series Applying Strategic Environmental Assessment: Good Practice Guidance for Development Co-operation. OECD, Paris.

Rwakakamba, M. (2013). 2013/2014 budget: bigger pie of agriculture sector budget should focus on dsip and local governments. Agency for Transformation (AfT), Kampala, Uganda. URL http://www.agencyft.org/20132014-budget-bigger-pie-of-agriculture-sector-budget-should-focus-on-dsip-and-local-governments/

Sayre, R., Beeching, J. R., Cahoon, E. B., Egesi, C., Fauquet, C., Fellman, J., Fregene, M., Gruissem, W., Mallowa, S., Manary, M., Maziya-Dixon, B., Mbanaso, A., Schachtman, D. P., Siritunga, D., Taylor, N., Vanderschuren, H. & Zhang, P. (2011). The BioCassava plus program: Biofortification of cassava for sub-Saharan Africa. *Annual Review of Plant Biology*, 62, 251–272. doi:10.1146/annurev-arplant-042110-103751.

UBOS & MAAIF (2010). Uganda census of agriculture 2008/2009. Volume IV: Crop area and production report. Uganda Bureau of Statistics (UBOS).

Ulleberg, I. (2010). *The role and impact of NGOs in capacity development: From replacing the state to reinvigorating education*. International Institute for Educational Planning, Paris, France.

USAID (2010). Market assessment and baseline study of staple foods. Country report - Uganda. United States Agency for International Development (USAID). URL http://pdf.usaid.gov/pdf_docs/PNADW642.pdf (last accessed: 01.08.2014).

USAID & EAT (2013). Institutional architecture for Food security policy change: Uganda. Enabling Agricultural Trade (EAT) project, United States Agency for International Development (USAID).

# 2

# Beyond good wood: Exploring strategies for small-scale forest growers and enterprises to benefit from legal and sustainable certification in Indonesia

Ahmad Maryudi [a,*], Ani A. Nawir [b], Purnomo Sumardamto [c], Dewi Ayu Sekartaji [d], Emma Soraya [a], Teguh Yuwono [a], Bowo Dwi Siswoko [a], Budi Mulyana [a], Nunuk Supriyatno [a]

[a]*Faculty of Forestry, Universitas Gadjah Mada (UGM), Yogyakarta, Indonesia*
[b]*Center for International Forestry Research (CIFOR), Bogor, Indonesia*
[c]*District Forest Service of Gunungkidul, Indonesia*
[d]*Ministry of Environment and Forestry, Indonesia*

**Abstract**

There is a growing trend towards small-scale forest operators engaging in certification initiatives in response to growing demands for certified timber. We aimed to analyse whether certified smallholders gain better access to timber markets and better financial incentives. We used three smallholder forest enterprises from Gunungkidul and Kulon Progo districts, Java, Indonesia as case studies of different managerial operations. Two were farmer cooperatives and the other was an external trading company in partnership with village-level farmer groups. Different certification schemes and managerial approaches were analysed whether such factors influenced the ability of smallholders to access certified timber markets and obtain improved offers for their timber. We found that smallholders find it challenging to enter and participate in certified timber markets and to obtain the promised premium prices. The obvious costs of certification and uncertain benefits are major challenges for smallholders. Even when market opportunities are present, certification alone is insufficient to tap into these markets. Certified forest enterprises need dedicated managers with adequate entrepreneurial skills to establish networks and contacts with potential buyers, and to actively seek information on what is specified by certified timber processors.

*Keywords:* smallholder growers, certification, market access, premium prices, entrepreneurial skills, market networks

## 1 Introduction

The past twenty years have witnessed a worldwide trend in certification initiatives, which specify how forests should be managed and timber produced. In general, certification is a process by which forest practices are assessed against a set of predefined standards agreed upon through independent audits (Viana, 1996; Rametsteiner & Simula, 2003; Cashore *et al.*, 2004; Durst *et al.*, 2006). The scope of forest and timber certification varies, from a wide range of environmental, social and economic aspects to schemes that focus more on legality (Maryudi, 2016). To encourage

* Corresponding author
Faculty of Forestry, Universitas Gadjah Mada (UGM),
Jln. Agro No. 1, Bulaksumur, Yogyakarta, 55281 Indonesia
Email: maryudi76@yahoo.com and ahmad_maryudi@ugm.ac.id

forest managers to implement desirable practices, certification promises market incentives of improved access and/or premium prices for certified products. Certification continues to gain wider support, reflected in the increase in the total area of certified forests. By May 2015, the area of certified forest worldwide, for two major schemes, Forest Stewardship Council (FSC) and Programme for the Endorsement of Forest Certification (PEFC), reached nearly 450 million ha (a tenth of the world's forest cover), including an estimated 7.5 million ha certified under both schemes (UNECE/FAO, 2015). This is a sharp increase from only six million ha in 1996 (UNECE/FAO, 2005). In addition to this, the number of Chain-of-Custody (CoC) certificates grew steadily to a total of 39,609 active CoC certificates in May 2015 (UNECE/FAO, 2015).

Certification, particularly the scheme provided by FSC, was not originally designed for small or non-industrial operations (Lindstrom *et al.*, 1999; Ricken-bach, 2002; Gulbrandsen, 2004; Butterfield *et al.*, 2005). In fact, certification was seen with great scepticism and was even opposed by small-scale operators in many countries, principally in Europe and North America (Lindstrom *et al.*, 1999; Rickenbach, 2002; Maryudi, 2005). This was largely due to potentially adverse economic consequences (Lindstrom *et al.*, 1999). In addition, certification was perceived as incompatible with smallholders with diverse goals and interests, un-like large scale and uniform forest operations (Ricken-bach, 2002). Nonetheless, there is a growing trend in small-scale forest operators practicing sustainable or legal forest production in response to demands for certi- fied/legal raw materials (Kandel, 2007; Wiersum *et al.*, 2013; Maryudi *et al.*, 2015; Fujiwara *et al.*, 2015). For example, FSC reported that in 2015 small-scale forestry made up 22% of all forest management certifi-cates (FSC, 2015). In Indonesia, there have been efforts to encourage smallholder farmers to adopt certification (Hinrichs *et al.*, 2008; Maryudi, 2009). In part, this could be facilitated by streamlined standards and pro-cedures exclusively designed for small-scale operations (Durst *et al.*, 2006; Overdevest, 2010; Wiersum *et al.*, 2013). The FSC, for example, now provides schemes for groups and Small or Low-Intensity Managed Forests (SLIMF). This trend may indicate increased confidence among small-scale operators that certification provides ways to unlock opportunities in timber markets, either improved market access or premium prices for certified products.

Demand for certified timber products are reported to be growing in eco-sensitive markets (UNECE/FAO, 2013). The public procurement policy on legally certified timber has also helped increase the growing interest in the market–based policy instruments (Cashore & Stone, 2012; Maryudi, 2016). Although evidence of a willingness to pay higher prices for legal/sustainable timber is unclear (Tiesl *et al.*, 2002; Jensen *et al.*, 2003; Anderson & Hansen, 2004; Owari *et al.*, 2006; Cai & Aguilar, 2013), there are at least strong signals from consumers' preference for such products (Hansmann *et al.*, 2006; Rickenbach & Overdevest, 2006; Brockmann *et al.*, 2012). In addition, export-oriented processing industries in timber producing countries are said to seek certified/legal timber (Harada & Wiyono, 2014; Maryudi *et al.*, 2015). Owari *et al.* (2006) argue that processing industries consider certification as important for maintaining market share and selling products. This may explain the increasing trend in small-scale forest operators practicing sustainable or legal forest production.

However, small-scale wood-processing industries prefer to buy logs through middle-men, because it is simpler and does not involve the complicated procedures of purchasing and transporting the logs, including unofficial payments, which frequently account for 15 to 20 % of the log prices (Triple Line Consulting, 2005). This supports the view that timber produced by small-scale tree growers fits the domestic market, where small-scale wood-processing industries will accept low-grade timber. This is a major concern for certified forest managers, particularly smallholders, who are struggling to capture market opportunities. While some have identified short-term financial benefits from engaging in certification (e.g. Harada & Wiyono, 2014; Maryudi *et al.*, 2015), others (Holding Anyonge & Roshetko, 2003; de Jong *et al.*, 2010; Perdana *et al.*, 2012; Foundjem-Tita *et al.*, 2013; Rohadi *et al.*, 2015) pointed out the difficulties faced by smallholders who wish to tap into potential markets. Maryudi *et al.* (2016) point out the importance of studies on the impacts of certification on smallholder forest operations in Indonesia. This is because tree planting and the sale of the subsequent timber contribute approximately 10–15 percent of the rural household income (Perdana *et al.*, 2012; Rohadi *et al.*, 2015). In this paper we aim to analyse whether certified smallholders gain better access to timber markets and even better financial incentives. Findings from this study will provide insights into how to best assist smallholders so that they can select the best and most appropriate strategies for their tree planting activities and marketing of the timber.

## 2 Smallholder tree growing in Indonesia: A brief overview

In the Indonesian context, smallholder tree growing (*Hutan Rakyat*) is defined as forest that is privately owned by a farmer with a minimum land size of 0.25 ha, and where a closed canopy of timber trees and/or other types of plants cover more than 50 % of the area; and/or in the first year a minimum of 500 trees per hectare are planted in the same area (Ministry of Forestry, 2009). Rural people across Indonesia have planted trees as part of their shifting cultivation, home gardening, farm forestry and meadow grazing (Michon & de Foresta, 1995). In most cases, smallholder tree growing in the country is far more diverse than industrial forestry in terms of species mix, age classes, and silvicultural operations. Planting systems vary and are decided by the individual farmer's preference, the socio-economic environment, and biophysical farm characteristics (e.g. soil types, slopes and farmland sizes). In many cases, a mixture of different trees and agricultural crops is the most popular arrangement as it produces seasonal crops to satisfy their daily needs (Awang *et al.*, 2007; Rohadi *et al.*, 2012). More trees are planted in areas less suited to agriculture and in extreme terrains (Buddhisatyarini, 2005). In recent years, intensive tree growing, including on fertile land, is not uncommon as rural people have more opportunities to earn an income from off-farm activities and are less reliant on subsistent farming (Sabastian *et al.*, 2014).

Initially, trees were grown to either improve the environment or to meet the owners' need for fuelwood and construction timber (Nawir *et al.*, 2007). The recent growing demand for timber in the processing industries appears to have further encouraged tree growing (Sabastian *et al.*, 2014; Nawir, 2013). It is estimated that standing timber stocks, in smallholder tree plantations, had reached 125 million $m^3$, 15 % of which was said to be ready for harvest (Shantiko *et al.*, 2013). Timber from smallholder plantations is increasingly sought by processing industries amidst the declining timber supply from state forests. For example, approximately 50 % of timber used by export-oriented furniture industries in Jepara – one of Indonesia's furniture centres – is sourced from smallholder plantations (Yovi *et al.*, 2009).

Rural farmers usually consider their tree plantations as an investment and insurance, while their daily livelihoods come from agricultural crops and off-farm earnings (Perdana *et al.*, 2012; Sabastian *et al.*, 2014). In most cases, trees are harvested intermittently only in hard times, particularly when farmers have no other liquid assets (Perdana *et al.*, 2012). Nonetheless, harvests have become more frequent in recent years. Many dimensions of rural life are changing fast and rural farmers spend more cash on electronic goods such as mobile phones, televisions, washing machines and motorbikes than they did in the past. In many cases, trees are harvested to satisfy these changing lifestyles. This raises concern about the future sustainability and timber supply from smallholder plantations as harvesting becomes more common.

## 3 Research methods

There are both sustainable forestry certification and legality verification for small forest operations in Indonesia. Sustainable certification is provided by two different bodies: 1) the Forest Stewardship Council (FSC) and 2) the Indonesian Ecolabel Institute (LEI), whereas the legality verification (known as *Sistem Verifikasi Legalitas Kayu*–SVLK) is of the Indonesian Ministry of Environment and Forestry (Maryudi, 2016). The study selected three smallholder enterprises from the two districts of Gunungkidul and Kulon Progo, the Special Region of Yogyakarta (in the south of Central Java), representing the different certification schemes, i.e. FSC, LEI and SVLK (see Maryudi *et al.*, 2015 for their details). The enterprises act as a parent group for smaller farmer groups, usually at the village level. The study focused on whether smallholders under certification are able to utilise the good credentials of their timber products. Smallholder tree growing have impressively expanded in the two districts. Particularly in Gunungkidul, smallholder tree growing is often regarded as one of Indonesia's reforestation success stories, and is often considered one of the most commercialised timber marketing hubs for local, national and international markets (Awang *et al.*, 2007; Nawir *et al.*, 2007).

The three case studies differ in their managerial operations (details in Table 1). Two are cooperatives of village-level farmer groups. The other is an external trading company in partnership with village-level smallholder groups that wish to engage in certification. The different management styles may influence the ability of the respective groups to find buyers for their timber. This study did not assess the demand side, i.e. whether a particular certification scheme might be preferred in the timber markets. Nonetheless, our aim was to test the hypothesis that adopting different certification schemes have impacts on market access. Studies show that consumers generally prefer certified timber products (Brockmann *et al.*, 2012) with a skewed preference for a particular certification scheme (Anderson

& Hansen, 2004; Hansmann *et al.*, 2006; Rickenbach & Overdevest, 2006).

Our research was conducted between April and October 2015. The primary data for the study was collected through semi-structured, in-depth interviews. Interviews were conducted with the managers, member farmers and a number of key actors in the value chains such as timber collectors, traders and industries. Those involved in the certification preparation of the respective enterprises were also interviewed. We also analysed documents, records and reports, particularly with regard to the costs of certification, timber sales and prices, and timber buyers.

## 4 Results

### *4.1 The case studies*

*Case study 1: The Koperasi Wana Manunggal Lestari (KWML) Cooperative*

In 2006, the Farm Forestry Working Group (*Kelompok Kerja Hutan Rakyat Lestari*), Universitas Gadjah Mada, and the district government of Gunungkidul facilitated the establishment of KWML, a cooperative of three farmer groups from three villages, with a total membership of 635 farmers, covering a forest area of 815 ha. KWML was one of the earliest management units of smallholder plantations in Indonesia. A committee, elected from the farmers, acts as manager of the whole forest. The cooperative regulates the harvests, by setting the annual allowable harvests for the whole group. It also has a role as a business unit, facilitating the members when marketing their timber. The cooperative and the supporting working group choose the certification provided by the national certification body (LEI), which exempts the cooperative from certification costs. In 2006, the LEI certificate initially covered 815 ha, but during a survey in 2012 the coverage was expanded to 1,153 ha. KWML was later declared eligible for SVLK-certification. The SVLK-certification was fully subsidized by central government through the *Multistakeholder Partnership Programme* funded by the British Government.

*Case study 2: The Koperasi Wana Lestari Menoreh (KWLM) Cooperative*

In 2007, Telapak (NGO) in collaboration with a local NGO (*Yayasan Bina Insan Mandiri*) and a local office of Credit Union established KWLM as a community logging programme. The programme later became a cooperative of smallholder farmers from 15 villages, with a total area of about eight thousand hectares. In 2011, the cooperative gained FSC certification for 210 hectares. Unlike KWML, this cooperative has a non-farmer manager with a university degree. The Humanist Institute for Cooperation (*Humanistisch Instituut voor Ontwikkelingssamenwerking*–HIVOS), a Dutch organisation for development, has covered the certification costs through Telapak. In 2015, KWLM also received SVLK certification, which was self-funded from revenues generated from timber sales as well as group savings. All members are obliged to sell their timber to the cooperative. The collaboration with the credit institution is crucial as the cooperative is able to provide soft loans for the members to delay their cuts when maximum annual cuts are reached. KWLM also works with non-certified smallholder forests to supply the demand for non-certified timber.

*Case study 3: Dipantara Forest Project*

In contrast to the previous two groups, Dipantara is a timber company operated by professionals and chaired by a former senior officer from the State Forest Company (Perhutani). Dipantara has a strong portfolio of supplying timber to processing industries. It does not have forests, instead it purchases timber from tree farmers to be sold to the processing industries. Dipantara collaborates with groups of smallholders at the village level. Viewing the potential certified timber markets, the company and 35 farmer groups have agreed to engage in FSC certification, which was achieved in 2012. The farmer groups have left the management responsibilities of the certification of their forests to the company, including paying the costs on their behalf. The company promised to purchase the farmers' timber at a premium price once the forests are certified. To prepare the certification, Dipantara worked with The Forest Trust (TFT), which has close links with both FSC and buyers of FSC-certified timber. Dipantara not only purchases certified timber from the farmers, but also continues its business with non-certified timber.

### *4.2 Costs of certification*

The cost of certification varied considerably across our three case studies. Sustainable and legal certification generally incurs two types of costs: 1) indirect costs for improving the forest management, and 2) direct costs for the certification audits (start up and surveillances). The indirect costs for improving forest management and to meet the required standards of the certification are rarely documented. In our case studies, investments were made to: 1) establish a registered farmer group cooperative, 2) create 'management contracts' (group

**Table 1:** *Summary of the case studies*

| *Managerial operator (Parent group)* | *Case study 1 KWML Cooperative* | *Case study 2 KWLM Cooperative* | *Case study 3 Dipantara Forest Project* |
|---|---|---|---|
| Type of managerial operation | • Cooperative of smallholder groups<br>• Manager and committee members elected from smallholder farmers | Cooperative of smallholder groups | • Trading company in partnership with smallholders<br>• Manager and committee are from the company |
| Profile of manager | Farmer with high-school education (Grade 12) | Non-farmer with a university degree | A former director at state-forest company |
| Certification | • LEI in 2006 (re-certification in 2012)<br>• SVLK in 2011 | • FSC in 2011<br>• SVLK in 2015 | FSC in 2012 |
| Members and forest area | 3 smallholder groups from 3 villages<br>• 635 farmers with 815 ha under 1st LEI certification in 2006 (1,153 ha after 2nd certification in 2012)<br>• 594 ha under 1st SVLK in 2011 (1,236 ha after 1st SVLK surveillance in 2012) | 18 smallholder groups (15 under certification)<br>• 1,083 farmers with 700 ha under FSC<br>• 1,341 farmers with 808 ha under SVLK | 100 smallholder groups (35 under certification)<br>• 330 ha under FSC (no information on the number of the farmers) |
| Annual sustainable cuts from certified forests | ≈ 700 $m^3$ | ≈ 500$m^3$ | ≈ 640 $m^3$ |

constitutions and rules) regulating the members, 3) improve the organisational capacity and forest operations, 4) map the forest, 5) establish management plans, and 6) conduct environmental assessment and monitoring. The certified entity (parent group) is required to organise the members and to ensure that each of its smallholder members follow the required standards. Each of our case studies revealed that it took more than a year before the enterprise could finally apply for certification.

It is difficult to calculate the costs for conducting each activity as the challenges faced by the farmer groups differed prior to the certification audits. In addition, the required actions and documentation standards also varied with the different certification bodies. The managers and the supporting NGOs estimated that the preparatory work might have cost them equal to or even more than the first certification audit. They suggested a conservative estimate of USD 5,000–10,000 per certification case depending on the number of farmer groups involved and the size of forests to be certified. The figure is slightly higher compared to other non-industrial forest certifications in other countries. For instance, the indirect certification costs of O'Neill Pine Company (OPC) in the US with 894 hectares were about USD 3,000 (see Butterfield *et al.*, 2005). The lower costs of OPC certification could be due to there being fewer members involved. There were only 47 forest owners involved in the certification, much lower than our case study enterprises.

The direct costs for the first audit and surveillance were even higher despite the simpler audit procedures compared to those for large enterprises (Table 2). The case study enterprises pay between USD 110 and 175 $ha^{-1}$ $year^{-1}$ for their FSC certification. As a comparison, the certification of Sample Forest in Guatemala of 750 hectares costs USD 21.33 $ha^{-1}$ $y^{-1}$ (see Molnar, 2003). To put these figures into perspective, the certification costs of large-scale forest operations in Indonesia range from USD 1.07 to USD 3.64 $ha^{-1}$ $y^{-1}$ (Ruslandi *et al.*, 2014). The costs of LEI certification and SVLK legality certification of KWML, which look comparably lower to FSC certification, are still higher than those of large operations.

KWML and KWLM only managed to cover the costs because of funding from donors, channelled through

**Table 2:** *Start-up costs of certification and surveillance audits for smallholder plantations*

| *Type of certification* | *Case study* | *Certification costs (× 1,000 USD)* | | *Type of surveillance* | *Validity period* | *Total cost for the whole validity period* (× 1,000 USD) |
|---|---|---|---|---|---|---|
| | | *Start-up certification* | *Surveillance audit* | | | |
| FSC sustainable certification | KWLM, Dipantara | 13 | 8 | Annual | 5 years | 37 |
| LEI sustainable certification | KWML | 8 | 0.5 | Annual | 5 years | 12 |
| SVLK legality verification | KWML, KWLM | 3 | 3 | Biannual | 10 years | 18 |

* Estimated costs for submitting annual reports to the certification body. No field surveillance conducted; surveillance was based on evaluation of the annual report

Source: interviews with the group managers and invoices from the certification bodies

tices and to prepare the certification documentation. In many cases, donors usually only provide short-term assistance, the certified groups are then left with the challenge of maintaining the certification. In the case of KWML, when it was unable to self-finance the recertification, its LEI certificate was suspended. It was only when further funding was available that KWML was able to recertify its forest operation. The financial benefits that offset the costs are crucial to the certified groups continuing the good wood business (discussed below). Both KWLM and Dipantara managers also suggested that the enterprises could reconsider certification if the expected benefits were minimal.

### *4.3 Timber sales, prices and market access*

Certification initiatives are based on the notion of market incentives provided to forest enterprises to implement improved practices. The incentives may include better market access and higher prices for certified timber products. The three case studies established contracts with export-oriented processing industries, which were willing to purchase the certified timber at higher prices compared to non-certified timber (Table 3). The NGOs assisting the respective enterprises connected the managers with the buyers and facilitated the initial contract. However, our research identified different approaches employed by the case study enterprises in utilising the good credentials of their timber.

KWML adopted a less 'pro-active' strategy compared to KWLM and Dipantara. It waited for the buyers to come to them, assuming that certified timber would be sought after as there were few certified smallholder forest operations. A limited network was said to be one of the main explanatory factors. The manager admitted to having limited entrepreneurial skills and said he was more familiar with farming activities. In addition, he could not commit fully to finding links to processing industries, as he had to dedicate most of his time to his farm. Initially, KWML relied on an early contract with a furniture company (*Jawa Furni Lestari*–JFL); attempts to broaden the network with more processing industries were unsuccessful. The manager said that there were rarely any real buyers, those that came to the cooperative never made a purchase. With no logs in the store due to member farmers' intermittent harvests, it is difficult for KWML to attract buyers, which prefer to check the logs before purchasing. The members will only harvest trees when transactions are secured as the cooperative lacks financial capital to buy logs from its members. The in-termittent tree harvests also contribute to KMWL's inab-ility to supply processing industries in regular volumes. The *Koperasi Hutan Jaya Lestari* Cooperative, an FSC-certified smallholder in Sulawesi, experienced a similar situation (Harada & Wiyono, 2014).

As a result, between 2007 and 2009, KMWL recorded only three transactions with a processing industry for total timber sales of about 300 $m^3$ (100 $m^3$ $y^{-1}$), much lower than its annual allowable cut amounting to 700 $m^3$. In the following year it recorded its lowest sales of certified timber of only 2 $m^3$ to JFL. Similarly, the sales of its legally verified timber are no better. In 2012, KWML managed only one sale of two legally verified teak logs of about 0.4 $m^3$ to JFL. The contracts with JFL, for both certified and legally verified timber, were totally abandoned by the end of that year. Individual KWML farmers are free to sell their timber if the cooperative is unable to find markets for certified timber. The farmers tend to sell their timber in non-certified timber markets where there is no premium price.

**Table 3:** *Price differences between certified and non-certified teak logs in each size class*

| *Log diameter* (in cm) | *Price differences* (%) | | |
|---|---|---|---|
| | KWML | KWLM | Dipantara |
| 10–15 | 14–27 | 20–30 * | not applicable |
| 16–19 | 8–20 | | 35 |
| 20–29 | 5 | | 29 |
| 30–39 | 3 | | 15 |

* Detailed information on different log diameter unavailable
Source: interviews with the managers and purchase notes

In contrast, KWLM and Dipantara have established more proactive strategies to find links to high value markets for their certified timber. Each of the enterprises employs a full time and highly qualified manager, who is tasked with creating business plans, and es-tablishing networks and contacts with potential buyers. The managers are actively engaged in business meet-ings and even use web-based information and technol-ogies. The KWLM manager, for instance, utilises the FSC-website to obtain contact details and information on FSC-certified processing industries from which the cooperative can learn about their products and their re-quired timber specifications. Similarly, the Dipantara manager uses his vast experience in timber selling as a director in the state forest company, which has nu-merous clients in the processing industries. He poin-ted out a major weakness in the way the state company sells its timber; it pre-determines the specifications of its logs, while different processing industries have dif-ferent requirements for their products. Dipantara uses these *niche markets* to its full advantage. Its contracts with processing industries have steadily increased from only one in 2008 to twelve in 2013.

Both KWLM and Dipantara have stored logs in contrast to KWML that only harvests trees when it has a buyer. This helps potential buyers to identify which logs they need. Both enterprises are able to buy trees from their members with loans from financial institutions. KWLM collaborates with a credit institution to provide micro credit for its members. The cooperative also accumulates financial capital from its timber sales. The premium price of its timber is divided equally between its members and the cooperative. To accumulate funds, both KWLM and Dipantara have become more integrated. They have more farmer groups than when they started, even from different districts, and they also sell non-certified timber. This strategy is said to be very important, it enables them to offer their members premium prices for certified timber, even when purchases from processing industries decline. As with any small-scale operation, both KWLM and Dipantara are still prone to competition with larger integrated companies. At the time of our research, the state forest company in the region offered massive discounts of up to 50 % on their timber. Understandably, many of their clients have abandoned their contracts and switched to the discounted, better quality timber.

### *4.4 Side benefits of engaging in certification*

Our case studies have also revealed that the small-scale enterprises have continued to maintain their certification of sustainable forestry and legal verification. While smallholders' continue to receive funding for maintaining and renewing their certification, the potential windows of opportunities from certification and legal verification may also explain the smallholders' continued certification. Certification has been promoted as an avenue to improve forest practices and governance. Experience shows that a number of stakeholders – governmental, non-governmental and donors – are in many cases committed to providing financial and technical assistance for improving the livelihoods of smallholders and to invest in capacity building and physical infrastructure.

To date, few smallholder groups have engaged in either certification or legal verification due to the continued concerns that the financial benefits hardly outweigh the costs. This is an opportunity for the currently certified groups to absorb the wide range of aid, as our case studies revealed. In most cases, the aid is framed in the context of rural development and empowerment of smallholder farmers. Farmer members were given free seedlings, fertilisers, farm equipment, livestock (cows, goats and sheep) and often, micro loans. The KWLM and Dipantara farmers even receive a share of the profits. As a cooperative, KWML and KWLM have a credit scheme for their respective members. In KWLM, the members have access to loans of more than USD 10,000. At the end of a financial year, they also receive dividends from the profits accumulated by the cooperative. In many cases, soft loans and dividends often serve as the main motivations for farmers to join a cooperative (see Harada & Wiyono, 2014).

Certification also provides opportunities for smallholders to increase their understanding of the requirements for managing their forests sustainably. Support from external stakeholders for the capacity building of farmers is notable in the case study villages. This includes training, cross-visits, internships and specialist consultations. In all cases, the farmer members are also trained to apply better silvicultural practices to improve the quality of the stand with the aim of obtaining better income from the wood. In most cases, smallholder tree farmers in Indonesia rarely apply appropriate silvicultural practices (singling, pruning, thinning and fertilising) and post-harvest management (Roshetko *et al.*, 2012). The poor silvicultural practices explain the smallholders low quality timber (Hardiyanto & Prayitno, 2007). The members are also trained to estimate the volume of harvested wood and associated prices. This is beneficial as in many cases farmers are often disadvantaged in transactions by traders who often conduct appraisals below the actual timber volume (Perdana *et al.*, 2012).

In addition to the funding received for the certification, KWML has received considerable assistance from the government (central and local) and donors since it was first certified in 2006. The most cited reason is that the cooperative demonstrates good forest management. For instance, in 2010, KWML received grants from the provincial government of nearly USD 4,000 as a start-up for a cattle business. In 2012, it also received a grant from the government of Japan for a training centre and office as well as a set of processing and drying machines. It is expected that these machines will support the co-operative's forest activities, not only planting but also processing industries. The group is also able to regularly generate income from providing timber milling and drying services. The district government has also sporadically provided the group with in kind (free seedlings, fertilisers etc.) and financial assistance.

## 5 Discussion

The core rationale that encourages farmer members to engage in certification is the promised premium prices for certified timber that exceeds the price offered when they sell non-certified timber to local traders. For the certified groups, the business of certification is only visible when the economic returns offset the costs. Our research has clearly shown that certification involves high transactional indirect and direct costs. Auld *et al.* (2008) argued that compared to integrated forestry, non-industrial operations face higher costs of preparing for, paying for and responding to certification audits. Small groups with intermittent harvests generally lead to uncertain supply and a high cost per unit reflected in a greater percentage of costs for maintaining their certification (Irvine, 2000; Nussbaum *et al.*, 2001). Our case studies revealed that due to the size of the forest, certification becomes an extremely expensive business for smallholder tree growers. Expanding the scope of forest operations, i.e. more farmers and forest area under certification might be an option; all of the case studies have pursued this strategy.

Scaling up of further adoption of these certification programmes might be challenging, as small-scale enterprises must rely on assistance and facilitation from external parties. This is shown in all of our case studies. Similar case studies have also found that smallholder farmer groups rely heavily on external donors for organising them and bearing the certification costs (Bass *et al.*, 2001; Hinrichs *et al.*, 2008). Analysts doubt such models can sustain since donors usually only provide short-term assistance. Such is clearly evidenced in the KWML case. Combined with the limited benefits from being certified, the cooperative gave up its good forest certification due to its lack of financial resources. Globally, decertification of small-scale operations is not uncommon. For instance, in 2009 FSC decertified 47 % of the earlier certifications of their small operations (Wiersum *et al.*, 2013). In contrast, suspension of certification of large operations is largely due to non-compliance to the standards (Forest Watch Indonesia, 2013). The high transaction costs appear to be the main issue for small-scale forestry. Our case studies revealed that whether smallholder groups remain in the 'good wood' business depends on their ability to self-finance and receiving worthwhile benefits from the certification.

While we did not assess the dynamics on the demand side, the case studies suggest that processing industries demand certified timber, particularly those with export orientations (see also Maryudi *et al.*, 2015). That KWML is unable to utilise the good credentials of its timber appears to be linked to its lack of market channels. This mirrors the common picture of smallholder farmers; they lack the capacity to understand market information and have weak linkages with market agents (Roshetko & Yuliyanti, 2002; Holding Anyonge & Roshetko, 2003; Roshetko *et al.*, 2008). In contrast, KWLM and Dipantara's success suggest that innovations and professionalism are useful for smallholder groups for tapping into the potential markets. The group managers actively searched for information regarding potential buyers of certified timber, what timber they specify and when they need the timber, and so on, using the internet and networking. There is also an increasing trend in the use of mobile phones and internet among smallholders that is said to provide avenues for timely and wider delivery of useful market information (Salami *et al.*, 2010). In Ecuadorian Amazon, smallholders obtain contracts with the buyers via phone calls or internet (Mejía & Pacheco, 2014). The use of mobile phones and internet in sub-Saharan Africa is also said to have had a positive impact on the way smallholders reach their customers (Aker, 2009).

## 6 Conclusion

Certification initiatives, both for sustainable forestry and legal verification, are promoted as a market-based policy instrument to promote improved forest management. The idea is that managers will be rewarded when they manage their forests wisely and sustainably. Smallholder tree growing in Indonesia is increasingly recognised as having good forest practices, as it produces a number of economic, ecological and social benefits for the farmers, the broader societies and the environment. Such is the main driver to encourage smallholder farmers to engage in certification to enjoy improved benefits from their tree growing activities. The more benefits smallholders obtain from tree growing, the more incentive for them to improve and sustain their productive base leading to an improved forest condition and environmental services.

There is a growing body of literature indicating growing market potential, particularly in Europe, for certified timber products. A number of companies, particularly do-it-yourself home improvement retailers, are entering certified wood markets as a response to a growing number of consumers willing to pay for forest products from sustainable and socially-responsible sources. Locally, there are emerging market opportunities for some timber industries, mostly for international markets looking for certified timber. However, our study has two major conclusions. First, we conclude that smallholders find it challenging to enter and participate in certified wood markets and to obtain the promised premium prices and remain vulnerable in the good wood business. The obvious costs of certification and the uncertain benefits have become major challenges for smallholders to become incorporated into the markets. While the streamlined procedures were introduced to cut the costs, the costs are still clearly prohibitive for smallholder groups in Indonesia. Expanding the scope of forest operations, i.e. aggregating smallholder forest resources, will not necessarily achieve a better economy of scale as the complexity and diversity of smallholder forests also tends to increase with scale.

Secondly, we also argue that even though the market opportunities are there for the taking if a forest is certified, being certified alone is insufficient to tap the market opportunities. Certified forest enterprises need dedicated managers with adequate entrepreneurial skills to establish networks and contacts with potential buyers and to actively search for information on the log specifications required. They may also need to work closely with processors of certified logs. In many cases, tree growers and their cooperative managers are more familiar with farming activities. In addition, they may not be able to fully focus on the marketing of their timber as much of their time is devoted to their own farming activities. Taking these points into consideration, we advocate the importance of building the capacity of smallholders, particularly the managers and committee members, in entrepreneurship and marketing skills. Smallholder groups may need support that links them with potential buyers. More importantly, they clearly need channels to certified markets. Partnerships with more established forest companies might also be a more viable option, particularly when smallholder farmers and the group managers are not prepared to or unable to self-finance the certification and to deal with the complexities of the timber markets. Established companies are usually more prepared and knowledgeable about promotion, negotiation, financing, payments and risks in the market. They may also bring technical, marketing and management expertise to the smallholders and their groups. The partnership must nonetheless be based on mutual benefits and shared values and risks between the participants. Other incentives of policy instruments might be more effective in increasing the comparative advantages of smallholder tree-growers in becoming more competitive commercially, compared to other plantation development and management strategies. For example, supporting robust and competitive enterprises at the management level by developing programmes for improving smallholders' business skills complemented by better silvicultural practices.

*Acknowledgements*

The project was funded by the Australian Centre for International Agricultural Research (ACIAR) (FST/2012/039). The local partners in Gunungkidul are Universitas Gadjah Mada (UGM) and the Farm Forestry Working Group (*Kelompok Kerja Hutan Rakyat Lestari*).

## References

Aker, J. C. (2009). Mobile Phones, Markets and Firms in Sub-Saharan Africa. *Private Sector & Development*, 4, 10–12.

Anderson, R. C. & Hansen, E. N. (2004). Determining consumer preferences for ecolabeled forest products: an experimental approach. *Journal of Forestry*, 102 (4), 28–32.

Auld, G., Gulbrandsen, L. H. & McDermott, C. L. (2008). Certification Schemes and the Impacts on Forests and Forestry. *Annual Review of Environment and Resources*, 33, 187–211.

Awang, S. A., Wiyono, E. B. & Sadiyo, S. (2007). *Unit Manajemen Hutan Rakyat: Proses Konstruksi Pengetahuan Lokal*. Banyumili Art Network, Yogyakarta.

Bass, S., Thornber, K., Markopoulos, M., Roberts, S. & Grieg-Gran, M. (2001). Certification's impacts on forests, stakeholders and supply chains. A report of the IIED project: Instruments for sustainable private sector forestry. International Institute for Environment and Development, London. Available at: http://pubs.iied.org/pdfs/9013IIED.pdf (last accessed: 26.11.2016).

Brockmann, K. L., Hemmelskamp, J. & Hohmeyer, O. (2012). *Certified tropical timber and consumer behaviour: The impact of a certification scheme for tropical timber from sustainable forest management on German demand*. Physica Verlag, Heidelberg.

Buddhisatyarini, T. (2005). *Ragam Pola Hutan Rakyat di Dlingo, Bantul. In: Awang, S.A. (Ed.). Petani, Ekonomi dan Konservasi: Aspek Penelitian dan Gagasan, Seri Bunga Hutan Rakyat, Pustaka Hutan Rakyat*. Debut Press, Yogyakarta, Indonesia.

Butterfield, R., Hansen, E., Fletcher, R. & Nikinmaa, H. (2005). *Forest Certification and Small Forest Enterprises: Key Trends and Impacts - Benefits and Barriers*. The Forest Trends, Washington DC.

Cai, Z. & Aguilar, F. X. (2013). Consumer stated purchasing preferences and corporate social responsibility in the wood products industry: A conjoint analysis in the US and China. *Ecological Economics*, 95, 118–127.

Cashore, B., Auld, G. & Newsom, D. (2004). *Governing through Markets: Forest certification and the emergence of non-state authority*. Yale University Press, New Haven.

Cashore, B. & Stone, M. W. (2012). Can legality verification rescue global forest governance? Analyzing the potential of public and private policy intersection to ameliorate forest challenges in Southeast Asia. *Forest Policy and Economics*, 18 (5), 13–22.

Durst, P., McKenzie, P. J., Brown, C. L. & Appanah, S. (2006). Challenges facing certification and ecolabelling of forest products in developing countries. *International Forestry Review*, 8 (2), 193–200.

Forest Watch Indonesia (2013). Certification Rift in Indonesia. Available at: http://fwi.or.id/english/publikasi/certification-rift-in-indonesia/ (last accessed: 12.11.2016).

Foundjem-Tita, D., Tchoundjeu, Z., Speelman, S., D'Haese, M., Degrande, A., Asaah, E., van Huylenbroeck, G., van Damme, P. & Ndoye, O. (2013). Policy and Legal Frameworks Governing Trees: Incentives or Disincentives for Smallholder Tree Planting Decisions in Cameroon? *Small-scale Forestry*, 12, 489–505.

FSC (2015). Market Info Pack: An overview of recent trends and current status of Forest Stewardship Council (FSC) certification. Available at: https://ic.fsc.org/preview.2015-fsc-market-info-pack.a-5067.pdf (last accessed: 11.01.2017).

Fujiwara, T., Awang, S. A., Widayanti, W. T., Septiana, R. M., Hyakumura, K. & Sato, N. (2015). Effects of National Community-Based Forest Certification on Forest Management and Timber Marketing: A Case Study of Gunung Kidul, Yogyakarta, Indonesia. *International Forestry Review*, 17 (4), 448–460.

Gulbrandsen, l. (2004). The emergence of non-state authority in forestry. *In:* Bostrom, M., Foellesdal, A., Klintman, M. & Soerensen, M. P. (eds.), *Politics consumerism: its motivations, power and conditions in the Nordic countries and elsewhere*. pp. 320–337, Nordic Council of Ministers, Oslo, Norway.

Hansmann, R., Koellner, T. & Scholz, R. W. (2006). Influence of consumers' socioecological and economic orientations on preferences for wood products with sustainability. *Forest Policy and Economics*, 8, 239–250.

Harada, K. & Wiyono (2014). Certification of a Community-based Forest Enterprise for Improving Institutional Management and Household Income: A Case from Southeast Sulawesi, Indonesia. *Small-scale Forestry*, 13 (1), 47–64.

Hardiyanto, E. B. & Prayitno, T. A. (2007). Present Utilization of Small-Diameter Teak Log from Community Teak Plantations in Java and Eastern Indonesia. Technical Report ITTO PPD 121/06 Rev. 2(I), International Tropical Timber Organization (ITTO), Yokohama, Japan. Available at: http://www.itto.int/files/itto_project_db_input/2592/Technical/PPD121-06%20%20Technical%20Report%20.pdf (last accessed: 26.11.2016).

Hinrichs, A., Muhtaman, D. R. & Irianto, N. (2008). *Forest Certification on Community Land in Indonesia*. Deutsche Gesellschaft für Technische Zusammenarbeit (GTZ) GmbH, Eschborn, Germany.

Holding Anyonge, C. & Roshetko, J. M. (2003). Farm-level timber production: orienting farmers towards the market. *Unasylva*, 54, 48–56.

Irvine, D. (2000). Certification and community forestry: Current trends, challenges and potential. Background paper for the World Bank/WWF Alliance Workshop on Independent Certification, Nov. 9–10, 1999, Washington, D.C., USA.

Jensen, K., Jakus, P. M., English, B. & Menard, J. (2003). Market participation and willingness to pay for environmentally certified products. *Forest Science*, 49 (4), 632–641.

de Jong, W., Cornejo, C., Pacheco, P., Pokorny, B., Stoian, D., Sagogal, C. & Louman, B. (2010). Opportunities and challenges for community forestry: Lessons from tropical America. *In:* Forests and society – Responding to global drivers of change. IUFRO World Series Volume 25, pp. 299–313, IUFRO, Vienna, Austria.

Kandel, P. N. (2007). Effects of forest certification towards sustainable community forestry in Nepal. *Banko Janakari*, 17 (1), 11–16.

Lindstrom, T., Hansen, E. & Juslin, H. (1999). Forest certification: the view from Europe's NIPFs. *Journal of Forestry*, 97 (3), 25–30.

Maryudi, A. (2005). Beberapa kendala bagi sertifikasi hutan rakyat. *Jurnal Hutan Rakyat*, 7 (3), 25–36.

Maryudi, A. (2009). *Forest Certification for Community-Based Forest Management in Indonesia: Does LEI Provide A Credible Option?*. Institute for Global Environmental Strategies, Japan.

Maryudi, A. (2016). Choosing legality verification as a policy instrument to tackle illegal logging in Indonesia. *Forest Policy and Economics*, 68, 99–104. doi: 10.1016/j.forpol.2015.10.010.

Maryudi, A., Nawir, A. A., Permadi, D. B., Purwanto, R. H., Pratiwi, D., Syofi'i, A. & Sumardamto, P. (2015). Complex regulatory frameworks governing private smallholder tree plantations in Gunungkidul District, Indonesia. *Forest Policy and Economics*, 59, 1–6. doi:10.1016/j.forpol.2015.05.010.

Maryudi, A., Nawir, A. A., Sekartaji, D. A., Sumardamto, P., Purwanto, R. H., Sadono, R., Suryanto, P., Soraya, E., Soeprijadi, D., Affianto, A., Rohman, R. & Riyanto, S. (2016). Smallholder farmers' knowledge of regulations governing the sale of timber and supply chains in Gunungkidul District, Indonesia. *Small-scale Forestry*, 1–13. doi:10.1007/s11842-016-9346-x.

Mejía, E. & Pacheco, P. (2014). Forest use and timber markets in the Ecuadorian Amazon. Center for International Forestry Research, Occasional Paper 111. Available at: http://www.cifor.org/publications/pdf_files/OccPapers/OP-111.pdf (last accessed: 26.11.2016).

Michon, G. & de Foresta, H. (1995). The Indonesian agro-forestry model: Forest resource management and biodiversity conservation. *In:* Halladay, P. & Gilmour, D. A. (eds.), *Conserving Biodiversity outside Protected Areas: The roles of traditional agroecosystems*. pp. 90–106, IUCN, Gland, Switzerland.

Ministry of Forestry (2009). Statistik kehutanan Indonesia (Forestry statistics of Indonesia) 2008, Ch. 3. Rehabilitasi lahan dan perhutanan sosial (Land Rehabilitation and Social Forestry). Available at: http://storage.jak-stik.ac.id/ProdukHukum/kehutanan/Statistik_Kehutanan_2008_RLPS.pdf (last accessed: 12.01.2017).

Molnar, A. (2003). *Forest Certification and Communities: Looking forward to the next decade*. The Forest Trends, Washington D.C.

Nawir, A. A. (2013). Commercial community tree-growing inside state forests: an economic perspective from eastern Indonesia. *International Forestry Review*, 15 (2), 200–217.

Nawir, A. A., Muniarti, Rumboko, L., Hiyama, C. & Gumartini, T. (2007). Portraits of rehabilitation projects in Indonesia: impacts and lessons learnt. *In:* Nawir, A. A., Murniati & Rumboko, L. (eds.), *Forest rehabilitation in Indonesia: Where to after three decades?*. pp. 113–175, Center for International Forestry Research (CIFOR), Bogor, Indonesia.

Nussbaum, R., Garforth, M., Scrase, H. & Wenban-Smith, M. (2001). *Getting small forest enterprises into certification: An analysis of current FSC accreditation, certification and standard-setting procedures identifying elements which create constraints for small forest owners*. ProForest, Oxford, England.

Overdevest, C. (2010). Comparing forest certification schemes: the case of ratcheting standards in the forest sector. *Socio-Economic Review*, 8, 47–76.

Owari, T., Juslin, H., Rummukainen, A. & Yoshimura, T. (2006). Strategies, functions and benefits of forest certification in wood products marketing: Perspectives of Finnish suppliers. *Forest Policy and Economics*, 9 (4), 380–391.

Perdana, A., Roshetko, J. M. & Kurniawan, I. (2012). Forces of competition: smallholding teak producers in Indonesia. *International Forestry Review*, 14 (2), 238–248.

Rametsteiner, E. & Simula, M. (2003). Forest certification-an instrument to promote sustainable forest management? *Journal of Environmental Management*, 67, 87–98.

Rickenbach, M. & Overdevest, C. (2006). More than Markets: Assessing Forest Stewardship Council (FSC) Certification as a Policy Tool. *Journal of Forestry*, 104 (3), 143–147.

Rickenbach, M. G. (2002). Forest certification of small ownerships: some practical challenges. *Journal of Forestry*, 100 (6), 43–47.

Rohadi, D., Herawati, T., Padoch, C. & Race, D. (2015). Making timber plantations an attractive business for smallholders. Center for International Forestry Research, Policy Brief No.114. Available at: http://www.cifor.org/publications/pdf_files/infobrief/5515-infobrief.pdf (last accessed: 26.11.2016).

Rohadi, D., Roshetko, J., Perdana, A., Blyth, M., Nuryartono, N., Kusumowardani, N., Pramono, A. A., Widyani, N., Fauzi, A., Sasono, J., Sumardamto, P. & Manalu, P. (2012). Improving economic outcomes for smallholders growing teak in agroforestry systems in Indonesia. Final Report: FST/2005/177. Australian Centre for International Agricultural Research (ACIAR). Available at: http://aciar.gov.au/files/node/14530/fr2012_11_improving_economic_outcomes_for_smallho_12552.pdf (last accessed: 26.11.2016).

Roshetko, J. M., Astho, A., Rohadi, D., Widyani, N., Manurung, G. S., Fauzi, A. & Sumardamto, P. (2012). Smallholder Teak Systems on Java, Indonesia: Income for families, timber for industry. *In:* Meyer, S. R. (ed.), *IUFRO Small-Scale Forestry Conference 2012: Science for Solutions Conference Proceedings*. pp. 162–167, IUFRO, Amherst, USA.

Roshetko, J. M., Snelder, D. J., Lasco, R. D. & van Noordwijk, M. (2008). Future Challenge: A Paradigm Shift in the Forestry Sector. *In:* Snelder, D. J. & Lasco, R. D. (eds.), *Smallholder Tree Growing for Rural Development and Environmental Services: Lessons from Asia*. pp. 453–486, Springer Verlag, Dordrecht, Netherlands.

Roshetko, J. M. & Yuliyanti (2002). Pemasaran untuk hasil-hasil wanatani di tingkat petani. *In:* Roshetko, J. M., Mulawarman, W. J., Santoso & Oka, I. N. (eds.), *Wanatani di Nusa Tenggara*. pp. 101–111, ICRAF and Winrock International, Bogor, Indonesia.

Ruslandi, Klassen, A., Romero, C. & Putz, F. E. (2014). Forest Stewardship Council certification of natural forest management in Indonesia: Required improvements, costs, incentives, and barriers. *In:* Katila, P., Galloway, G., de Jong, W., Pacheco, P. & Mery, G. (eds.), *Forest under Pressure: Local Response to Global Issues*. ch. 15, pp. 255–273, IUFRO, Tampere, Finland.

Sabastian, G., Kanowski, P., Race, D., Williams, E. & Roshetko, J. (2014). Household and farm attributes affecting adoption of smallholder timber management practices by tree growers in Gunungkidul region, Indonesia. *Agroforestry Systems*, 88 (2), 257–268.

Salami, A., Kamara, A. B. & Brixiova, Z. (2010). Smallholder Agriculture in East Africa: Trends, Constraints and Opportunities. African Development Bank, Working Papers Series No 105. Available at: http://www.afdb.org/fileadmin/uploads/afdb/Documents/Publications/WORKING%20105%20%20PDF%20d.pdf (last accessed: 26.11.2016).

Shantiko, B., Purnomo, H. & Irawati, R. H. (2013). *Furniture, timber and forest ecosystem service value chains*. Center for International Forestry Research (CIFOR), Bogor, Indonesia.

Tiesl, M. F., Peavey, S., Newman, S., Buono, J. & Hermann, M. (2002). Consumer reactions to environmental labels for forest products: a preliminary look. *Forest Products Journal*, 52 (1), 44–50.

Triple Line Consulting (2005). Sawn Timber Supply Chain Study. Final Report. International Finance Corporation Program for Eastern Indonesia SME Assistance (IFC-PENSA), Jakarta.

UNECE/FAO (2005). Forest Products Annual Market Review 2004-2005. Available at: http://www.unece.org/forests/fpamr2005.html (last accessed: 12.11.2016).

UNECE/FAO (2013). Forest Products Annual Market Review 2012-2013. Available at: http://www.unece.org/forests/fpamr2013.html (last accessed: 12.11.2016).

UNECE/FAO (2015). Forest Products Annual Market Review 2014-2015. Available at: http://www.unece.org/forests/fpamr2015.html (last accessed: 12.11.2016).

Viana, V. M. (1996). *Certification of Forest Products: Issues and Perspectives*. Island Press, Washington DC.

Wiersum, K. F., Humphries, S. & van Bommel, S. (2013). Certification of community forestry enterprises: experiences with incorporating community forestry in a global system for forest governance. *Small-scale Forestry*, 12 (1), 15–31.

Yovi, E. Y., Bahruni & Nurrochmat, D. R. (2009). Sources of timber and constraints to the timber acquisition of Jepara's small-scale furniture industries. *Jurnal Manajemen Hutan Tropika*, 15 (1), 32–40.

# 3

# Optimising contributions of goat farming to household economic success and food security in three production systems in Ethiopia

Tatek Woldu [a,b,*], André Markemann [a], Christoph Reiber [a], Philipp C. Muth [a], Anne Valle Zárate [a]

[a] *University of Hohenheim, Institute of Agricultural Sciences in the Tropics (Hans-Ruthenberg-Institute), Department of Animal Breeding and Husbandry in the Tropics and Subtropics, Stuttgart, Germany*

[b] *Jimma University College of Agriculture and Veterinary Medicine, Jimma, Ethiopia*

**Abstract**

The study aims to analyse factors affecting contributions of goat farming to household economic success and food security in three goat production systems of Ethiopia. A study was conducted in three districts of Ethiopia representing arid agro-pastoral (AAP), semi-arid agro-pastoral (SAAP) and highland mixed crop-livestock (HMCL) systems involving 180 goat keeping households. Gross margin (GM) and net benefit (NB1 and NB2) were used as indicators of economic success of goat keeping. NB1 includes in-kind benefits of goats (consumption and manure), while NB2 additionally constitutes intangible benefits (insurance and finance). Household dietary diversity score (HDDS) was used as a proxy indicator of food security. GM was significantly affected by an off-take rate and flock size interaction ($P<0.001$). The increment of GM due to increased off-take rate was more prominent for farmers with bigger flocks. Interaction between flock size and production system significantly ($P<0.001$) affected both NB1 and NB2. The increment of NB1 and NB2 by keeping larger flocks was higher in AAP system, due to higher in-kind and intangible benefits of goats in this system. Effect of goat flock size as a predictor of household dietary diversity was not significant ($P>0.05$). Nevertheless, a significant positive correlation ($P<0.05$) was observed between GM from goats and HDDS in AAP system, indicating the indirect role of goat production for food security. The study indicated that extent of utilising tangible and intangible benefits of goats varied among production systems and these differences should be given adequate attention in designing genetic improvement programs.

*Keywords:* dietary diversity, economic success, goats, intangible benefits, off-take rate

## 1 Introduction

In developing countries, huge goat resources are present (Aziz, 2010) and the demand for meat products is strongly increasing (Narrod *et al.*, 2011). Thus, goat farming could play a considerable role in improving the livelihoods of poor African farmers (Peacock, 2005). Ethiopia's estimated goat population was about 25 million in 2013, accounting for 7.2 % and 2.6 % of the African and global goat population, respectively (FAOSTAT, 2015). Among ruminants, goats are less numerous as compared to cattle and sheep in Ethiopia; however, the sheep to goat ratio decreased from 1.29 to 1.06 within the last 20 years (FAOSTAT, 2015). The country is home to genetically diverse goat populations that are widely distributed across all agro-ecologies (Hassen *et al.*, 2012).

* Corresponding author
University of Hohenheim, Department of Animal Breeding and Husbandry in the Tropics and Subtropics, Institute of Agricultural Sciences in the Tropics (Hans-Ruthenberg-Institute), Garbenstr. 17, 70599 Stuttgart, Germany
Email: tatekwbelete@yahoo.com

Despite the huge genetic diversity and valuable contributions of goats to the livelihoods of farmers in rural areas, the sector has been given low research and development attention at global (Aziz, 2010) and national (Solomon *et al.*, 2014) levels. This is mainly due to an inadequate recognition of the contributions goats make to the livelihoods of the poor, resulting in underutilisation of the diverse goat genetic resources (Aziz, 2010). Community-based breeding programs (CBBPs) are considered as a promising tool for livestock genetic improvement under smallholder tropical conditions (Mueller *et al.*, 2015). Presently, a research project is underway to improve goat productivity in Ethiopia and Cameroon by CBBPs (ILRI, 2013). This research paper is part of the recent initiative in Ethiopia.

Improved knowledge on the economic value and roles of goats that influence the overall benefits for smallholders will help in designing optimised breeding programs that consider both, tangible and intangible benefits (Kosgey *et al.*, 2004). It was reported by a number of studies that intangible benefits, such as finance and insurance, comprise a sizable portion of the overall benefits of livestock in different parts of Africa (Ayalew Kebede, 2000; Kosgey *et al.*, 2004; Moll, 2005). Even though considerable attention was given in valuing intangible benefits of small ruminants so far, the economic value of such benefits across production systems along with farmers' strategies to exploit them was not adequately investigated.

The different contributions of goats to smallholder families include their role in improving household food security. A number of studies reported a significant association between dietary diversity and the nutritional status of children in developing countries (Moursi *et al.*, 2008). Likewise, dietary diversity was also reported to be correlated with caloric intake, even though the strength of relationship varies among different studies (Maxwell *et al.*, 2014). The objectives of the present study were to analyse factors affecting contributions of goat farming to household economic success and household dietary diversity as a proxy for food security in three largely differing production systems of Ethiopia.

## 2 Materials and methods

### 2.1 *Description of the study area and production systems*

The study was conducted in three districts of Ethiopia, namely Abergele, Konso and Meta Robi, representing arid agro-pastoral (AAP), semi-arid agropastoral (SAAP) and highland mixed crop-livestock (HMCL) systems, respectively. The AAP system is characterised by a dry and hot climate with annual precipitation ranging from 300 to 496 mm with average daily minimum and maximum temperatures of 21 and 41°C, respectively. Crop farming is practiced around homestead areas with seasonal movements of livestock during feed shortage periods. Abergele goat types are the most predominant goats in this district (Hassen *et al.*, 2012). In the SAAP system, the climate is semi-arid with a daily average minimum and maximum temperature of 12 and 33°C, respectively, while mean annual rainfall ranges from 400 to 1000 mm. Farmers in this system practice agro-pastoralism with some periodic movement of satellite goat flocks. Woyto-Guji goats are the most predominant breeds in the area (Tucho, 2004). The HMCL system is characterised by settled farming with high integration of crop and livestock. The area receives an average annual rainfall of 1100 mm and the daily annual temperature ranges between 15 and 32°C. The climate is conducive for crop farming. The central highland goat breed is widely reared in this system (Tucho, 2004).

### 2.2 *Sampling and data collection*

The study is part of an ongoing goat CBBP (ILRI, 2013) which is being implemented in five districts. For the purpose of this study, three districts and two villages from each district were selected based on diverse agroecologies and production systems, potential of the areas for goat production, and accessibility. In each district, two villages were selected based on advice from key informants from the district's office of Agriculture and Rural Development. Farmers, who owned at least five goats, were identified from the list of farmers in collaboration with development agents and village administrators. Systematic random sampling was used in the last step to select 30 households from the pre-selected farmers, i.e. 60 households per district and a total of 180 households for the study. In addition, in each village a few farmers were put on a waiting list. Three households from Abergele and two from Meta Robi, which were sampled for data collection, but had very few or no goats were replaced by households from the waiting list.

Data were collected between December 2013 and February 2014 by using a semi-structured questionnaire, which captured socio-economic and demographic variables, livestock holdings, income generated and costs incurred by the major agricultural enterprises including livestock, crops and off-farm activities within the last 12 months, number of livestock slaughtered for meat consumption and amount of home-produced and consumed crops.

### 2.3 *Household dietary diversity*

Based on a 24-hour recall method (FAO, 2013), farmers were asked to describe the type of food consumed by members of the household during the previous day. Mixed meals were described by each ingredient. The food items consumed were grouped into 12 food categories including cereals, legumes (pulses and peanut), vegetables, white tubers (potato, sweet potato), fruits (domestic and wild), meat (beef, poultry, sheep and goat), fish, oil and fat, sweets (sugar and honey), milk and milk products, eggs and spices. In each district, two enumerators, who could speak the local language were recruited and trained to assist during data collection.

### 2.4 *Income and costs*

The income from goat production comprises cash revenues (CR) from the sale of kids, bucks, does and castrated goats. CR from the sale of other livestock species was also calculated. Sales of dairy products (mainly butter) were also considered for estimating CR from cattle, which was not the case for goat milk, because farmers in the study area did not sell or process goat milk. Sheep milk was neither consumed nor marketed in any of the study areas.

In-kind benefits included goat meat, milk and manure. The monetary value of goat meat consumption was estimated by multiplying the number of goats slaughtered per year with the average yearly price of goats during the study period. Average lactation milk off-take was estimated based on average milk off-take (346 ml/day) and lactation length (12 weeks) of Abergele goats (Alubel, 2015). The milk off-take was multiplied by the average price of milk during the study period. Manure was valued by estimating the daily dry matter faecal output of goats by using a regression formula developed by Fernández-Rivera *et al.* (1995) cited by Ayalew Kebede (2000). The average nitrogen and phosphorus contents of the goats' faecal dry matter reported by Schlecht *et al.* (1997) and Somda *et al.* (1995) cited by Ayalew Kebede (2000) were used to calculate the annual nitrogen and phosphorus outputs. The unit price of nitrogen and phosphorus was derived from the average price of diammonium phosphate (DAP) and urea during the study period.

Intangible socio-economic benefits of goat production, i.e. financial (F) and insurance (I) functions of goats, were estimated. The financial benefit of a goat flock per household was valued by the following equation:

$$F_i = \gamma P_i \tag{1}$$

Where $F_i$ is the financial benefit of a goat flock in the $i^{th}$ household, $\gamma$ is the opportunity cost of alternative financial sources, such as costs to obtain formal or informal credits (Moll, 2005), and $P_i$ is the monetary value of the goat flock (number of goats owned × market price of goats) of the $i^{th}$ household in the year 2012. Interest rate of micro-finances in the study area (0.10) was used to estimate $\gamma$.

The insurance value of goats was estimated by the equation suggested by Moll (2005):

$$I_i = \alpha (P_i + P_i^*)/2 \tag{2}$$

Where Ii is the insurance value of the goat flock of the $i^{th}$ household, $P_i$ and $P_i^*$ are the average monetary values of the goat flock of the $i^{th}$ household in the years 2012 and 2011, respectively, and $\alpha$ is the insurance function. The size of $\alpha$ is usually determined based on existing alternative insurance systems. Guesstimates criteria based on climatic condition as suggested by Moll (2005) were implemented. Considering the annual rainfall and temperature in the study sites, insurance factors of 0.05, 0.075 and 0.1 were assigned for the HMCL, SAAP and AAP systems, respectively.

The major variable costs of goat production included veterinary costs, feed and hired labour costs for herding. Veterinary costs comprised costs for vaccination, deworming and medication, while feed costs included expenses for purchased feedstuffs used for supplementation. Since browsing is the major source of feed for goats in the study area, costs for supplementation from own sources were ignored. Hired labour cost included the wage payment and/or the monetary value of in-kind payments given for the herders. Fixed costs such as depreciation of housing and machineries were not considered in the study, because goats are mainly housed in simple fenced barns, caves (e.g. AAP system) or in the main house together with the family members in some cases.

The economic parameters were calculated by using the following equations (in ETB household$^{-1}$ year$^{-1}$):

$$GM = CR - VC \tag{3}$$

$$NB1 = (CR + BC + BM) - VC \tag{4}$$

$$NB2 = (CR + BC + BM + F + I) - VC \tag{5}$$

Where: *GM* is the gross margin (not including in-kind and intangible benefits of goats), *CR* are cash revenues,

*VC* are variable costs, *NB1* is the net benefit including in-kind benefits of goats, *BC* is the benefit of consuming goat products, *BM* is the benefit of using manure, *NB2* is the net benefit including in-kind and intangible benefits of goats, *F* is the financial function, and *I* is the insurance function. All the economic parameters are given in ETB (Ethiopian birr) whereby 1 USD $\approx$ 19 ETB in 2012.

### 2.5 Data analysis

The contribution of goat farming to household income was assessed by the proportion of gross margin (GM) generated from goats to all other household income sources. Goat flock sizes (TLU), off-take rates, costs and economic efficiency parameters were not normally distributed. Hence, the Wilcoxon-Mann-Whitney test was employed for detecting significant differences between production systems. The P-values were estimated by using Monte-Carlo simulation methods due to the presence of tied observations in the data set.

A linear mixed model with villages as random effect was used to analyse the effects of production system, use of veterinary services, supplementation of goats before selling, flock size, off-take rate (percentage of total sales of goats per annual average flock size) and fecundity (total number of kids born per total number of mating does) (Rosa *et al.*, 2007) on the economic success of goat keeping. At first, fixed effects and all possible two-way interactions between factors were screened by backward selection procedure of GLMSELECT procedure in SAS (SAS Institute Inc., 2011), whereas factors showing minimum contribution to model variation were removed based on Schwarz Bayesian criteria (SBC). Finally, all factors involved in significant interactions and the random village effect entered the linear mixed model. The normality of residuals and the homogeneity of error variance were tested. The final reduced models employed were the following:

$$y_{ijk} = \beta_0 + S_i + \beta_1 t_{ijk} + \beta_2 f_{ijk} + \beta_3 t_{ijk} f_{ijk} + \beta_{4i} f_{ijk} + z_j + \varepsilon_{ijk} \quad (6)$$

With $y_{ijk}$ = *NB1* and *NB2* of the $k^{th}$ household, $\beta_0$ = intercept, $\beta_1$–$\beta_4$ = regression coefficients, $S_i$ = effect of production system ($i$ = AAP, SAAP, HMCL), $t_{ijk}$ = off-take rate treated as a continuous variable, $f_{ijk}$ = flock size treated as a continuous variable, $t_{ijk} f_{ijk}$ is interaction between offtake rate and flock size, $\beta_{4i} f_{ijk}$ interaction between $i^{th}$ production system and flock size, $z_j$ = the random effect of village, ( $j$ = 1, 2, 3, 4, 5, 6 ) and $\varepsilon_{ijk}$ = random error term.

$$y_{jk} = \beta_0 + \beta_1 t_{jk} + \beta_2 f_{jk} + \beta_3 t_{jk} f_{jk} + z_j + \varepsilon_{jk} \quad (7)$$

$y_{jk}$ = *GM* for the $k^{th}$ household and the variables as previously explained.

The food categories consumed by the household were summarised into terciles of lower (0–3), medium (4–5) and higher (6–7) diversity, following the procedure suggested by Swindale & Paula (2006). An ordered logit model was fitted to analyse effects of socio-economic variables to predict terciles of households' dietary diversity.

$$\begin{aligned} y^*_{ij} = \; & \beta_0 + \beta_1 x_{1ij} + \beta_2 x_{2ij} + \beta_3 x_{3ij} + \\ & \beta_4 x_{4ij} + \beta_5 x_{5ij} + \beta_6 x_{6ij} + \beta_7 x_{7ij} \end{aligned} \quad (8)$$

Where $y^*_{ij}$ = latent dietary diversity terciles of the $j^{th}$ household, $\beta_0$ = intercept, $\beta_1$–$\beta_7$ = coefficients of regression, $x_{1ij}$ = production system of the $j^{th}$ household ($i=0$ for AAP, $i=1$ for SAAP and $i=2$ for HMCL), $x_{2ij}$ = gender of the $j^{th}$ household head ($i=1$ for male, and 0 for female ), $x_{3ij}$ = literacy of the $j^{th}$ household head ($i=1$ for literate, and 0 for illiterate), $x_{4ij}$ = family size of the $j^{th}$ household ($i=0$ for $\leq 7$, and 1 for $> 7$), $x_{5ij}$ = cultivated land of the $j^{th}$ household ($i=0$ for $< 1$ ha, $i=1$ for 1–2 ha and $i=2$ for $> 2$ ha), $x_{6ij}$ = livestock holding in tropical livestock units (TLU)[1] of the $j^{th}$ household ($i$ = 0 for $\leq 9$ and 1 for $> 9$), $x_{7ij}$ = goat (TLU) holding of the $j^{th}$ household ($i=0$ for $\leq 1.7$ and 1 for $> 1.7$). The Pearson's correlation coefficient was used to determine the association between income from goat production and household dietary diversity in the three production systems. All analyses were carried out using SAS version 9.3 (SAS Institute Inc., 2011).

## 3 Results

### 3.1 Household characteristics

Goat owners in the study area indicated that 91 % of the households were male-headed with a mean household size of 6.9 (2 to 14) persons. The literacy rate among the household heads was 23.7 % and similar across production systems. The average TLU owned per household was 7.3, of which small ruminants accounted for 32.5 %. Goats accounted for 23.9 % of the total TLU and 74.2 % of the total small ruminants. In the AAP system, small ruminants represented the majority of the total TLU (48.9 %) followed by cattle (39.2 %). In contrast, the proportion of cattle was higher than small ruminants in both, HMCL (72.0 % vs 21.5 %) and SAAP (68.0 % vs 26.0 %) systems. The average goat flock

[1] Conversion factors used were 0.7, 0.5, and 0.1 for cattle, donkey and small ruminants, respectively (Jahnke, 1982)

size per household was significantly (P<0.001) different among production systems. It was highest in the AAP (27.3), followed by the SAAP (16.5) and HMCL (8.6) systems.

### *3.2 Contribution of goats to household economy*

Cattle provided 44.8 % of household GM, representing the biggest contributor, while goats contributed 23.2 % and 30.9 % to the total GM and livestock GM of the surveyed households, respectively (Figure 1). On average, goats provided a 3.4 and 1.6 times higher GM in the AAP system than sheep and cattle, respectively. However, the contribution of goats to household GM was 2.5 and 3.5 times lower than for cattle in the HMCL and SAAP systems, respectively. It is worth noting that goats contributed more than sheep to household GM in all production systems.

The estimated monetary values of goat benefits to the households in each production system are presented in Table 1. The highest benefit from keeping goats in the surveyed households were from live sales of goats, followed by intangible benefits, manure and milk consumption. Economic benefits from goat meat consumption were the lowest in all production systems. Only 12 % of the surveyed households slaughtered goats at least once per year, mainly as a sacrifice during holidays and social events such as weddings and remembrance days. Goat milk was found to be an important commodity for home consumption only in the AAP system, contributing 20.3 % to the total value of goat benefits, whereas in the SAAP and HMCL systems, goat milk was neither consumed nor marketed.

The goat marketing strategies of farmers differed across production systems (Figure 2). In the HMCL system, goat kids of less than one year were sold most frequently (54.2 %), followed by mature males (30.5 %), while does (10.2 %) and castrated goats (5.1 %) had a lower share of sales. In contrast, almost an equal proportion of kids, bucks and does were sold in the AAP system. The average annual off-take rate for live sale of goats was significantly (P<0.01) higher in HMCL system (21.5 %) than AAP (11.7 %) and SAAP (10.0 %) system, while no significant difference were detected between the AAP and SAAP systems.

Costs and economic efficiency of goat keeping across the production systems is presented in Table 2. The total variable costs varied significantly between production systems (P<0.05). Veterinary expenses accounted for the biggest share of total variable costs in the SAAP (68.5 %) and HMCL (71.7 %) systems, whereas it was significantly lower in the AAP (13.7 %) system. Hired labour costs accounted for the biggest share (68.1 %) of the total variable costs in the AAP system. Feed costs were not significantly different among production systems. Only 18.0 % of all farmers purchased additional supplements for goats. The major feedstuffs purchased were crop residues, mainly used for the fattening of goats. On average, 5.0 %, 16.7 % and 25.0 % of the farmers in the HMCL, AAP and SAAP system, respectively, had a negative GM, while a positive NB1 and NB2 was obtained for all of the surveyed farmers.

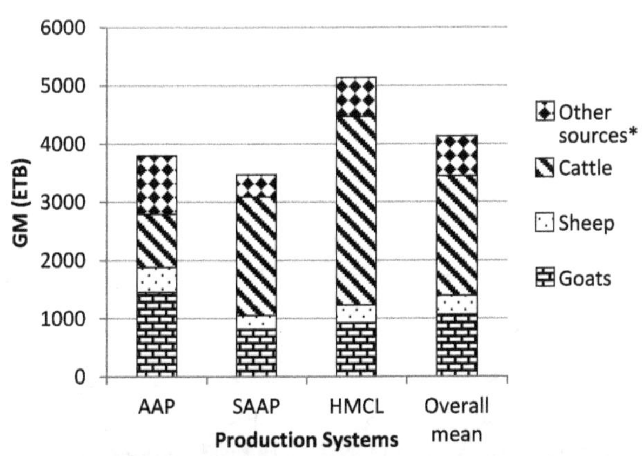


**Fig. 1:** *Contribution of livestock to the household gross margin (GM) in arid agro-pastoral (AAP), semi-arid agro-pastoral (SAAP) and highland mixed crop-livestock (HMCL) systems of Ethiopia. ETB= Ethiopian birr, 1 USD ≈ 19 ETB in 2012. Other sources include income from sale of poultry and honey bee products.*

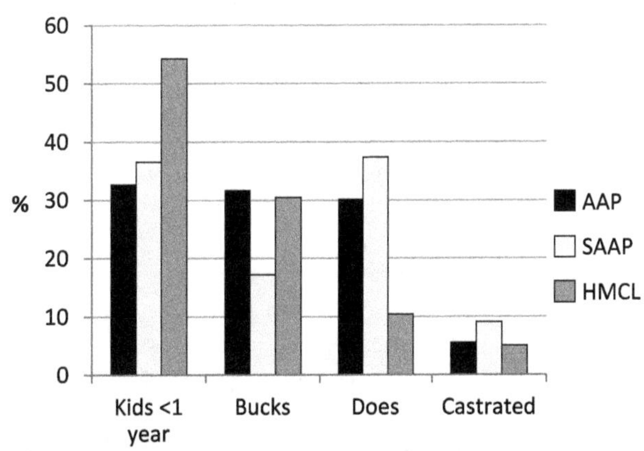


**Fig. 2:** *Proportion of goat types sold within one year in arid agro-pastoral (AAP), semi-arid agro-pastoral (SAAP) and highland mixed crop livestock (HMCL) systems in Ethiopia.*

**Table 1:** *Estimated goat benefits (GV) from live sales, meat and milk consumption, manure and intangible functions to the households in three production systems of Ethiopia in the year 2012*

| *Benefits (GV)* | *Production systems* | | | | | |
|---|---|---|---|---|---|---|
| | *AAP* | | *SAAP* | | *HMCL* | |
| | *Value* (ETB) | *% of total* | *Value* (ETB) | *% of total* | *Value* (ETB) | *% of total* |
| Live sales | 1645.0 | 32.4 | 920.2 | 38.4 | 1014.7 | 57.8 |
| Milk | 1029.1 | 20.3 | 0.0 | 0.0 | 0.0 | 0.0 |
| Meat | 200.8 | 4.0 | 192.0 | 8.0 | 176.9 | 10.0 |
| Manure | 599.6 | 11.8 | 368.4 | 15.4 | 166.5 | 9.5 |
| Financial | 1064.2 | 21.0 | 608.0 | 25.4 | 264.5 | 15.1 |
| Insurance | 532.1 | 10.4 | 305.0 | 12.7 | 132.2 | 7.5 |
| Total | 5070.8 | | 2393.6 | | 1754.8 | |

AAP=arid agro-pastoral, SAAP=semi-arid agro-pastoral, HMCL=highland mixed crop livestock, GV=gross value. ETB= Ethiopian birr, 1 USD ≈ 19 ETB in 2012.

**Table 2:** *Variable costs and economic efficiency of goat rearing in three production systems of Ethiopia in the year 2012*

| *Parameters* (ETB) | *Production systems* | | | | | | *P-value* * |
|---|---|---|---|---|---|---|---|
| | *AAP* | | *SAAP* | | *HMCL* | | |
| | *Mean* | *Median* | *Mean* | *Median* | *Mean* | *Median* | |
| *Variable Costs* | | | | | | | |
| Feed costs | 34.2 | $0.0^{a}$ | 18.4 | $0.0^{a}$ | 12.1 | $0.0^{a}$ | 0.17 |
| Veterinary costs | 25.9 | $0.0^{a}$ | 71.4 | $35.0^{b}$ | 61.6 | $6.0^{a}$ | < 0.01 |
| Hired labour costs | 128.5 | $0.0^{a}$ | 14.5 | $0.0^{a}$ | 12.2 | $0.0^{a}$ | 0.09 |
| Total variable costs | 188.6 | $27.0^{b}$ | 104.3 | $48.0^{b}$ | 85.9 | $6.0^{a}$ | 0.05 |
| *Economic Efficiency* | | | | | | | |
| Total variable costs/goat | 8.4 | $1.5^{a}$ | 7.4 | $3.3^{b}$ | 9.8 | $1.4^{ab}$ | 0.04 |
| GM/goat | 71.1 | $61.2^{a}$ | 68.6 | $34.8^{a}$ | 180.9 | $84.9^{b}$ | 0.01 |
| NB1/goat | 149.0 | $132.4^{b}$ | 109.3 | $73.2^{a}$ | 233.7 | $130.0^{b}$ | < 0.01 |
| NB2/goat | 208.2 | $188.1^{b}$ | 163.9 | $130.1^{a}$ | 277.9 | $192.1^{b}$ | < 0.01 |

AAP=arid agro-pastoral, SAAP=semi-arid agro-pastoral, HMCL=high land mixed crop-livestock, GM= Gross margin, NB1= comprise GM and in-kind benefits NB2= comprise GM, in-kind and intangible socio-economic benefits,
[abc] Medians with different superscripts within a row differ significantly (P<0.05),
* Estimated by Monte Carlo simulation method, ETB= Ethiopian birr, 1 USD ≈19 ETB in 2012.

### *3.3 Factors affecting economic success of goat keeping*

The interaction between off-take rate and flock size significantly affected GM and revealed a positive regression coefficient (Table 3). Thus, the increase in GM due to an increase in flock size depended on off-take rates and vice versa. For instance, the increment in GM through increasing the flock size was more pronounced for farmers, who had off-take rates >12% than compared to those with lower off-take rates. The interaction between production system and flock size significantly affected both NB1 and NB2 (Table 3). For instance, in the AAP system, an increase in flock size by only one head caused a rise in NB1 and NB2 by 45 and 95 ETB, respectively, while in the HMCL system the increment was only 4 and 56 ETB, respectively. Moreover, increasing flock size by one head in the AAP system resulted in a 2.5 and 1.3 times higher NB1 and NB2, respectively, than in the SAAP system. As illustrated in Figure 3, the rise in NB2 with increased flock sizes followed a different pattern among production systems. In the AAP system, NB2 continuously increased nearly up to a flock size of 50 heads, while the curve started to flatten thereafter. In contrast, the NB2 curve started to flatten at smaller flock sizes in the other production systems (Figure 3).

### *3.4 Food security contribution of goats*

The diets of the surveyed household members were composed of cereals, spices, grain legumes and vegetables. Consumption of animal products in the study areas was low and constituted only a small fraction of the diet (Figure 4). Milk, meat and egg products were only consumed by 21.0 %, 10.0 % and 3.0 % of the household members, respectively. When considering the pro-

**Table 3:** *Factors affecting gross margin (GM) and net benefits (NB1 and NB2) of goat farms in the year 2012*

| *Parameters* | *Coefficient (β)* | *SE* | *P-value* |
|---|---|---|---|
| *GM* | | | |
| Intercept | 8.55 | 66.09 | 0.897 |
| Flock size | -13.09 | 2.71 | < 0.001 |
| Off-take | 10.07 | 2.92 | 0.001 |
| Off-take*Flock size | 6.26 | 0.27 | < 0.001 |
| *NB1* | | | |
| Intercept | 72.91 | 224.81 | 0.767 |
| Production systems | | | |
| AAP | 170.41 | 309.66 | 0.523 |
| SAAP | -28.98 | 306.69 | 0.925 |
| HMCL | Reference | | |
| Flock size | 4.45 | 7.33 | 0.545 |
| Flock*Production system | | | |
| Flock size*AAP | 40.59 | 7.47 | < 0.001 |
| Flock size*SAAP | 13.44 | 8.18 | 0.103 |
| Flock size*HMCL | Reference | | |
| Off-take rate | 11.52 | 3.52 | 0.001 |
| Off-take rate *Flock size | 6.46 | 0.32 | < 0.001 |
| *NB2* | | | |
| Intercept | -39.65 | 164.35 | 0.825 |
| Production systems | | | |
| AAP | 315.27 | 219.09 | 0.152 |
| SAAP | 64.67 | 214.35 | 0.763 |
| HMCL | Reference | | |
| Flock size | 56.38 | 7.85 | < 0.001 |
| Flock size*Production system | | | |
| Flock size*AAP | 38.64 | 7.98 | < 0.001 |
| Flock size*SAAP | 14.59 | 8.74 | 0.096 |
| Flock size*HMCL | Reference | | |
| Off-take rate | 13.02 | 3.74 | 0.001 |
| Flocks*off-take rate | 6.68 | 0.34 | < 0.001 |

AAP=arid agro-pastoral, SAAP=semi-arid agro-pastoral, HMCL=high land mixed crop-livestock, GM= gross margin, NB1= includes GM and in-kind benefits NB2= includes GM, in-kind and intangible socio-economic benefits.

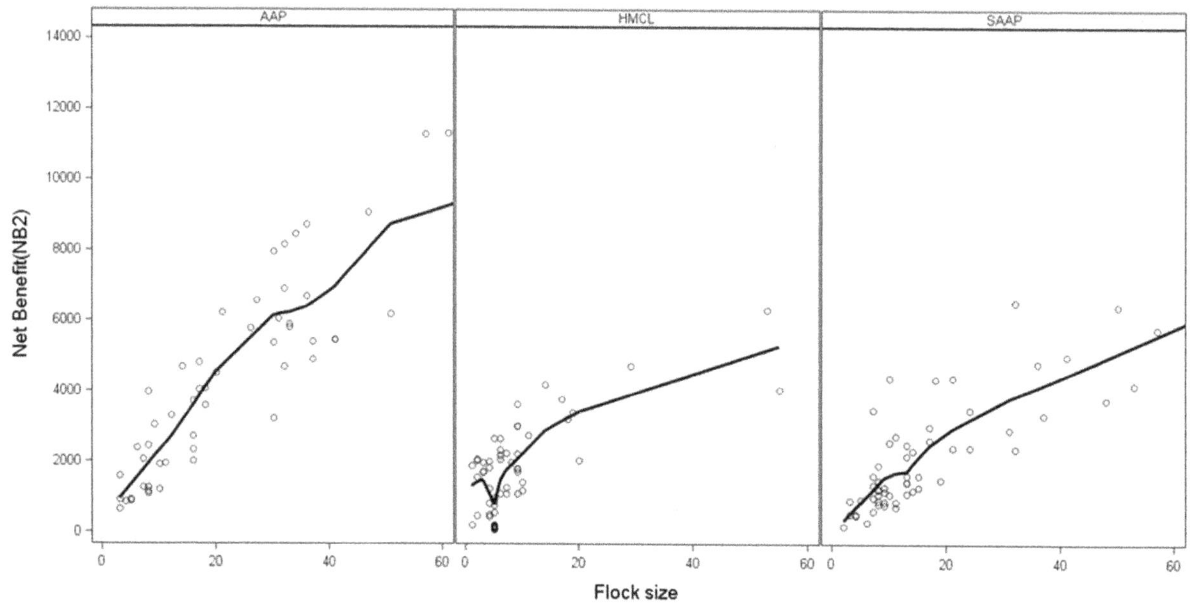


**Fig. 3:** *Trends in net benefits (including in-kind and intangible benefits of goats, NB2) with increasing flock sizes in arid agro-pastoral (AAP), semi-arid agro-pastoral (SAAP) and highland mixed crop-livestock (HMCL) systems of Ethiopia.*

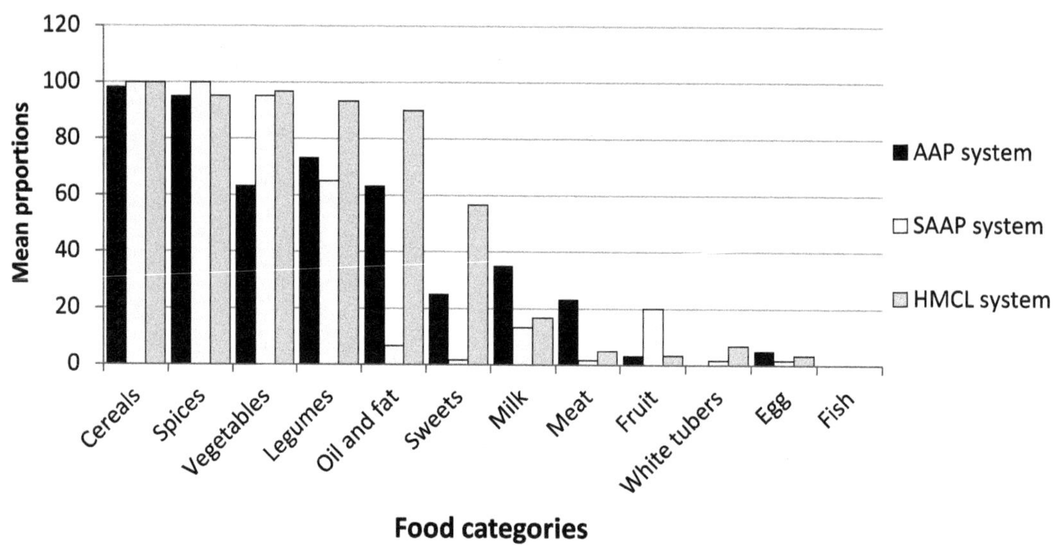


**Fig. 4:** *Proportion of food categories consumed by household members in arid agro-pastoral (AAP), semi-arid agro-pastoral (SAAP) and highland mixed crop-livestock (HMCL) systems of Ethiopia.*

duction systems separately, milk consumption by household members was higher in the AAP (35.0 %) as compared to the SAAP (13.6 %) and HMCL (16.7 %) systems. About 25.0 % of the total households, who consumed milk in the AAP system, reported that the source of milk was from goats, while cow's milk was the sole source of milk in SAAP and HMCL systems. Goat milk in the AAP region was consumed mainly by children, who are responsible for herding the goats. Only 13.0 % of the total households, who consumed meat as part of their diet, used their own goats as a source of meat.

The average dietary diversity score (DDS) of the surveyed households was 4.9 (Range: 2 to 8). The highest average diversity score was 5.7 in the HMCL system, followed by 4.9 in the AAP and 4.1 in the SAAP system. The ordered logit analysis showed that production system and gender of household head significantly affected household dietary diversity, while literacy, family size, livestock holding, area of cultivated land and goat flock holding were not significant (Table 4). Households in the HMCL system had a six times higher chance of being in the upper DDS terciles as compared to the

**Table 4:** *Effect of socio-economic characteristics on household dietary diversity score (HDDS)*

| *Parameters* | *Lower HDDS* (%) | *Medium HDDS* (%) | *Upper HDDS* (%) | *Odds ratio (P-value)* |
|---|---|---|---|---|
| *Production systems* | | | | |
| AAP | 13.3 | 6.7 | 13.3 | |
| SAAP | 24.4 | 4.4 | 4.4 | 0.2 (0.00) |
| HMCL | 2.2 | 10.0 | 21.1 | 6.1 (0.00) |
| *Gender* | | | | |
| Female | 7.2 | 1.1 | 1.1 | |
| Male | 32.7 | 20.0 | 37.8 | 5.0 (0.01) |
| *Literacy* | | | | |
| Illiterate | 33.9 | 15.0 | 29.4 | |
| Literate | 28.2 | 28.2 | 43.6 | 1.2 (0.61) |
| *Family size* | | | | |
| ≤ 7 | 27.2 | 13.3 | 22.2 | |
| > 7 | 12.7 | 7.8 | 16.6 | 1.7 (0.13) |
| *Cultivated land* | | | | |
| < 1 ha | 13.4 | 2.2 | 7.8 | |
| 1–2 ha | 10.6 | 8.4 | 16.7 | 1.2 (0.71) |
| > 2 ha | 16.2 | 10.1 | 14.5 | 0.6 (0.33) |
| *Livestock TLU* | | | | |
| ≤ 9 | 26.1 | 13.3 | 26.1 | |
| > 9 | 13.8 | 7.8 | 12.8 | 2.0 (0.50) |
| *Goat TLU* | | | | |
| ≤ 1.7 | 27.7 | 11.7 | 26.1 | |
| > 1.7 | 12.2 | 9.4 | 12.8 | 1.3 (0.55) |

HDDS=Household dietary diversity score, TLU= Tropical livestock unit, conversion factor of 0.7, 0.5, and 0.1 for cattle, donkey and small ruminants, respectively (Jahnke, 1982).

**Table 5:** *Pearson correlation coefficient between economic success of goat keeping and HDDS in three production systems of Ethiopia*

| | *Production systems* | | | | | |
|---|---|---|---|---|---|---|
| | *AAP* | | *SAAP* | | *HMCL* | |
| | *HDDS* | *P-value* | *HDDS* | *P-value* | *HDDS* | *P-value* |
| GM | 0.26 | 0.04 | –0.03 | 0.82 | –0.27 | 0.04 |
| NB1 | 0.33 | 0.01 | 0.02 | 0.86 | –0.26 | 0.05 |
| NB2 | 0.32 | 0.01 | 0.05 | 0.69 | –0.24 | 0.06 |

AAP = arid agro-pastoral, SAAP = semi-arid agro-pastoral, HMCL = highland mixed crop-livestock, HDDS = Household dietary diversity score, GM = gross margin, NB1 = includes GM and in-kind benefits NB2 = includes GM, in-kind and intangible socio-economic benefits.

AAP system. Male-headed households had five times higher chances of consuming more diversified diets than female-headed households. A significant positive correlation was detected between HDDS and GM, NB1 and NB2 from goats in the AAP system (Table 5), while correlations were either not significant (P>0.05), or negative in the SAAP and HMCL systems, respectively.

## 4 Discussion

Net benefits from goat production were positive for almost all farmers in the present study, which was mainly due to low variable costs. The net benefit reported in this study would probably be slightly reduced by inclusion of family labour and fixed costs. The lower proportion of feed costs and the relatively higher proportion of veterinary costs observed in the HMCL system is in agreement with Legesse *et al.* (2010) who reported that veterinary costs accounted for a great share (60 %) of small ruminant production under similar production conditions. In contrast, reports from Kenya (Ogola *et al.*, 2010) and Jordan (Al-Khaza'leh *et al.*, 2015) stated feed costs as major expenses of smallholder goat production. Moreover, the high proportion of veterinary expenses observed in the SAAP and HMCL systems were in agreement with Netsanet (2014) who reported that diseases such as contagious caprine pleuropneumonia (CCPP), trypanosomiasis, internal and external parasites are the major constraints of goat production in the same study areas. The higher economic efficiency in terms of GM per goat observed in the HMCL system is probably due to lower total variable costs per goat and better market accessibility and subsequently higher selling prices of goats in this system as compared to the other systems.

The higher goat off-take rate observed in the highland areas than in agro-pastoral production systems could be a reason for absence of significant differences in GM among production systems despite the differences in flock size. Moreover, increasing flock size at a low off-take rate did hardly influence GM, mainly due to high VC to maintain larger flock sizes. Still, farmers in agro-pastoral systems (AAP and SAAP) continued to keep larger flock sizes at low off-take rates, deliberately foregoing economic gain in terms of GM, even though adequate goat markets are accessible. Kosgey *et al.* (2004) also argued that pastoralists in tropical environments continue to build larger flock sizes despite the net financial losses.

On the contrary, when in-kind and intangible benefits are considered in the evaluation of economic success, farmers in the AAP system attained an increased NB1 and NB2 by keeping larger flock size mainly due to utilisation of more products from goats, such as milk, as well as the higher insurance and financial benefits of goats in this production system. This implies that intangible benefits of goats are effectively exploited in AAP and SAAP systems through keeping larger flock sizes. Barrett *et al.* (2004) also observed that pastoralists in northern Kenya and southern Ethiopia keep larger flock sizes for sociocultural reasons and to reduce risks during drought periods rather than increasing off-take rates. Lack of responsiveness of goat sales to changes in price was also reported in Botswana (Seleka, 2001). In contrast, the relatively higher off-take rate and the tendency of farmers to sell more growing kids in the HMCL system indicates that the major purpose of keeping goats in this system is generation of cash income through increased commercial off-take. In the AAP system, where in-kind and intangible benefits of goats are highly valued, benefits in goat production could therefore be optimised by the incorporation of adaptive traits, such as fertility and disease resistance, in goat breeding objectives. However, a thorough investigation is required in cost and benefits of a higher disease resistance (Bishop & Morris, 2007). Whereas, targeting reproduction traits such as improving fertility rate could be a better option in the HMCL system. Although, disease resistance/tolerance abilities of goat breeds in the investigated systems were not yet studied, the higher twining rate (46.9 %) of central highland goats in HMCL system (Netsanet, 2014) than Abergele (4.0 %) and Woyito Guji (15.1 %) goats (Alubel, 2015; Netsanet, 2014) depicts the potential of selecting central highland goats for improved reproductive efficiency.

The average HDDS observed in this study (4.9) is similar to the dietary diversity score of 4.6 reported by Mersha (2014) for mixed crop-livestock systems of Ethiopia, but higher than the average dietary diversity score of 2.7 for Borana pastoral communities (Megersa *et al.*, 2014). This variation in dietary diversity is mainly due to the dependence of pastoralists on cereals and milk as the main source of their diet (Villa *et al.*, 2011; Megersa *et al.*, 2014), while additional foodstuffs including legume pulses, vegetables, oil and fats are consumed in the mixed crop-livestock systems. The higher probability of households consuming diversified food diets in the HMCL system than in the other systems is mainly due to better access of the households to diverse foodstuffs and a higher GM from agricultural activities (Figure 1).

Contrary to other findings (Demeke *et al.*, 2011; Megersa *et al.*, 2014) the number of livestock owned in general and goats in particular were not determinant factors of household dietary diversity, this is probably due to a limited direct contribution of livestock products to food diets in the study area (Figure 4). Nevertheless, the significant positive correlation (P<0.05) between GM and HDDS in the AAP system could indicate that cash income generated from goat sales is used to purchase other foodstuffs to diversify diets. This point to an indirect function of goat keeping to possibly increasing dietary diversity and thus, household food security. The negative correlation between income from goats and HDDS in the HMCL system could partly be explained by the observation that goats played a less important role in determining HDDS of farmers in this system, contributing only 9.0 % to the total GM (Figure 1). Furthermore, farmers owning a higher number of goats were relatively poorer, because better-off farmers kept more cattle and depended on crop production as a major source of household income.

The cultural habit of consuming goat milk and its exclusive use for nourishing children and the elderly in the AAP system indicates the potential of improving nutritional status of children by improving goat milk production through improved management of the available feed resources and genetic improvement of goats for milk production. In contrast, consumption of goat milk is considered as a cultural taboo in the HMCL system. This implies that goat traits to be included in defining breeding objectives should also consider the culture and norms of the society. The higher dietary diversity of male-headed households compared to female-headed households could be an indicator of gender to be an important predictor of food security. This is mainly because female headed households are mostly single households; as a result the endowment with household family labour is severely affected.

## 5 Conclusions

The farmers' strategies to utilise tangible and intangible benefits of goats were found to be different among production systems. Thus, during the design and implementation of goat genetic improvement programs, differences in marketing strategies of farmers across production systems, as well as their priorities in utilising tangible and intangible benefits should be taken into consideration. Intangible benefits of goats should be considered in defining goat breeding objectives in agro-pastoral systems, while in mixed crop-livestock systems more attention should be given improving reproductive efficiency to increase the number of marketable goats and optimise benefits from goat farming. Since the current profitability of goat keeping by smallholders relies on low variable costs; a cost-benefit analysis would be suitable which considers the cost and benefits of any intervention.

### Acknowledgements

The research work has been financially supported by the Dr. Hermann Eiselen PhD Grant from the Foundation *fiat panis* and the International Livestock Research Institute (ILRI), through the ILRI-BecA project "Harnessing genetic diversity for improved goat productivity." This publication is an output of the scholarship from the Food Security Center, University of Hohenheim, which is supported by the German Academic Exchange Service (DAAD) and German Federal Ministry for Economic Cooperation and Development (BMZ). This paper also benefited from the insightful comments of Dr. Haile, A., Dr. Rischkowsky, B., and Dr. Dessie T. and we are grateful for the comments and suggestions received.

### References

Al-Khaza'leh, J., Reiber, C., Al Baqain, R. & Valle Zárate, A. (2015). A comparative economic analysis of goat production systems in Jordan with an emphasis on water use. *Livestock Research for Rural Development*, 27 (5). URL `http://www.lrrd.org/lrrd27/5/khaz27081.html` (last accessed: 5 June 2015).

Alubel, A. (2015). *On-farm phenotypic characterization and performance evaluation of Abergelle and Central highland goats as an input for designing community based breeding programs*. Master's thesis, Haramay University, Ethiopia.

Ayalew Kebede, W. (2000). *Do Small holder farmers benefit more from crossbred (Somali X Anglo-Nubian) than from indigenous goats?*. Ph.D. thesis, Georg-August University of Göttingen, Cuvillier Verlag, Göttingen.

Aziz, A. M. (2010). Present status of the world goat populations and their productivity. *Lohmann Information*, 45 (2), 42–52.

Barrett, C. B., Bellemare, M. F. & Osterloh, S. M. (2004). Household-level livestock marketing behaviour among Northern Kenyan and Southern Ethiopian pastoralists. Cornell University, Available at SSRN.

Bishop, S. C. & Morris, C. A. (2007). Genetics of disease resistance in sheep and goats. *Small Ruminant Research*, 70, 48–59.

Demeke, A. B., Keil, A. & Zeller, M. (2011). Using panel data to estimate the effect of rainfall shocks on smallholders food security and vulnerability in rural Ethiopia. *Climate Change*, 108, 185–206.

FAO (2013). Guidelines for measuring household and individual dietary diversity. Nutrition and consumer protection division, Food and agriculture organization of the United Nations (FAO). Rome, Italy. , pp: 53.

FAOSTAT (2015). Food and agricultural organization of the United Nations, statistical division. URL `http://faostat3.fao.org/browse/Q/QA/E` (last accessed: 7 March 2015).

Fernández-Rivera, S., Williams, T. O., Hiernaux, P. & Powell, J. M. (1995). Fecal excretion by ruminants and manure availability for crop production in semi-arid West Africa. *In:* Powell, J. M., Fernández-Rivera, S., Williams, T. O. & Renard, C. (eds.), *Livestock and Sustainable Nutrient Cycling in Mixed Farming Systems of sub-Saharan Africa. Volume II: Technical Papers*. pp. 149–169, International Livestock Centre for Africa, Addis Ababa, Ethiopia.

Hassen, H., Lababidi, S., Rischkowsky, B., Baum, M. & Tibbo, M. (2012). Molecular characterization of Ethiopian indigenous goat populations. *Tropical Animal Health and Production*, 44 (6), 1239–1246.

ILRI (2013). Harnessing the genetic diversity of goats to improve productivity in Africa. BecA-ILRI Hub, ILRI, University of Dschang. Poster prepared for the ILRI APM 2013, Addis Ababa, 15-17 May 2013. Nairobi, ILRI, Kenya. URL `http://www.slideshare.net/ILRI/apm2013-beca-goatdiversity` (last accessed: 20 October 2015).

Jahnke, H. E. (1982). *Livestock production systems and livestock development in tropical Africa*. Kieler Wissenschaftsverlag Vauk, Kiel, Germany.

Kosgey, I. S., Van Arendonk, J. A. M. & Baker, R. L. (2004). Economic values for traits in breeding objectives for sheep in the tropics: impact of tangible and intangible benefits. *Livestock Production Science*, 88, 143–160.

Legesse, G., Siegmund-Schultze, M., Abebe, G. & Valle Zárate, A. (2010). Economic performance of small ruminants in mixed-farming systems of Southern Ethiopia. *Tropical Animal Health and Production*, 42, 1531–1539.

Maxwell, D., Vaitla, B. & Coates, J. (2014). How do indicators of household food insecurity measure up? An empirical comparison from Ethiopia. *Food Policy*, 47, 107–116.

Megersa, B., Markemann, A., Angassa, A. & Valle Zárate, A. (2014). The role of livestock diversification in ensuring household food security under a changing climate in Borana, Ethiopia. *Food Security*, 6, 15–28.

Mersha, S. Y. (2014). *Contribution of cattle and dairy products to smallholder farming systems, household income and nutrition in southern Ethiopia*. Ph.D. thesis, Institute of Animal Production in the Tropics and Subtropics, Universität Hohenheim, Stuttgart, Germany.

Moll, A. J. H. (2005). Costs and benefits of livestock systems and the role of market and non-market relationships. *Agricultural Economics*, 23, 181–193.

Moursi, M. M., Arimond, M., Dewey, K. G., Trèche, S., Ruel, M. T. & Delpeuch, F. (2008). Dietary diversity is a good predictor of the micronutrient density of the diet of 6- to 23-month-old children in Madagascar. *Journal of Nutrition*, 138, 2448–2453.

Mueller, J. P., Rischkowsky, B., Haile, A., Philipsson, J., Mwai, O., Besbes, B., Valle Zárate, A., Tibbo, M., Mirkena, T., Duguma, G., Solkner, J. & Wurzinger, M. (2015). Community-based livestock breeding programs: essentials and examples. *Journal of Animal Breeding and Genetics*, 132, 155–168.

Narrod, C., Tiongco, M. & Scott, R. (2011). Current and predicted trends in the production, consumption and trade of live animals and their products. *Revue scientifique et technique*, 30 (1), 31–49.

Netsanet, Z. (2014). *On-farm phenotypic characterization and performance evaluation of central highland and woyto-goji goat types for designing community based breeding strategies in Ethiopia*. Master's thesis, Haramaya University, Ethiopia.

Ogola, T. D. O., Nguyo, W. K. & Kosgey, I. S. (2010). Economic contribution and viability of dairy goats: implications for a breeding program. *Tropical Animal Health and Production*, 42, 875–885.

Peacock, C. (2005). Goats–A pathway out of poverty. *Small Ruminant Research*, 60, 179–186.

Rosa, H. J. D., Ventura, P. F. M., Sousa, J. T., Moreira, D., Borba, A. E. S. & Rego, O. A. (2007). Productive and reproductive performance of Romney Marsh and Merino Branco sheep in the Azores under different feeding regimens. *Small Ruminant Research*, 67, 296–301.

SAS Institute Inc. (2011). Base SAS® 9.3 procedure guide. SAS Institute Inc., Cary, NC, USA.

Schlecht, E., Fernández-Rivera, S. & Hiernaux, P. (1997). Timing, size and nitrogen concentration of fecal and urinary excretions in cattle, sheep and goats: Can they be exploited for better manuring of croplands? *In:* Renard, G., Neef, A., Becker, K. & Oppen, V. M. (eds.), *Proc. Soil Fertility Management in West African Land Use Systems. Niamey, Niger, 4–8 March 1997*. pp. 361–367, Margraf Verlag, Weikersheim, Germany.

Seleka, T. B. (2001). Determinants of short-run supply of small ruminants in Botswana. *Small Ruminant Research*, 40, 203–214.

Solomon, A. K., Mwai, O., Grum, G., Haile, A., Rischkowsky, B., Solomon, G. & Dessie, T. (2014). Review of goat research and development projects in Ethiopia. ILRI project report. Nairobi, Kenya: International Livestock Research Institute.

Somda, Z. C., Powell, J. M., Fernández-Rivera, S. & Reed, J. (1995). Feed factors affecting nutrient excretion by ruminants and the fate of nutrients when applied to soil. *In:* Powell, J. M., Fernández-Rivera, S., Williams, T. O. & Renard, C. (eds.), *Livestock and Sustainable Nutrient Cycling in Mixed Farming Systems of Sub-Saharan Africa. Volume II: Technical Papers*. pp. 227–243, International Livestock Centre for Africa, Addis Ababa, Ethiopia.

Swindale, A. & Paula, B. (2006). Household Dietary Diversity Score (HDDS) for Measurement of Household Food Access: Indicator Guide (v.2). Washington, D.C.: FHI 360/FANTA.

Tucho, T. A. (2004). *Genetic characterization of indigenous goat populations of Ethiopia using Microsatellite DNA Markers*. Ph.D. thesis, Karnal (Haryana), India: National Dairy Research Institute (Deemed University).

Villa, K. M., Barrett, C. B. & Just, D. R. (2011). Whose fast and whose feast? Intra-household asymmetries in dietary diversity response among East African pastoralists. *American Journal of Agricultural Economics*, 93, 1062–1081.

# 4

# Agricultural microcredit and technical efficiency: The case of smallholder rice farmers in Northern Ghana

Benjamin Tetteh Anang [a,b,*], Stefan Bäckman [a], Timo Sipiläinen [a]

[a]*Department of Economics and Management, University of Helsinki, FI-00014, Finland*

[b]*Department of Agricultural and Resource Economics, Faculty of Agribusiness and Communication Sciences, University for Development Studies, Tamale, Ghana*

**Abstract**

In the current study, we compared technical efficiency of smallholder rice farmers with and without credit in northern Ghana using data from a farm household survey. We fitted a stochastic frontier production function to input and output data to measure technical efficiency. We addressed self-selection into credit participation using propensity score matching and found that the mean efficiency did not differ between credit users and non-users. Credit-participating households had an efficiency of 63.0 percent compared to 61.7 percent for non-participants. The results indicate significant inefficiencies in production and thus a high scope for improving farmers' technical efficiency through better use of available resources at the current level of technology. Apart from labour and capital, all the conventional farm inputs had a significant effect on rice production. The determinants of efficiency included the respondent's age, sex, educational status, distance to the nearest market, herd ownership, access to irrigation and specialisation in rice production. From a policy perspective, we recommend that the credit should be channelled to farmers who demonstrate the need for it and show the commitment to improve their production through external financing. Such a screening mechanism will ensure that the credit goes to the right farmers who need it to improve their technical efficiency.

*Keywords:* microcredit, propensity score matching, selection bias, smallholder farming, stochastic frontier analysis, technical efficiency

## 1 Introduction

Majority of Ghanaian smallholder farmers operate less than 2 hectares of land (Seini & Nyanteng, 2005). As a result of limited use of capital and low adoption of production technologies, yields and incomes among smallholders are generally low. The participation of Ghanaian smallholders in the formal financial sector is limited by lack of collateral, perceived high risk of lending, and high transaction cost of loans (Boniphace *et al.*, 2015; UNCTAD, 2015) while statistics attest the fact that the demand for financial services for rural people remains largely unmet (Zeller & Sharma, 1998; UNDP, 2004). Commercial banks are not interested in lending to rural households due to lack of individual collateral (Phillip *et al.*, 2008). According to Anang *et al.* (2015), some lenders may consider farm households without adequate capital endowment too poor and not creditworthy thus limiting their access to credit. Dittoh (2006) identified access to credit as the main concern of Ghanaian small-scale farmers. According to some researchers, the lack of access to credit (and other financial services) by smallholder farmers has implications for agricultural development, farm efficiency and productivity (Owusu-Antwi, 2010; Boniphace *et al.*, 2015). Liquidity constraints therefore impact agricultural growth and productivity in northern Ghana.

* Corresponding author
Email: benjamin.anang@helsinki.fi

Credit is a key component of financial services and fundamental in all aspects of production, including agricultural production. Agricultural production and financial decisions by farm households are interrelated and this has led to a growing research interest in order to understand this interrelationship (Barry & Robinson, 2001). A large body of literature highlights the important role of credit in agricultural production (Chaovanapoonphol *et al.*, 2005; Ruben & Kolk, 2005; Dittoh, 2006; Komicha & Öhlmer, 2007; Martey *et al.*, 2015). These studies portray the key role of agricultural credit in technical efficiency and productivity of farm households. The role of credit in raising both the technical and allocative efficiency of agricultural production has been attested to by Chaovanapoonphol *et al.* (2005).

Farm households need credit to purchase external inputs, contract wage labour, acquire food and non-food items, invest in education, etc. (Ruben & Kolk, 2005). Access to credit also enables farmers to adopt more capital-intensive methods of production to improve their level of technical efficiency (Hazarika & AIwang, 2003). Alene & Hassan (2006) and Komicha & Öhlmer (2007) also indicate that the capacity of farmers to adopt improved production technologies can be constrained by resource limitations including credit constraints.

Capital market imperfections as a result of asymmetric information and problems of incentive compatibility have been identified as the cause of credit constraint encountered by borrowers (Stiglitz & Weiss, 1992; Blancard *et al.*, 2006). Alene & Hassan (2006) attest to credit market imperfections as common phenomena in developing countries due to poorly developed infrastructure, weak institutional environment and less competitive market situation.

Credit affects farm production both directly and indirectly. Directly, credit affords producers the purchasing power to acquire essential production inputs and carry out long-term investments. On the other hand, credit affects production indirectly through its effect on farmers' risk behaviour (Guirkinger & Boucher, 2005). For example, farmers who are credit constrained are more likely to invest in activities that are less risky and less productive. As indicated by Komicha & Öhlmer (2007), this risk behaviour can affect farmers' choice of technology and adoption decisions with implications for technical efficiency of the producers. Lack of credit can therefore serve as a binding constraint that limits investment in productivity-enhancing technologies and production inputs and limits the household's ability to reduce vulnerability (Owusu-Antwi, 2010).

It is evident from the foregoing that lack of credit can serve as a critical factor limiting productivity and efficiency of production of farm households. Recent studies on the effect of credit on efficiency, especially technical efficiency of production include Ayaz & Hussain (2011) who investigated the effect of institutional credit on the production efficiency of Pakistani farmers. The authors found credit to have a positive impact on technical efficiency. Pinheiro (1992) however found no effect of credit on technical, allocative and economic efficiency of farmers in Dominican Republic, while Chaovanapoonphol *et al.* (2005) found credit to reduce technical inefficiency of rice farmers in Thailand.

Studies on the efficiency of rice production in Ghana include Abdulai & Huffman (2000), Seidu *et al.* (2004), Al-Hassan (2008), and Martey *et al.* (2015). The results from these studies show high variability in the estimates of technical efficiency even for the same ecological zone. On the other hand, it is quite typical that average efficiencies may differ due to the method and sample used. The results from these studies however highlight considerable inefficiency of production which calls for measures that will improve the level of technical efficiency of Ghanaian farmers.

On the effect of credit on technical efficiency of Ghanaian farmers, we found very limited studies, which necessitated the current study. The few studies include Martey *et al.* (2015) who found a positive effect of credit on technical efficiency of maize producers in northern Ghana. Abdallah (2016) also investigated agricultural credit and technical efficiency of maize farmers in Ghana and found a positive effect of credit on efficiency.

To the best of our knowledge, there is no study that directly assesses the effect of credit on technical efficiency of rice production in Ghana. In most of the previous efficiency studies, a credit dummy has been included in the inefficiency effects model to explain the effect of credit on efficiency. The limitation of these approaches is that the selection bias arising from access to credit or credit participation is ignored which may lead to biased estimates of the impact of credit.

The credit impact assessment on technical efficiency requires that the researchers control for factors which influence participation in credit. One of the innovative approaches used by many researchers to account for selection bias, as in the case of credit-programme participation, is propensity score matching (PSM). Among the recent applications of PSM in agriculture are Mayen *et al.* (2010), Abdoulaye & Sanders (2013) and Abate *et al.* (2014). This sample selection method reduces the selection bias in programme participation, and therefore helps to obtain less unbiased estimates of the impact of an intervention or programme.

The current study therefore employs the propensity score matching technique to assess the effect of participation in microcredit on technical efficiency of smallholder farmers in northern Ghana. The participation in credit means that the household actually received credit from a particular source for the purpose of farming. By definition, microcredit refers to a limited amount of credit offered to poor people usually without collateral. The average loan received by the respondents in the current study suggests that the credit is micro in nature.

## 2 Materials and methods

### 2.1 *Theoretical background*

Economic theory stipulates that economic agents aim at output maximisation given the quantity of inputs and existing technology. This means that given fixed input levels, the producer must produce on or very close to the production frontier. Producers however differ in their ability to produce efficiently, that is, on the production frontier. Thus, with the same set of inputs, some producers will produce more output than others.

Different methods of estimating efficiency exist in the economic literature. The approaches can be categorised into parametric, semi-parametric and non-parametric ones (Chakraborty *et al.*, 1999). Unlike the parametric approach, the non-parametric method assumes no functional form. The parametric approach often employs stochastic frontier analysis (SFA) while the non-parametric approach typically employs data envelopment analysis (DEA). The stochastic frontier approach attributes deviations from the production frontier to inefficiency and random errors whereas the deterministic approach attributes all errors to inefficiency (Coelli *et al.*, 2005). The productive efficiency literature also distinguishes between technical, allocative, and economic efficiencies (see Khan & Saeed, 2011). We focus on technical efficiency in this study.

### 2.2 *The stochastic frontier model*

A firm is technically efficient in production if it is able to achieve maximum output, with given level of inputs and production technology. The stochastic frontier model assumes that maximum output may not be realised from a given set of inputs because of inefficiency. This model can be used to estimate efficiency and its determinants using either a two-step or a one-step procedure. The two-step procedure has been criticised for its theoretical inconsistency (see Kumbhakar *et al.*, 1991; Reifschneider & Stevenson, 1991), hence we apply the one-step procedure proposed by Battese & Coelli (1995) to estimate the parameters of the stochastic production frontier and inefficiency effects model using maximum likelihood estimation. The stochastic frontier production function is defined as follows:

$$Y_i = \exp(X_i\beta + V_i - U_i) \qquad (1)$$

where $Y_i$ is rice output, $X_i$ is a vector of inputs, $V_i$ is a symmetric error term indicating the effects of pure random factors on production, $U_i$ is a one-sided error term indicating the effects of inefficiency and $\beta$ is a vector of parameters to be estimated. Technical efficiency (TE) is computed as the ratio of the observed output $Y_i$ to the frontier output $Y_i^*$.

$$TE_i = \frac{Y_i}{Y_i^*} = \frac{\exp(X_i\beta + V_i - U_i)}{\exp(X_i\beta + V_i)} = \exp(-U_i) \qquad (2)$$

where $0 \le TE \le 1$. The technical inefficiency effects, $U_i$, are obtained by truncation (at zero) of the normal distribution with mean $\mu_i$ and variance $\sigma_i^2$ such that:

$$U_i = \delta_0 + \sum_{n=1}^{N} \delta_n Z_{ni} \qquad (3)$$

where $Z_i$ represents a vector of farm-specific independent variables and $\delta$ is a vector of unknown coefficients of the farm-specific inefficiency effects. We used the statistical software package Stata version 14 for the frontier analysis.

The study also conducted tests of the null hypotheses for the parameters in the production function and inefficiency effects model using the generalised likelihood-ratio test statistic defined as:

$$\lambda = -2\{\ln[LL_0/LL_1]\} \qquad (4)$$

where $LL_0$ is the likelihood function under the null hypothesis and $LL_1$ is the likelihood function under the alternative hypothesis. For the test of functional form, the test statistic $\lambda$ has approximately a Chi-squared or a mixed Chi-squared distribution. The difference between the number of parameters in the null and alternative hypothesis represents the degrees of freedom. For the inefficiency model, the critical values for $\lambda$ are derived from Kodde & Palm (1986).

### 2.3 *Propensity score matching and self-selection*

Selection bias arises if the participation in credit by households is not random. Non-randomness in participation may arise if certain individuals are unable to participate or certain individuals decide not to participate. The failure to account for selection bias in credit participation is likely to lead to a biased estimate of the impact of credit.

The present study employs a matching approach (propensity score matching or PSM) to address the problem of self-selection. Matching models are a special case of selection models formulated on the assumption that conditioning on observable variables eliminates (or significantly reduces) sample selection bias (Heckman & Navarro-Lozano, 2004). Matching models create the condition of an experiment in which the treatment condition (i.e. participation in credit *versus* non-participation) is randomly assigned and provides a causal link between the treated group (e.g. credit participants) and the outcome of interest (i.e. technical efficiency).

The basic idea of the PSM method is to match observations of farmers with credit (the treated) and those without credit (the untreated) according to their predicted propensity of credit participation (Rosenbaum & Rubin, 1983; Heckman *et al.*, 1998). Rosenbaum & Rubin (1983) defined the propensity score as the conditional probability of receiving a treatment based on pre-treatment characteristics. It is expressed as

$$p(X) = \Pr\{L=1 \mid X\} = E\{L \mid X\} \tag{5}$$

where $L = \{0, 1\}$ represents the treatment indicator variable (e.g. participation in credit), $E$ is the expectation sign (expected value) and $X$ is a vector of pre-treatment characteristics such as farm and household characteristics.

We used the estimated propensity scores to obtain an estimate of the average treatment effect on the treated (*ATT*) which measures the effect of microcredit on participants. It is assumed that farmers have two potential technical efficiency outcomes, $Y$, given the participation status ($L$) such that $Y = Y_0$ if $L = 0$ and $Y = Y_1$ if $L = 1$. The average treatment effect (*ATE*) is represented by $ATE = E(Y_1 - Y_0)$. The average treatment effect on the treated (*ATT*), which is our variable of interest is given as $ATT = E\big((Y_1 - Y_0) \mid L=1\big)$. The *ATT* can further be expressed as $ATT = E(Y_1 \mid L=1) - E(Y_0 \mid L=1)$.

### 2.4 *Empirical production frontier and probit models*

The two most commonly used functional forms in efficiency analysis are the Cobb-Douglas and translog specifications. We conducted a formal test of the functional form and the Cobb-Douglas form was preferred above the translog specification. The current study therefore used the Cobb-Douglas production function in equation (6) to estimate efficiency of rice production in northern Ghana.

$$\ln Y_i = \beta_0 + \sum_{k=1}^{3} \beta_k D_{ki} + \sum_{j=1}^{6} \beta_j \ln X_{ji} + V_i - U_i \tag{6}$$

where $Y_i$ represents rice output of the $i^{\text{th}}$ farmer and $j$ is the $j^{\text{th}}$ input used in rice production. $D_{ki}$ is the $k^{\text{th}}$ intercept dummy variable where $D_1$ is an irrigation dummy, $D_2$ is a location dummy and $D_3$ is a cropping intensity dummy; ln = natural logarithm; $X_1$ = total land used for rice production; $X_2$ = total labour in man-days; $X_3$ = quantity of seed planted; $X_4$ = quantity of inorganic fertiliser applied; $X_5$ = other variable costs; $X_6$ = farm capital. $V_i$ and $U_i$ are as previously defined.

The technical inefficiency effect Ui is a linear function of socio-economic and management factors as defined in equation (7).

$$U_i = \delta_0 + \sum_{n=1}^{14} \delta_n Z_{ni} \tag{7}$$

where $\delta_n$ is the coefficient of the explanatory variables. $Z_i$ represents farmer and household characteristics accounting for inefficiency in production. The $Z_i$ variables included in the inefficiency model include the gender of the household head, age and its squared value, household size, contact with extension agents, educational status, association membership, participation in off-farm work, specialisation in rice production, distance to the nearest market, regional dummy, access to irrigation, participation in microcredit and herd ownership.

We specified the probit model for participation in credit as an index function, with an unobserved continuous variable ($L_i^*$) as follows:

$$L_i^* = \gamma Z_i + e_i \tag{8}$$

$$L_i = \begin{cases} 1 & \text{if } L_i^* > 0 \\ 0 & \text{if } L_i^* \leq 0 \end{cases}$$

where $L_i$ = participation in credit (equals 1 for participants, 0 otherwise) and $e_i$ is the random error term in the probit model. The explanatory variables included in the model are the gender, educational status, age, household size, total farm size, household income, access to irrigation (dummy), value of farm capital, adoption of improved variety (HYV dummy), association membership, distance to market, contact with extension (dummy), region (dummy), and awareness of lending/microfinance institutions (MFIs dummy). We used Stata version 14 to analyse credit participation and the propensity score.

### 2.5 *Sampling and study area*

The data used for the study came from a farm household survey conducted during the 2013/2014 farming season in northern Ghana. Northern Ghana is made up of three administrative Regions: Upper East, Upper West and Northern Region. Northern Ghana produces the bulk of the country's rice hence the choice of the location. The study involved 300 smallholder rice farmers distributed across northern Ghana.

We used a multi-stage stratified random sampling technique to select the respondents. First, we purposively selected two Regions, namely the Upper East and Northern Regions because of their contribution to domestic rice production and the presence of irrigation schemes for rice cultivation. After that, we selected three irrigation schemes based on size and geographical location. They included the Botanga Irrigation Scheme in the Northern Region and the Vea and Tono irrigation schemes in the Upper East Region. Next, we selected at random five communities within the catchment area of each irrigation scheme. Finally, we stratified the farmers into irrigators and non-irrigators, and selected equal number of respondents from each group. The study used a semi-structured questionnaire to solicit responses related to rice production, input and output quantities and prices, and whether the household participated in microcredit and the amounts borrowed.

## 3 Results

### 3.1 *Characteristics of the respondents*

Table 1 shows the descriptive statistics of the variables used in the study. About 40 percent of the sampled farmers participated in credit. As shown in the table, 104 credit users were matched to the non-credit users in the sample. In addition, credit users produced more rice and had higher household income than non-users. Credit participants also used more inputs in production with the exception of expenditure on other inputs.

Farmers in the Northern Region reported higher participation in credit while credit users had higher participation in farmer-based organisations. On the other hand, household size, educational status, age and herd size did not differ between credit participants and non-participants. Contrary to our *a priori* expectation, participants in credit devoted less land to rice cultivation while adoption of high-yielding varieties (HYV) was lower for credit users. Furthermore, one-third of credit users double-cropped their fields compared to one-quarter of non-users.

The amount of loan received by the respondents is shown in Table 2. Majority of the respondents took very small loans not exceeding GH₵200 with very few taking loans exceeding GH₵600. The average loan size was GH₵246.

The source of the credit included rural banks, government-subsidised credit targeted at poverty alleviation, non-governmental organisations working with farmers, farmers' cooperatives, relatives and moneylenders. Majority of the credit was collateral-free while subsidised credit from government sources, non-governmental organisations and farmers' cooperative had very low interest rates and limitations in terms of loan size. Very few farmers used credit from commercial sources. The loans were used primarily to finance land preparation and hiring in labour as well as the purchase of farm inputs notably fertilisers, chemical sprays and seeds. Majority of the farmers were credit-constrained as the loan amounts offered fell below the amount they actually requested. It was observed that farmers were reluctant to borrow from commercial sources which offer larger loan amounts. The lack of collateral and the high interest rates compared to the alternative credit sources, may account for this behaviour. Most of the non-commercial sources provided only limited amount of credit which may be due to the large number of applicants.

### 3.2 *Propensity score matching analysis*

We present the probit estimates of the credit propensity equation in Table 3. The model had a good fit as indicated by the pseudo-$R^2$, the percentage of correct predictions and the Chi-squared value.

Several variables included in the model had a significant effect on credit market participation. Female farmers and rain-fed producers were more likely to participate in credit, just as households with higher income and contact with extension agents. In addition, total household assets was positively related to credit participation while farmers who were aware of the presence of lending institutions in the area as well as farmers located in the Northern Region were more likely to participate in credit. However, contrary to our expectation we found participation in credit to increase with distance to the nearest market, while farmers who planted traditional varieties were also more likely to participate in credit.

**Table 1:** *Descriptive statistics of the variables used in the study*

| *Variable* | *Credit-users* (N = 121) | | *Non-credit users* (N = 179) | | *Matched credit users* [a] (N = 104) | | *t-test* [b] |
|---|---|---|---|---|---|---|---|
| | *Mean* | *SD* | *Mean* | *SD* | *Mean* | *SD* | |
| Output (kg) | 1864 | 2285 | 1502 | 1962 | 1530 | 1661 | 1.466 |
| Household income (Cedi) [c] | 2796 | 2403 | 2073 | 1678 | 2467 | 2176 | 3.070*** |
| Land area under rice (ha) | 0.95 | 0.76 | 0.79 | 0.62 | 0.83 | 0.56 | 1.996** |
| Labour (man-days) | 69.9 | 45.6 | 60.4 | 44.5 | 64.4 | 33.9 | 1.802* |
| Seed (kg) | 186 | 179 | 139 | 134 | 172 | 164 | 2.617*** |
| Inorganic fertiliser (kg) | 317 | 370 | 275 | 321 | 292 | 360 | 1.047 |
| Other costs (Cedi) | 190 | 205 | 183 | 179 | 170 | 188 | 0.297 |
| Farm capital (Cedi) | 150 | 175 | 114 | 132 | 143 | 167 | 2.032** |
| Total household assets (Cedi) | 728 | 1226 | 493.4 | 1001 | 636 | 1018 | 1.817* |
| Cropping intensity (1=double) | 0.33 | 0.47 | 0.25 | 0.44 | 0.33 | 0.47 | 1.494 |
| Sex (1=Male) | 0.75 | 0.43 | 0.81 | 0.40 | 0.74 | 0.44 | −1.079 |
| Years of formal education | 3.93 | 5.48 | 3.94 | 5.27 | 4.01 | 5.57 | −0.021 |
| Age (years) | 41.9 | 12.0 | 40.7 | 12.5 | 41.4 | 12.4 | 0.841 |
| Household size (number) | 10.3 | 6.0 | 9.20 | 7.89 | 10.1 | 6.02 | 1.331 |
| Total land area | 7.05 | 5.94 | 4.79 | 4.09 | 5.98 | 3.86 | 3.895*** |
| Access to irrigation (1=Yes) | 0.50 | 0.50 | 0.50 | 0.50 | 0.49 | 0.50 | 0.117 |
| Adopt improved variety (1/0) | 0.59 | 0.49 | 0.72 | 0.45 | 0.60 | 0.49 | −2.429** |
| Group membership (1=Yes) | 0.75 | 0.43 | 0.60 | 0.49 | 0.73 | 0.45 | 2.794*** |
| Share of land under rice (%) | 38.9 | 20.7 | 49.7 | 26.9 | 39.2 | 20.6 | −43.735*** |
| Herd ownership (1=Yes) | 0.37 | 0.49 | 0.31 | 0.46 | 0.37 | 0.48 | 1.060 |
| Distance to market (km) | 7.49 | 4.42 | 8.21 | 4.21 | 7.91 | 4.58 | −1.414 |
| Extension contact (1=Yes) | 0.72 | 0.45 | 0.58 | 0.50 | 0.69 | 0.46 | 2.551** |
| Regional dummy (1=Northern) | 0.45 | 0.50 | 0.25 | 0.44 | 0.40 | 0.49 | 3.734*** |
| Awareness of MFIs [d] (1=Yes) | 0.92 | 0.28 | 0.74 | 0.44 | 0.90 | 0.30 | 3.884*** |

***, ** and * stand for statistical significance at the 1, 5 and 10 percent level, respectively.[a] The subsample of credit participating farms matched to non-participating farms on the basis of the estimated likelihood or propensity of participating in credit.[b] The test of mean difference between the unmatched groups.[c] GH₵1 = US$0.26. [d] MFI means microfinance institution.

**Table 2:** *Amount of loan received by respondents*

| *Loan size* (GH₵)* | *Frequency* | *Percentage* | *Cumulative* (%) |
|---|---|---|---|
| 1–200 | 69 | 57.0 | 52.0 |
| 201–400 | 30 | 24.8 | 81.8 |
| 401–600 | 15 | 12.4 | 94.2 |
| 601–800 | 5 | 4.1 | 98.4 |
| 801–1000 | 2 | 1.7 | 100.0 |
| Total | 121 | 100.0 | 100.0 |

*1GH₵ is equivalent to US$0.26.

**Table 3:** *Probit results of the determinants of access to agricultural microcredit*

| *Variable* | *Coefficient* | *Std. Error* | $P>\|z\|$ |
|---|---|---|---|
| Sex | −0.666*** | 0.225 | 0.003 |
| Education | 0.017 | 0.016 | 0.298 |
| Age | 0.008 | 0.007 | 0.281 |
| Household size | −0.023 | 0.014 | 0.103 |
| Land | 0.022 | 0.022 | 0.320 |
| Household income | 0.189* | 0.112 | 0.092 |
| Access to irrigation | −0.344* | 0.188 | 0.068 |
| Total household assets | 0.115** | 0.048 | 0.016 |
| Improved variety adoption | −0.455** | 0.183 | 0.013 |
| Group membership | 0.128 | 0.193 | 0.507 |
| Distance to market | −0.038* | 0.021 | 0.071 |
| Extension contact | 0.710*** | 0.210 | 0.001 |
| Regional dummy | 1.337*** | 0.243 | 0.000 |
| Awareness of MFIs | 1.094*** | 0.258 | 0.000 |
| Constant | −1.490*** | 0.518 | 0.004 |

***, ** and * stand for statistical significance at the 1, 5 and 10 percent level, respectively. Number of observations = 300, Log-likelihood = −158.7, Wald Chi$^2$ (14) = 87.22, Prob > Chi (2) = 0.000, Pseudo $R^2$ = 0.216, Percentage correctly predicted = 71.3.

We used the estimates of the probit model to obtain a propensity score (the predicted probability of participation in credit) for each farm after which each credit-participant was matched to a non-participant with similar propensity score. The propensity score matching technique produced a subsample of 283 matched farms comprising 104 credit participants and 179 non-participants. We used this new sub-sample to estimate the production frontier. We ensured that the matched samples were within the common support region to ensure the robustness of the matching. As indicated earlier, the common support region indicates values of the propensity scores where the treated (credit users) and untreated units (non-credit users) can be found. Without a common support, suitable matches are unlikely to be obtained. We present a plot of the treated and untreated units after the matching in Figure 1. The plot shows the propensity scores on the x- axis with the matched treated units above the horizontal line and untreated units below the horizontal line.

To check the robustness of the propensity score matching, a balancing test of the matched sample was performed and the results are reported in Table 4. The balancing test is a test of the mean equality of the covariates for credit users and non-users before and after the matching. The results indicate that the unmatched sample does not to satisfy the balancing property as the

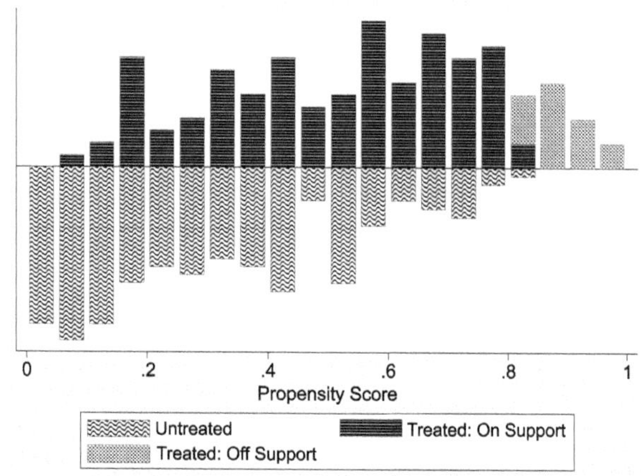


**Fig. 1:** *Distribution of common support region by treatment status (Note: The treated are the credit users)*

two groups are comparable in only 6 out of the 14 covariates. The matched sample however showed no systematic differences in the observed covariates between credit users and non-users thus justifying the validity and robustness of the matching.

### 3.3 *Tests of hypotheses*

We present the results of the tests of hypotheses regarding the functional form and inefficiency effects model in Table 5. From the results, we adopt the Cobb-Douglas functional form and reject the null hypothesis of no inefficiency effects in the specified model implying that the traditional average response model is not an appropriate representation of the data. The result of the second assumption indicates that the variables included in the inefficiency effects model jointly measure production inefficiency of the respondents.

### 3.4 *Estimation of technical efficiency and its determinants*

We present in Table 6 the maximum-likelihood estimates of the parameters of the Cobb-Douglas stochastic frontier and inefficiency models based on the PSM subsample. All the conventional inputs maintained a positive sign in line with our *a priori* expectation. Furthermore, all the conventional inputs apart from capital and labour had a significant effect on rice production. This shows that the size of farm, seed, fertiliser and other costs positively influence the output of smallholder rice producers in northern Ghana. The intercept dummies included in the model to account for shifts in the production function were statistically significant. Irrigators as well as producers in the Northern Region and households who double-cropped their land had a higher production frontier.

**Table 4:** *Balancing test of matched sample*

| *Variable* | *Unmatched Sample* | | | *Matched Sample* | | |
|---|---|---|---|---|---|---|
| | *Mean* | | *Diff: P>\|t\|* | *Mean* | | *Diff: P>\|t\|* |
| | *Treated* | *Control* | | *Treated* | *Control* | |
| Sex | 0.752 | 0.804 | 0.281 | 0.740 | 0.762 | 0.718 |
| Education | 3.926 | 3.939 | 0.984 | 4.010 | 3.975 | 0.963 |
| Age | 41.93 | 40.72 | 0.401 | 41.41 | 40.57 | 0.619 |
| Household size | 10.32 | 9.196 | 0.184 | 10.06 | 9.508 | 0.578 |
| Land | 7.047 | 4.790 | 0.000 | 5.976 | 6.296 | 0.628 |
| Household income | 2796 | 2073 | 0.002 | 2467 | 2491 | 0.934 |
| Access to irrigation | 0.504 | 0.497 | 0.907 | 0.490 | 0.466 | 0.723 |
| Total household assets | 728.0 | 493.4 | 0.070 | 636.2 | 653.8 | 0.910 |
| Improved variety adoption | 0.587 | 0.721 | 0.016 | 0.596 | 0.604 | 0.907 |
| Group membership | 0.752 | 0.598 | 0.006 | 0.731 | 0.647 | 0.193 |
| Distance to market | 7.492 | 8.207 | 0.158 | 7.909 | 7.915 | 0.992 |
| Extension contact | 0.719 | 0.575 | 0.011 | 0.692 | 0.652 | 0.534 |
| Regional dummy | 0.455 | 0.251 | 0.000 | 0.404 | 0.412 | 0.903 |
| Awareness of MFIs | 0.917 | 0.743 | 0.000 | 0.904 | 0.919 | 0.698 |

Note: The treated are the credit users.

**Table 5:** *Generalised likelihood-ratio tests of hypotheses*

| *Null hypothesis* | *LR statistic* ($\lambda$) | *Critical value* * | *Decision* |
|---|---|---|---|
| Production function is Cobb-Douglas | 30.6 | 32.7 | Accept $H_0$ |
| No inefficiency effects: $H_0 : \delta_0 = \delta_1 = \ldots = \delta_{12} = \gamma = 0$ | 59.9 | 25.7 | Reject $H_0$ |
| Inefficiency model does not explain inefficiency: $H_0 : \delta_0 = \delta_1 = \ldots = \delta_{12} = \gamma = 0$ | 55.0 | 24.4 | Reject $H_0$ |

* We obtained critical values for the inefficiency model from Kodde & Palm (1986)

A 1 % increase in land area increased output by 0.31 % while a 1 % increase in labour and seed increased output by 0.15 % and 0.16 % respectively. In addition, a 1 % increase in fertiliser, other costs and capital increased output by 0.07 %, 0.09 % and 0.01 % respectively. Land had the highest effect on output followed by seed and labour. Capital had the least effect on out-put while the sum of the coefficients of the input vari-ables, which is a measure of economies of scale, was 0.79. The result implied diminishing returns to scale in rice production.

The inefficiency effects model in Table 6 shows that male farmers recorded higher efficiency than female farmers. Efficiency increased with age but later decreased with the progression in age. Thus, there is an increase in efficiency with age but at a decreasing rate. Farmers who allocated a greater portion of their land to rice cultivation as well as herd owners were also more efficient in production. In addition, efficiency was higher for farmers in the Northern Region and for users of irrigation. Finally, efficiency decreased with the educational level of the household head but increased with the distance to the nearest market.

**Table 6:** *Maximum likelihood estimation results of the stochastic frontier production function and inefficiency effects model*

| *Variable* | *Parameter* | *Coefficient* | *Standard Error* | *p–value* |
|---|---|---|---|---|
| Constant | $\beta_0$ | −0.170 | 0.118 | 0.149 |
| Cropping intensity dummy | $\beta_{01}$ | 0.259 ** | 0.102 | 0.011 |
| Irrigation dummy | $\beta_{02}$ | 0.257 ** | 0.103 | 0.012 |
| Regional dummy | $\beta_{03}$ | 0.581 *** | 0.107 | 0.000 |
| Land | $\beta_1$ | 0.313 *** | 0.104 | 0.003 |
| Labour | $\beta_2$ | 0.146 | 0.097 | 0.135 |
| Seed | $\beta_3$ | 0.160 ** | 0.068 | 0.019 |
| Fertiliser | $\beta_4$ | 0.067 ** | 0.026 | 0.010 |
| Other costs | $\beta_5$ | 0.089 *** | 0.030 | 0.003 |
| Capital | $\beta_6$ | 0.010 | 0.031 | 0.747 |
| *Inefficiency model* | | | | |
| Constant | $\delta_0$ | 4.710 *** | 1.540 | 0.002 |
| Participation in credit | $\delta_1$ | −0.039 | 0.265 | 0.882 |
| Sex of household head | $\delta_2$ | −1.138 *** | 0.340 | 0.001 |
| Age of household head | $\delta_3$ | −0.113 * | 0.067 | 0.090 |
| Age squared | $\delta_4$ | 0.001 * | 0.001 | 0.077 |
| Household size | $\delta_5$ | −0.019 | 0.023 | 0.391 |
| Extension contact | $\delta_5$ | −0.119 | 0.296 | 0.688 |
| Years of formal education | $\delta_7$ | 0.049 ** | 0.025 | 0.049 |
| Association membership | $\delta_8$ | −0.383 | 0.264 | 0.146 |
| Share of land under rice | $\delta_9$ | −0.013 ** | 0.007 | 0.041 |
| Distance to nearest market | $\delta_{10}$ | −0.064 ** | 0.032 | 0.044 |
| Herd ownership | $\delta_{11}$ | −0.818 ** | 0.367 | 0.026 |
| Regional dummy | $\delta_{12}$ | −0.799 | 0.580 | 0.168 |
| Irrigation dummy | $\delta_{13}$ | −0.849 ** | 0.382 | 0.026 |
| Off-farm work | $\delta_{14}$ | 0.062 | 0.248 | 0.802 |
| *Variance parameters* | | | | |
| Sigma-squared | $\sigma^2$ | 0.893 *** | 0.057 | 0.000 |
| Gamma | $\gamma$ | 0.719 *** | 0.012 | 0.000 |
| Log likelihood function | $\lambda$ | −239.6 | | |
| Returns to scale | | 0.785 | | |

***, ** and * stand for statistical significance at the 1, 5 and 10 percent level, respectively.

### *3.5 Difference in technical efficiency between credit participants and non-participants*

The estimated mean technical efficiency for the PSM subsample was 63.0 % (SE 0.019) for credit users and 61.7 % (SE 0.016) for non-users. The means were not statistically different. We used nearest-neighbour matching to estimate the propensity score and the average treatment effect on the treated (ATT). The result indicates a non-significant effect of microcredit on smallholders' technical efficiency (ATT of 0.013, SE 0.031) which is consistent with the results obtained using the PSM subsample.

## 4 Discussion

### *4.1 Propensity score analysis*

The study indicates that women farmers are more likely to take part in microcredit. This result is supported by Jazairy *et al.* (1992) who found female borrowers to be more creditworthy. Akudugu (2012) also found that women were more likely to demand credit than men in the Upper East Region of Ghana. Furthermore, increasing farmers' awareness of the presence of lending institutions promotes their participation in credit programmes. The result agrees with Gaih & Thapa (2006) who reported that lack of awareness is a factor excluding some groups from microfinance. The results of the study also highlight the positive effect of extension contact on smallholders' participation in credit. Contact with extension agents enhances farmers' knowledge about the presence of lending institutions and the sources of credit thereby facilitating their participation in microcredit programmes. The result is consistent with Muhongayire *et al.* (2013) who found extension contact to enhance farmers' participation in formal credit in rural Rwanda.

The greater participation of households with higher income (and larger total household assets) in microcredit suggests that wealth status could affect smallholders' participation in rural credit programmes. As indicated by Anang *et al.* (2015), some lenders may consider poor households as risky borrowers thus constraining their participation in microcredit. The reasons behind the effects of market distance, geographical location and choice of rice variety on credit participation were not obvious. Finally, the lower participation rate of irrigators in credit indicates that agricultural microcredit may be seasonal and less available during the dry (off) season when irrigation farming is mostly practiced. The reason may also be that irrigators get more frequent returns from farming and therefore face less liquidity constraints.

### *4.2 Technical efficiency and its determinants*

The intercept dummy variables included in the production function suggest that participation in irrigation, cropping intensity and location of the farm (Northern Region = 1) shifted the production frontier upwards. In other words, irrigators, farmers who double-cropped their farms and farmers in the Northern Region operate on a higher production frontier which implies higher productivity.

All the conventional inputs had a positive effect on rice output in line with the monotonicity assumption of production functions. The area of land under rice production had the highest impact on rice output compared to the other variable inputs as shown by the estimated output elasticity with respect to land. Capital had the least effect on output and this may be due to the low use of capital inputs by the farmers. The study also revealed diminishing returns to scale in rice production. Thus increasing all inputs by 1 % will increase rice output by 0.79 %.

The result of the study also indicates women's lower efficiency of production relative to men, which is consistent with Abdulai *et al.* (2013) and Donkoh *et al.* (2013). Many researchers have recognised the important role of women as agricultural producers. However, gender inequality in access to production technology in many developing countries means that women farmers are often disadvantaged which can adversely affect their level of efficiency. Women also face other challenges that have negative impact on their technical efficiency. As shown by Abdulai *et al.* (2013), women's domestic and economic roles tend to affect their technical efficiency in farming.

Technical efficiency of production also tends to increase with the age of the household head. The result suggests that older farmers who are likely to be more experienced in farming utilise resources more efficiently in production. However, with progression in age, productivity begins to decrease as farmers become less energetic. The result agrees with Taiwo *et al.* (2014) who found that efficiency in cassava production in Nigeria increased with age but declined as farmers became very old. If the household head is older, there is the likelihood that the family labour may increase as the children become older. However, this may not be the case in the situation where the older children out-migrate. Participation in off-farm work may also decrease family labour for farming activities. Hence, there is the likelihood that family labour is getting less in the current study area and thus having adverse effect on efficiency as the household heads grow older.

Farmers' technical efficiency also increased with participation in farmers' organisations which is consistent with our *a priori* expectation. Farmers belonging to a farmers' group benefit from economies of scale, the sharing of production information, and access to production inputs and agricultural extension service, thus enhancing their efficiency in production. The result is consistent with Shehu *et al.* (2010) who reported that association membership enables yam farmers in Nigeria to access loans and productive inputs, which are easier to obtain collectively than individually. Idiong (2007) also found membership of farmers' association to increase the technical efficiency of Nigerian cocoyam farmers due to information sharing among members.

Farmers who allocated a greater proportion of their land to rice cultivation were more efficient in production because of specialisation. The result is in line with classical economic theory which views specialisation as an important determinant of efficiency. The study also highlighted the importance of draught animals (animal traction) in smallholder production and efficiency. Households having cattle were more efficient in production because the use of draught animals (cattle) enabled timely and more efficient farm operations.

Education, which is an important part of human capital, improves the quality of labour (Hyuha *et al.*, 2007). Education is therefore expected to improve the technical efficiency of farmers. The lower efficiency level of educated farmers in the current study may be due to the fact that educated farmers are more likely to find jobs outside the farm sector, which may interfere with the time they allocate to farming activities. Donkoh *et al.* (2013) and Asante *et al.* (2014) found similar effect of education on the efficiency of smallholder rice production in Ghana.

The distance to the nearest local market exerted a positive influence on technical inefficiency contrary to our *a priori* expectation. This shows that farmers living further away from the local market are more efficient in production. The longer distance to markets is likely to affect the timely acquisition of farm inputs to carry out farm operations which can affect technical efficiency. The result of our study agrees with Martey *et al.* (2015) who found that the technical efficiency of Ghanaian maize farmers increased with an increase in the distance to the local market.

Irrigation users also had higher efficiency of production than non-irrigators. Access to irrigation enables farmers to maximise the use of other inputs such as fertiliser due to the availability of water throughout the farming season. The result is consistent with other research findings (Makombe *et al.*, 2007; Mariano *et al.*, 2011).

### *4.3 Effect of credit on technical efficiency*

The main objective of the study was to compare the mean efficiency of credit users and non-users. We found the mean efficiency for credit users to be statistically not different from non-users. Furthermore, the result of the average treatment effect on the treated which measured the impact of credit on participants in microcredit programme showed no significant impact of credit on technical efficiency. A possible reason for the insignificant effect of microcredit on technical efficiency may be the small size of credit as shown in Table 2. By relaxing the liquidity constraints of farmers, credit helps producers to hire in labour and buy other production inputs that may enhance their technical efficiency. The small amount of credit to the respondents in the current study may therefore be insufficient in augmenting their technical efficiency. Hence increasing the loan size given to farmers could improve technical efficiency of rice production in northern Ghana.

## 5 Conclusion

The study investigated the effect of microcredit on technical efficiency of smallholder rice production in northern Ghana using cross-sectional data from 300 farm households. The study involved the estimation of a credit participation model and a Cobb-Douglas production function. We controlled for self-selection using propensity score matching and found that efficiency did not differ between credit-participating and non-participating farms although it was slightly higher in the credit-participating group. Controlling for self-selection using the PSM approach to match farmers based on their observed characteristics ensured that we obtained a more reliable comparison of technical efficiency for both participants and non-participants. The insignificant effect of credit on technical efficiency may be due to the small size of loans. From the results of the study, we conclude that credit should be channelled to farmers who demonstrate the need for it and show the commitment to improve their production through external financing. Such a screening mechanism will ensure that credit goes to the right farmers who have need for it to improve their technical efficiency. Credit institutions may also consider providing credit in kind rather than in cash to make inputs readily available to farmers as well as minimise the possibility of channelling the credit into other uses. Finally, improving access to irrigation and enabling intensification of production are possible options to improve productivity of rice farmers in the study area.

**Acknowledgements**

The authors express gratitude to the Wienco Chair at the Faculty of Agriculture, University for Development Studies in Tamale, Ghana, as well as the Nordic Africa Institute in Uppsala, Sweden for providing financial support towards the fieldwork and data collection.

## References


Abate, G. T., Francesconi, G. N. & Getnet, K. (2014). Impact of agricultural cooperatives on smallholders' technical efficiency: Empirical evidence from Ethiopia. *Annals of Public and Cooperative Economics*, 85 (2), 257–286.

Abdallah, A.-H. (2016). Agricultural credit and technical efficiency in Ghana: is there a nexus? *Agricultural Finance Review*, 76 (2), 309–324.

Abdoulaye, I. D. & Sanders, J. H. (2013). A matching approach to analyse the impact of new agricultural technologies: Productivity and technical efficiency in Niger. Paper presented at the Agricultural & Applied Economics Association's 2013 AAEA & CAES Joint Annual Meeting, Washington, DC, August 4-6, 2013.

Abdulai, A. & Huffman, W. (2000). Structural adjustment and economic efficiency of rice farmers in Northern Ghana. *Economic Development and Cultural Change*, 48 (3), 503–520.

Abdulai, S., Nkegbe, P. K. & Donkoh, S. A. (2013). Technical efficiency of maize production in Northern Ghana. *African Journal of Agricultural Research*, 8 (43), 5251–5259.

Akudugu, M. A. (2012). Estimation of the Determinants of Credit Demand by Farmers and Supply by Rural Banks in Ghana's Upper East Region. *Asian Journal of Agriculture and Rural Development*, 2 (2), 189–200.

Al-Hassan, S. (2008). Technical efficiency of rice farmers in Northern Ghana. AERC Research Paper 178, African Economic Research Consortium, Nairobi.

Alene, A. D. & Hassan, R. M. (2006). The efficiency of traditional and hybrid maize production in eastern Ethiopia: An extended efficiency decomposition approach. *Journal of African Economies*, 15 (1), 91–116.

Anang, B. T., Sipiläinen, T., Bäckman, S. & Kola, J. (2015). Factors influencing smallholder farmers' access to agricultural microcredit in Northern Ghana. *African Journal of Agricultural Research*, 10 (24), 2460–2469. doi:10.5897/AJAR2015.9536.

Asante, B. O., Wiredu, A. N., Martey, E., Sarpong, D. B. & Mensa-Bonsu, A. (2014). NERICA adoption and impacts on technical efficiency of rice producing households in Ghana: Implications for Research and Development. *American Journal of Experimental Agriculture*, 4 (3), 244–262.

Ayaz, S. & Hussain, Z. (2011). Impact of institutional credit on production efficiency of farming sector: a case study of District Faisalabad. *Pakistan Economic and Social Review*, 49 (2), 149–162.

Barry, P. J. & Robinson, L. J. (2001). Agricultural Finance: Credit, Credit Constraints and Consequences. *In:* Gardner, B. & Rausser, G. (eds.), *Handbook of Agricultural Economics*. Amsterdam, Elsevier.

Battese, G. E. & Coelli, T. J. (1995). A model for technical inefficiency effects in a stochastic frontier production function for panel data. *Empirical Economics*, 20 (2), 325–332.

Blancard, S., Boussemart, J., Briec, W. & Kerstens, K. (2006). Short- and long-run credit constraints in French agriculture: A directional distance function framework using expenditure-constrained profit functions. *American Journal of Agricultural Economics*, 88 (2), 351–364.

Boniphace, N. S., Fengying, N. & Chen, F. (2015). An analysis of smallholder farmers' socio-economic determinants for inputs use: A case of major rice producing regions in Tanzania. *Russian Journal of Agricultural and Socio-Economic Sciences (RJOAS)*, 2 (38), 41–55.

Chakraborty, K., Biswas, B. & Lewis, W. C. (1999). Technical Efficiency in Stochastic Production Frontier: A Simultaneous Equation Approach. Economic Research Institute Study Papers. Paper 160.

Chaovanapoonphol, Y., Battese, G. E. & Chang, H. (2005). The impact of rural financial services on the technical efficiency of rice farmers in the Upper North of Thailand. Contributed Paper at the Annual Conference of the Australian Agricultural and Resource Economics Society at Coffs Harbour, 9-11 February 2005.

Coelli, T. J., Rao, D. S. P., O'Donnell, C. J. & Battese, G. E. (2005). *An introduction to efficiency and productivity analysis*. (2nd ed.). USA: Springer.

Dittoh, S. (2006). Effective aid for small farmers in Sub-Saharan Africa: southern civil society perspective. Ghana case study. Canadian Food Security Policy Group, Canadian International Development Agency (CIDA).

Donkoh, S. A., Ayambila, S. & Abdulai, S. (2013). Technical efficiency of rice production at the Tono Irrigation Scheme in Northern Ghana. *American Journal of Experimental Agriculture*, 3 (1), 25–42.

Gaih, R. & Thapa, G. (2006). A methodology for assessment of the impact of microfinance on empowerment and vulnerability. International Fund for Agricultural Development (IFAD). URL `http://www.ifad.org/operations/projects/regions/pi/paper/2.pdf` (last accessed: 25.04.2015).

Guirkinger, C. & Boucher, S. (2005). Credit constraints and productivity in Peruvian agriculture. Department of Agricultural and Resource Economics, University of California-Davis.

Hazarika, G. & Alwang, J. (2003). Access to credit, plot size and cost inefficiency among smallholder tobacco cultivators in Malawi. *Agricultural Economics*, 29, 99–109.

Heckman, J., Ichimura, H., Smith, J. & Todd, P. (1998). Characterizing selection bias using experimental data. *Econometrica*, 66 (5), 1017–1098.

Heckman, J. & Navarro-Lozano, S. (2004). Using matching, instrumental variables, and control functions to estimate economic choice models. *The Review of Economics and Statistics*, 86, 30–57.

Hyuha, T. S., Bashaasha, B., Nkonya, E. & Kraybill, D. (2007). Analysis of profit efficiency in rice production in Eastern and Northern Uganda. *African Crop Science Journal*, 15 (4), 243–253.

Idiong, I. C. (2007). Estimation of farm level technical efficiency in small-scale swamp rice production in Cross River State of Nigeria: A Stochastic Frontier Approach. *World Journal of Agricultural Sciences*, 3 (5), 653–658.

Jazairy, I., Mohiudden, A. M. & Pannuccio, T. (1992). *The state of world poverty*. Rome: IFAD.

Khan, H. & Saeed, I. (2011). Measurement of Technical, Allocative and Economic Efficiency of Tomato Farms in Northern Pakistan. International Conference on Management, Economics and Social Sciences (ICMESS'2011) Bangkok, December, 2011.

Kodde, D. A. & Palm, F. C. (1986). Wald criteria for jointly testing equality and inequality restrictions. *Econometrica*, 54 (5), 1243–1248.

Komicha, H. H. & Öhlmer, B. (2007). Influence of Credit Constraint on Technical Efficiency of Farm Households in Southeastern Ethiopia. International Conference on African Development Archives. Paper 125.

Kumbhakar, S. C., Ghosh, S. & McGuckin, J. T. (1991). A Generalized Production Frontier Approach for Estimating Determinants of Inefficiency in U.S. Dairy Farms. *Journal of Business and Economic Statistics*, 9, 279–286.

Makombe, G., Kelemework, D. & Aredo, D. (2007). A comparative analysis of rain-fed and irrigated agricultural production in Ethiopia. *Irrigation and Drainage*, 21, 35–44.

Mariano, M. J., Villano, R. & Fleming, E. (2011). Technical efficiency of rice farms in different agro climatic zones in the Philippines: An application of a stochastic metafrontier model. *Asian Economic Journal*, 25 (3), 245–269.

Martey, E., Wiredu, A. N. & Etwire, P. M. (2015). Impact of credit on technical efficiency of maize producing households in Northern Ghana. Selected Paper presented at the Centre for the Study of African Economies (CSAE) Conference 2015, University of Oxford, March 22-24, 2015.

Mayen, C. D., Balagtas, J. V. & Alexander, C. E. (2010). Technology adoption and technical efficiency: Organic and conventional dairy farms in the United States. *American Journal of Agricultural Economics*, 92 (1), 181–195. doi:10.1093/ajae/aap018.

Muhongayire, W., Hitayezu, P., Mbatia, O. L. & Mukoya-Wangia, S. M. (2013). Determinants of Farmers' Participation in Formal Credit Markets in Rural Rwanda. *Journal of Agricultural Sciences*, 4 (2), 87–94.

Owusu-Antwi, G. (2010). The analysis of the rural credit market in Ghana. *International Business and Economics Research Journal*, 9 (8), 45–56.

Phillip, D., Nkonya, E., Pender, J. & Oni, O. A. (2008). Constraints to increasing agricultural productivity in Nigeria. Nigeria Strategy Support Program. Brief No. 4.

Pinheiro, A. E. (1992). *An Econometric Analysis of Farm-level Efficiency of Small Farms in the Dominican Republic*. Master's thesis, University of Connecticut.

Reifschneider, D. & Stevenson, R. (1991). Systematic departures from the frontier: a framework for the analysis of firm inefficiency. *International Economic Review*, 32, 715–723.

Rosenbaum, P. R. & Rubin, D. B. (1983). The central role of the propensity score in observational studies for casual effects. *Biometrika*, 70 (1), 41–55.

Ruben, R. & Kolk, H. (2005). Credit use, factor substitution and rural income distribution: A study on maize farmers in Occidental Honduras. *Investigacion Economica*, 64 (251), 13–32.

Seidu, A., Sarpong, D. B. & Al-Hassan, R. (2004). Allocative efficiency, employment and rice production risk: An analysis of smallholder paddy farms in the Upper East Region of Ghana. *Ghana Journal of Development Studies*, 1 (2), 142–163.

Seini, A. W. & Nyanteng, K. V. (2005). Smallholders and structural adjustment in Ghana. *In:* Djurfeldt, G., Holmen, H., Jirström, M. & Larsson, R. (eds.), *The African food crises: Lessons from the Asian Green Revolution*. UK: Biddles Ltd, King's Lynn.

Shehu, J. F., Iyortyer, J. T., Mshelia, S. I. & Jongur, A. A. U. (2010). Determinants of yam production and technical efficiency among yam farmers in Benue State, Nigeria. *Journal of Social Sciences*, 24 (2), 143–148.

Stiglitz, J. E. & Weiss, A. (1992). Asymmetric information in credit markets and its implications for macroeconomics. *Oxford Economic Papers*, 40, 694–724.

Taiwo, O., Dayo, O. S. & Bolariwa, K. O. (2014). Technical efficiency analysis of cassava production in Nigeria: Implication for increased productivity and competitiveness. *Research Journal of Agriculture and Environmental Management*, 3 (11), 569–576.

UNCTAD (2015). The role of smallholder farmers in sustainable commodities production and trade. United Nations Conference on Trade and Development. URL `http://unctad.org/meetings/en/SessionalDocuments/tdb62d9_en.pdf` (last accessed: 21.01.2015).

UNDP (2004). United Nations Development Programme. Building inclusive financial services in Africa.

Zeller, M. & Sharma, M. (1998). *Rural finance and poverty alleviation*. Washington DC. International Food Policy Research Institute.

# 5

# Participatory Rural Appraisal for Diagnostic Analysis of spate irrigation systems in Raya Valley, Ethiopia

Giulio Castelli *, Elena Bresci

*Department of Agricultural, Food and Forestry Systems, University of Florence, Italy*

**Abstract**

Spate irrigation is a complex and unique form of water management, which represent the main source of irrigation water in semi-arid river catchments. Water is diverted from seasonal rivers by using diversion structures made by stones, earth and brushwood, located within the river bed. The modernisation of spate irrigation realised in Raya Valley (northern Ethiopia) resulted in disappointing performances. One of the main reasons for this failure was the poor consideration of the characteristics of seasonal catchments and local communities' needs and preferences. Local farmers, who showed a deep knowledge of the river system, were involved only at the level of consultation. The aim of this research was to develop a participatory Diagnostic Analysis (DA) for a traditional non-modernised spate irrigation system in Raya Valley, in order to involve local farmers within the development process, and to build a solid knowledge basis for effective improvements. A Participatory Rural Appraisal (PRA) of the Harosha spate irrigation system was undertaken. PRA techniques focusing on spatial, temporal, socio-economical and spatiotemporal aspects of the system were performed with local farmers in order to identify and rank main problems and constraints to development. Farmers recognised the need of more resistant diversion structures and gabion walls for the stabilisation of the river bank. The involvement of farmers also helped to highlight that not only irrigation-related problems, but also flood-related problems threaten agricultural production and rural livelihoods. Rather than an irrigation system approach, an approach integrating irrigation development and flood risk mitigation is suggested for framing future development strategies.

*Keywords:* participatory rural appraisal, indigenous knowledge, rural development, water harvesting, Ethiopia, Tigray, arid climates

## 1 Introduction

Spate irrigation is a unique and ancient form of water management typical for ephemeral river catchments (namely wadis) in arid climates. The technique is based on the diversion of seasonal wadi flash floods with the use of diversion bunds, built within the river bed during dry periods. Spate irrigation has been practiced since 3000 BC, and today it covers around 3 million hectares of irrigated land around the world in areas distributed in arid and semi-arid zone of Near East, Africa, South and Central Asia and Latin America (Van Steenbergen *et al.*, 2010). In these contexts, spate irrigated agriculture is often not recognised as part of the formal irrigation sector, while it is usually one of the main sources of livelihood for the poorest sector of society.

In arid areas, wadi represents the only source of runoff available, and erratic and scarce rainfalls could not sustain agriculture. Spate systems allow turning flash floods into productive water for irrigation and other agricultural activities, while the deposition of fine sediments

* Corresponding author
Giulio Castelli, Via San Bonaventura 13, 50145 Firenze, Italy
Email: giulio.castelli@unifi.it

suspended in diverted water contributes to soil fertility. In addition to this, local populations have developed great wisdom in spate irrigation systems construction and they have reached effective management strategies and water rights systems, which allow coping with the unpredictability typical of wadi flows (Van Steenbergen *et al.*, 2010).

Despite its relevance for rural livelihoods and the potential as strategy for water management in arid climates, spate irrigation has been neglected in the technical literature (ibid.). Only in the last 20 years, governments, development agencies and NGOs have started recognising the relevance of spate systems for rural livelihoods and development in arid areas and began to implement modernisation programmes (Mehari *et al.*, 2011).

Spate irrigation in Ethiopia has developed relatively more recently than in other countries, due to the increasing food demand caused by population growth (Van Steenbergen *et al.*, 2010). In the northern arid Tigray region, the regional government has made strong efforts to improve traditional irrigation systems in the last 15 years, mainly focusing on the rural area of Raya Valley (Kidane, 2009; Van Steenbergen *et al.*, 2011) where spate irrigation has been practised for centuries (Kidane, 2009; Yazew *et al.*, 2014). In spate systems of Raya Valley, structural problems represent the main constraint. Diversion structures, which are built as spur-shape discharge separators using local material, such as earth, brushwood and stones, are usually washed away by most powerful spate flows. This leads to a lack of irrigation water and to a high need of labour for diversions maintenance and reconstruction after floods. A recent modernisation process based on the construction of new diversion structures and improvement of the channel systems, used more resistant material, such as concrete and gabions (Kidane, 2009; Yazew *et al.*, 2014).

Despite high investments, the result of this modernisation process was disappointing. Most of the interventions were dominated by an engineering approach, applying a design strategy typical for irrigation systems of permanent rivers. Farmers' ideas and preferences, their specific technical knowledge of spate systems management, their own well adapted institutional system and knowledge of local environment and hydrology were not considered and incorporated in the design. As a result, most of the modernised systems stopped to operate due to technical problems, related to a wrong assumption of river system hydrology, in particular sedimentation, and to institutional problems related to new operation and management strategies (Kidane, 2009; Erkossa *et al.*, 2014; Yazew *et al.*, 2014; Libsekal *et al.*, 2015). On the other hand, research showed that, despite structural problems, traditional farmer managed spate irrigation systems are performing better than modernised ones, due to the technical knowledge and experience of the farmers, who have been using the technique for centuries. For effective improvements in spate irrigation systems, the involvement of farmers in the planning and design phase is required (Kidane, 2009; Erkossa *et al.*, 2014; Yazew *et al.*, 2014; Libsekal *et al.*, 2015).

The aim of this work was to develop a participatory Diagnostic Analysis (DA) of a traditional spate system, in order to identify with local farmers the main problems and to set the ground for the design of appropriate technical solutions.

The DA concept is expressed as the "appraisal and analysis of existing irrigation systems with the objective to identify problems and to define the causes or constraints, underlying these problems" (Falciai, 1996). DA was conceived as the first part of a four-phases development model (Clyma *et al.*, 1977) but it has then been used in other development methodologies (Dedrick *et al.*, 2000; Bruscoli *et al.*, 2001). The technique is based on the analysis and identification of existing problems in order to develop appropriate solutions (Bresci & Letterio, 2007).

Participatory Rural Appraisal (PRA) was adopted for system diagnostic, in order to describe and identify major problems. PRA was defined by Chambers (1994) as "family of approaches and methods to enable local (rural or urban) people to express, enhance, share and analyse their knowledge of life and conditions, to plan and to act", and can be used as a tool for participatory diagnostic analysis of rural systems (Bruscoli *et al.*, 2001; Bresci & Letterio, 2007). The key concept in PRA is that local people are creative, capable of carrying their own analysis, identifying problems and constraints, planning and eventually taking actions. Researchers and field workers should act as facilitators and help local people to carry on their own system analysis. PRA involves a series of methods, which can be used and adapted to each case study, ranging from simple spatial representations of the study area (participatory mapping) to matrix ranking of different options. The information generated with PRA is shared and discussed with local farmers, providing a consistent ground for planning future development (Chambers, 1994).

## 2 Materials and methods

### 2.1 *Study area*

The analysis was carried out for Harosha spate irrigation system, located in the south of Raya Valley, in Harele *tabia* (municipality). Harosha wadi flows from west highlands of southern Raya Valley to the eastern valley (Fig. 1), where it ends spreading the flows in many channels, namely its distributary system (Nichols, 2009).

The spate irrigation system is located in the distributary system of the wadi. The first diversion structure (D1) is located upstream of the first division of the distribution system, on the left side (Fig. 2). Three further diversions (D2–D4) are on the left side of the northern branch of the river, before the road to Addis Ababa. Downstream the bridge, six smaller diversions are located on the left side of the northern branch and, more downstream, three on the right side. Harele villages are built on the right side of the northern branch. On the right bank, a gabion wall is protecting the villages from floods. The analysis was focused on the command area of the 4 upstream diversions, identified as Diversion 1 (D1) to Diversion 4 (D4), starting from upstream. A scheme of the area is shown in Figure 2.

The rainy season covers the months of June, July and August, with the possibility of early rains during April and May. The mean annual rainfall is 724 mm, the mean annual potential evapotranspiration about 1752 mm y$^{-1}$ (Hagos, 2010).

### 2.2 *Participatory Rural Appraisal*

PRA techniques were organised according to the classification proposed by Tesfai & de Graaff (2000), adding a fourth category: techniques focusing on spatial, temporal, socio-economical and spatio-temporal aspects of spate irrigation system. PRA was carried out from April to June 2014 with the help of a local interpreter for both oral and written communication. The analysis was carried out mainly with male farmers in the field, as women are usually carrying out household or non-agricultural work. Women farmers who are living alone were involved in a final meeting for problems discussion.

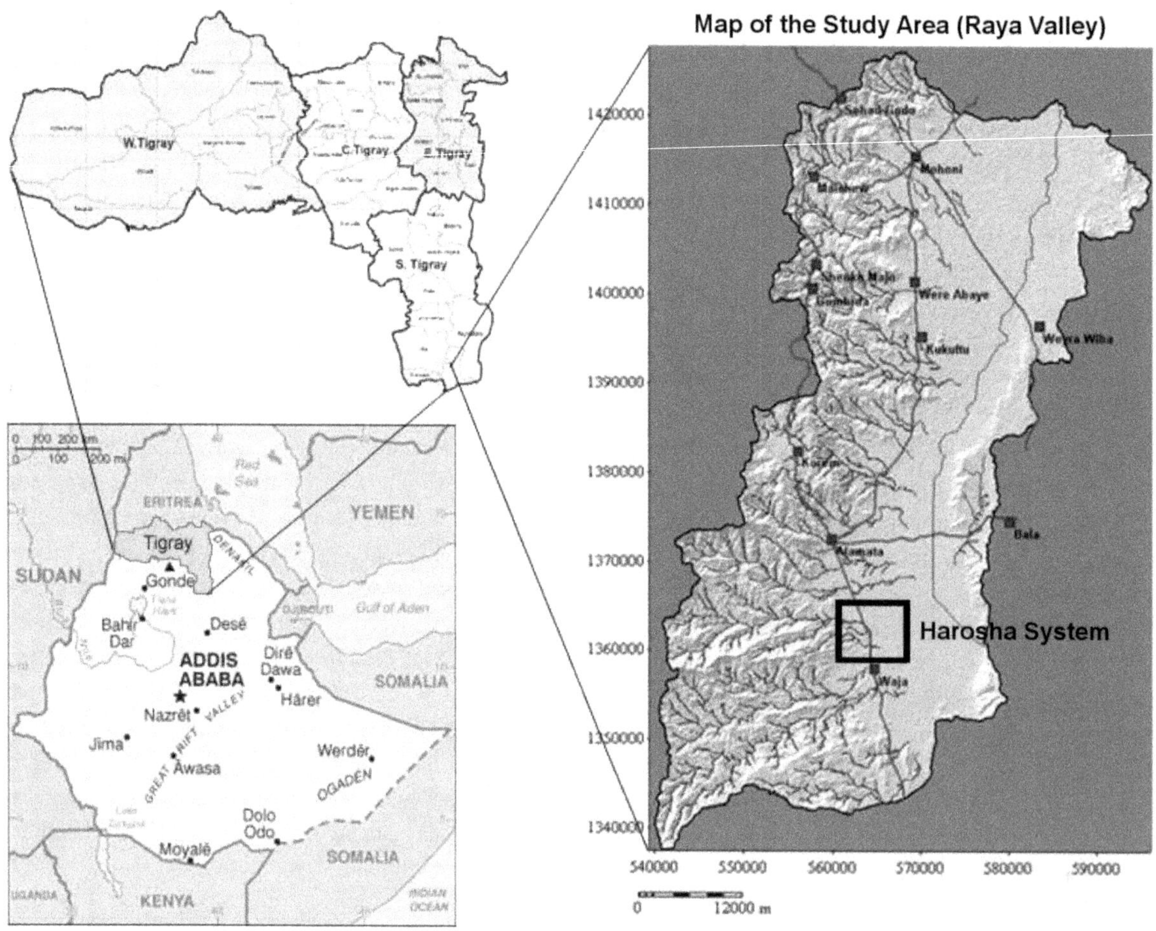


**Fig. 1:** *Location of Harosha system, adopted from Hagos (2010)*

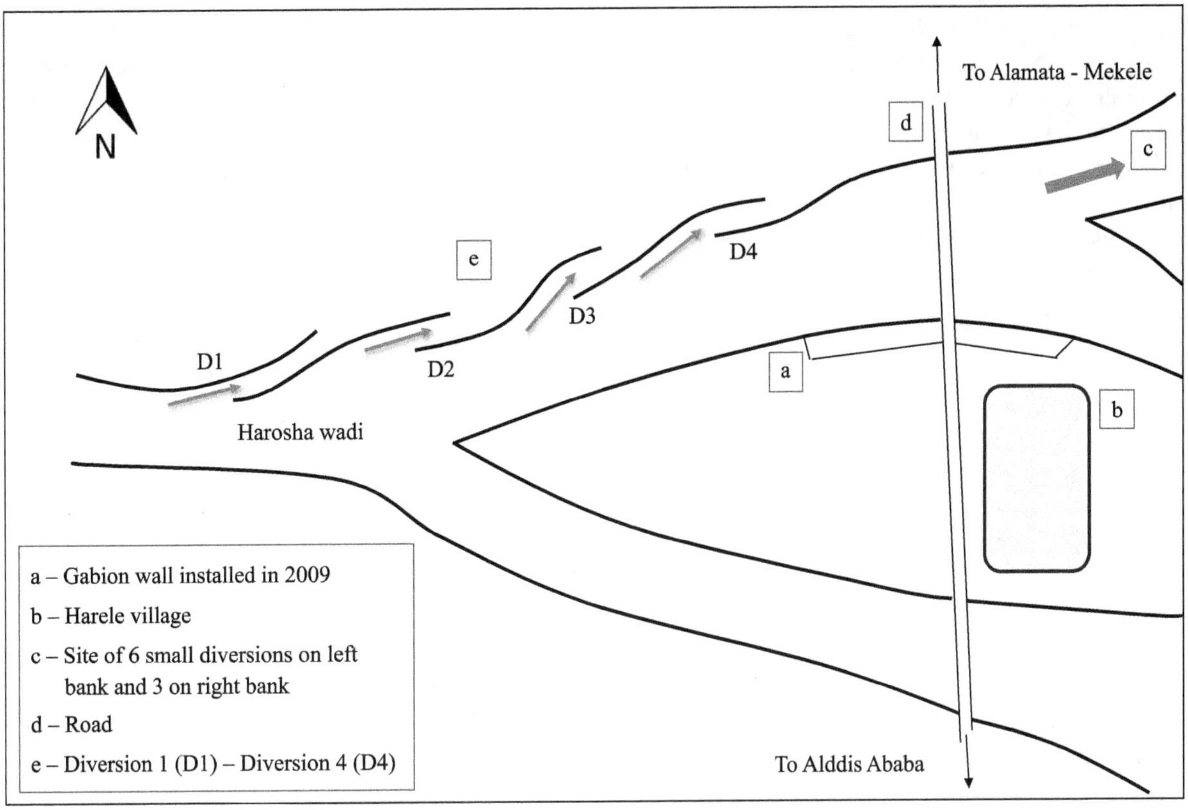


**Fig. 2:** *Scheme of study area and diversion sites*

*2.3 Techniques focusing on spatial aspects*

*Participatory maps:* Participatory mapping of the system was realised asking a group of four farmers to draw a sketch map of the command area. The map helped to obtain detailed information on the spatial characteristics of the system, to identify the main structures and it was used for planning the following PRA activities.

*Field walks:* Field walks were realised to acquire information about the system operation and water management at field level. The walk started with one or two farmers and interviews were done with other farmers met during the walks.

*Transects:* Transect walks were realised with groups of three or four farmers in order to obtain information about the differences found along one precise path in the system. The transect paths were made along canals for understanding the influence of the distance of each field from the diversion structure on water availability and sediment loads.

*Structure analysis:* Diversion structures were object of a more detailed analysis. A structured questionnaire about the characteristics of a diversion was realised with farmer representatives of the area. Design discussion: In a final meeting, to which all farmers were invited, a discussion on possible technical solutions was undertaken, as part of PRA analysis. The discussion involved about forty farmers.

*2.4 Techniques focusing on temporal aspects*

*Trend lines:* Trend analysis was realised to evaluate, year by year, the river water availability, irrigation water availability for the farmers, damage to diversions, erosion and crop production considering the last 5 years. Trend lines were realised for the command area of each diversion, with groups from three to four farmers. Values ranging from 1 to 5 were used for the representation of the considered variables, in order to facilitate the discussion and comparison of the values throughout the years.

*Seasonal calendar:* The calendar of farming activities throughout the year helped to understand farming and irrigation practices. Calendars were realised with the same groups of participatory maps.

*Semi Structured Interviews (SSIs) on hydrology:* SSIs about the hydrology of the wadi were realised in order to obtain information on the river system and how to organise a structured hydrological analysis, utilising living memory. Two to three SSIs were realised for the command area of each diversion.

*Hydrological analysis:* Wadis are commonly characterised by a lack of hydrological information, making hydrological modelling difficult. Most experienced farmers were involved for an analysis of the water levels in the wadi, selecting a cross section in which the level identification was simple for the farmers. Discharges were calculated using slope-area method (Van Steenbergen *et al.*, 2010). The following data were analysed:

- Maximum flow level within living memory
- Average of the maximum yearly levels within living memory
- Level of "high", "medium" or "low" flow during the years according to farmers' experience.

The number of occurrences of each of the above mentioned flow levels was defined, considering the cases of a dry, normal and wet year. The total flow time and the peak time were also analysed.

As the selected cross section was located in the northern branch of the system, the discharge calculated with slope area method was corrected with the following formula (Eq. 1), for obtaining the discharge of the full basin:

$$Q_i^* = f \cdot Q_i(y) \tag{1}$$

where

- $Q_i^*$ is the full catchment discharge occurring before the first division on the wadi distributary system, for the level $i$ ($m^3\,s^{-1}$)
- $Q_i$ is the discharge for level i, calculated with slope area method from the depth $y$ ($m^3\,s^{-1}$)
- $f$ is the full discharge factor, calculated as (Eq. 2):

$$f = \frac{L_1 + L_2}{L_1} \tag{2}$$

in which $L_1$ and $L_2$ are the width of northern and southern branches (Fig. 2).

### 2.5 *Techniques focusing on socio-economical aspects*

*Interviews:* One or two farmer representatives for each diversion command area and government officials from the agricultural bureau of Harele were interviewed.

*SSIs:* SSIs on socio-economical aspects were realised with farmers, considering history of the system, problems in the system, management structures (WUAs, representatives, rules and regulation), off-farm activities (what, when, how much, income). Nine SSIs were carried out with single farmers and in small groups (2–3 people).

*Ranking of problems:* The problems identified in the PRA activity were ranked in a meeting, just before the design discussion, to which all farmers were invited. The ranking methodology was organised providing "problem sheets" in which a problem was written and drawn (for farmers who cannot read). The sheets were used to visualise the ranking. The ranking was agreed with a free discussion, without the use of discussion tools such as pair-wise ranking. Figure 3 shows problem sheets.

**Fig. 3:** *Problem sheets used during PRA*

### 2.6 *Techniques focusing on spatio-temporal aspects*

*Crop production analysis:* Georeferenced crop production trend lines were realised by asking farmers to rank from 1 to 5 the crop production of their fields for the last 5 years. Each trend line was associated to a GPS point taken on the field, in order to analyse how crop production trends vary with the position in the system. 28 trends were realised.

## 3 Results

### 3.1 *Spatial aspects*

#### 3.1.1 *Characteristics of the command areas*

Participatory maps, field walks and transect walks were used to gain detailed information about the command area. The area covered around 70 ha, with roughly

150 households owning irrigated lands. Each family holds one or more plots in different parts of the system with an average plot size of 75 × 50 m. Each diversion deviates water to a primary canal, which delivers irrigation water to a determined area, namely the command area. In the command areas of D2, D3 and D4 primary canals convey water to secondary canals that deliver water to fields. In the D1 command area, two secondary canals are present, then water is conveyed to the fields through tertiary canals. Field canals convey water to each plot from the secondary or tertiary canals. The canal system is managed using micro diversion bunds (Fig. 4) placed inside the canals which are built and breached to direct water according to the irrigation order (Table 1).

The primary canal of D2 is being lengthened by farmers in order to extend the command area, allowing more farmers to have water. Farmers reported that there is generally no problem with the water delivery.

### *3.1.2 Diversion structures*

Diversion structures in traditional spate systems are usually conceived as structures built with local materials like earth, stones and wood, in order to be easily repaired and adapted by the farmers to the morphodynamical evolution of the river. On the other hand, fixed concrete structures may result out of place from one to the next year, due to the continuous shifts of wadi beds (Van Steenbergen *et al.*, 2010; Yazew *et al.*, 2014). Structure analysis and field walks revealed that diversion structures present in Harosha system area characterised by the typical "spur-type" design, as described by Van Steenbergen *et al.* (2010), namely bunds, parallel to the flow, used for deviating a part of the discharge into a diversion canal.

**Table 1:** *Main characteristics of the command areas under the respective diversion structures (D1–D4)*

| *Diversion* | *Number of parcels in the command area* | *Command area [ha]* | *Number of secondary canals* | *Number of tertiary canals* |
|---|---|---|---|---|
| D1 | 100 | 25 | 2 | 20 |
| D2 | 80 | 20 | 15 | – |
| D3 | 50 | 15 | 10 | – |
| D4 | 40 | 10 | 10 | – |

**Fig. 4:** *Micro diversion bund at field level*

The shape and the design of diversion structures are reported to be effective. In particular, the solution of using multiple diversions allow the farmer to be capable of irrigating a part of their land even if a diversion got broken, as they own land in more than one command area. Despite the advantages of traditional diversions, the main problem of diversion structures is that they are too often broken or washed away by flows.

To alleviate the heavy burden for reconstruction to farmers, local government has financed the use of gabions for reinforcing earthen diversion structures after floods in 2010. D3 and D4 were reinforced with gabions, which were installed by the farmers, and showed higher resistance to wadi ephemeral flows.

The following local techniques were described by farmers:

- *Upward slope of diversion canals:* farmers build diversion canals with upward slope in order to slow down water velocity and facilitate sedimentation in the first part of the channel. The system is working, but the upward slope reduced the amount of water diverted. Farmers suggested a refined design for optimising the balance between sedimentation and diversion efficiency.
- *Fuse structures:* fuse structures are used for diversion management. During high flows, farmers amount a pile of earth at the intake point, in order to avoid excessive flow in the diversion canal. During dry periods, farmers extend the diversion spur with a smaller earthen bund, in order to intercept low flows.

#### *3.1.3 River bank collapse*

In 2009, local government built a gabion wall for flood protection on the right side of the northern branch of the distributary system, in front of diversion structures sites (position "a" in Fig. 2). Gabion walls reduced the available section for water flows. As a result, the left bank of the channel collapsed, causing a reduction of cultivable lands. A loss of 13 ha in 3 years was reported. Farmers observed that protection walls should be built more distant from the wadi bed.

### *3.2 Temporal aspects*

Temporal analysis revealed that flood protection structures were built in 2009, in 2010 abundant rains led to good crop production in the whole command area, but high flows washed away all the diversions. In 2011, diversions were rebuilt, reinforced with the use of gabions, financed by local government. 2013 was reported to be the driest year of the last decade.

Specific trend lines realised for each diversion (Fig. 5a to 5e) show that more water is available for D1 which is located upstream of the first subdivision of the river. This results also in higher water availability for farmers, leading to a better crop production, and higher erosion of the left bank. Trends 5b and 5d showed how D3 and D4 have been more resistant, leading to better water availability for farmers throughout the years.

Figure 6 shows the seasonal calendar of farming activities. For ploughing activity, most farmers used animal traction. Livestock rearing is found in all households; especially cattle and small ruminants. Some farmers have off-farm activities such as being civil worker in a neighbouring city.

Table 2 shows the results of the hydrological analysis realised considering local people's experience of seasonal and historical floods. Observed water depths were used as input for the discharge analysis.

Table 3 shows the results of the discharge calculation. The discharge of wadi flow events may range on average from 22 to 194 $m^3 s^{-1}$, with peaks up to 750 $m^3 s^{-1}$, making it extremely difficult to design structures capable to abstract low water flows without being damaged by peak flows.

### *3.3 Socio-economical aspects*

One or two farmer representatives are elected as *Abo-Mais* (father of the river) for each diversion, and they stay in charge for life. Their duties are to organise system operation and maintenance, direct water delivery, and apply and collect penalties.

Farmers get water rights according to their contribution to maintenance works. Old and ill farmers, and female farmers alone, receive water even without any work contribution. If female farmers participate to maintenance works, they get water first, regardless of the time of contribution. If there is enough discharge, all tertiary canals are opened at the same time and water is delivered according to a list for each canal. If the discharge is not sufficient to supply all the canals together, water is delivered according to an aggregated order, considering all the canals as a whole. Penalties are applied if a farmer is not respectful of irrigation rules and may range from 50 to 100 Birrs (2.20–4.40 Euros). The penalties paid are used for buying building materials.

### *3.4 Spatio-temporal aspects*

The crop production analysis trends, realised at single household plot level, confirmed the information gathered with trend lines made at diversion level: 2010 was reported as the most productive year, due to high

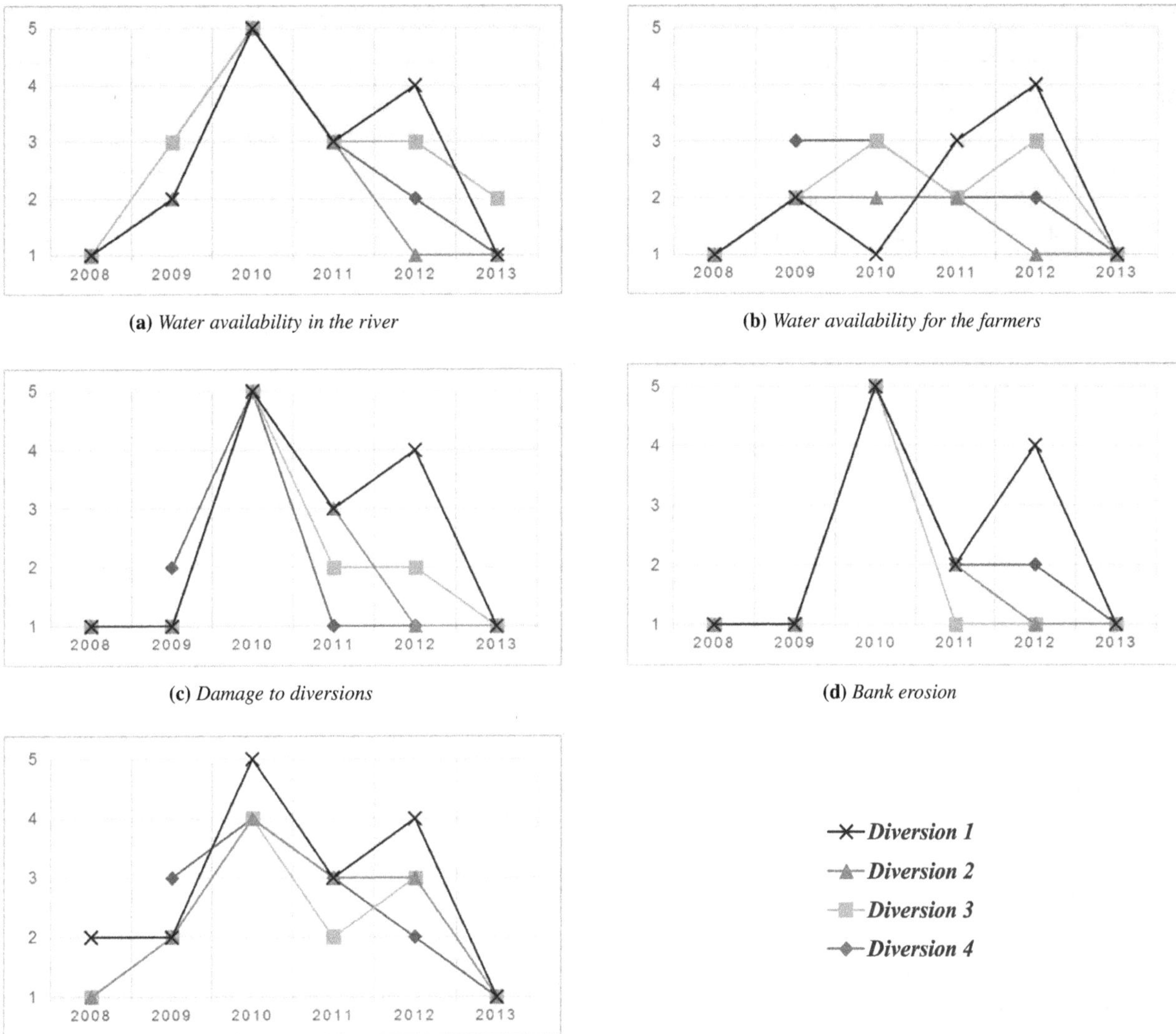


(a) *Water availability in the river*

(b) *Water availability for the farmers*

(c) *Damage to diversions*

(d) *Bank erosion*

(e) *Crop production*

**Fig. 5:** *Trend lines of the four diversions on a scale from 1 (low) to 5 (high)*

| *Activity* | JAN | FEB | MAR | APR | MAY | JUN | JUL | AUG | SEP | OCT | NOV | DEC |
|---|---|---|---|---|---|---|---|---|---|---|---|---|
| Diversion building | ■ | | | | | ■ | | | | | | |
| Ploughing | | ■ | ■ | | | | | | | | | |
| Seeding (maize, sorghum) | | | | ■ | ■ | | | | | | | |
| Seeding (teff) | | | | | | ■ | | | | | | |
| Maintenance works on diversion | | | | | | | ■ | ■ | ■ | | | |
| Removing of weeds | | | | | | | | | ■ | | | |
| Harvesting (teff, sorghum) | | | | | | | | | | ■ | | |
| Harvesting (maize) | | | | | | | | | | | ■ | |
| Threshing | | | | | | | | | | | | ■ |

**Fig. 6:** *Seasonal calendar of farmer activities*

**Table 2:** *Hydrological analysis results*

| *Type of flow* | *Observed water depth* [cm] | *Duration* | | *Number of occurrences* | | |
|---|---|---|---|---|---|---|
| | | *to peak* [hr] | *to end* [hr] | *dry year* | *average* | *wet year* |
| Maximum flow level within living memory | 400 | – | – | – | – | – |
| Mean level of the yearly maximum within living memory | 220 | – | – | – | – | – |
| High | 180 | 2 | 8 | 1 | 2 | 6 |
| Medium | 100 | 1 | 5 | 2 | 3 | 8 |
| Low | 50 | 0.5 | 2 | 3 | 4 | 12 |

**Table 3:** *Application of the slope area method to the Harosha catchment*

| *Level* | *y* [m] | *Name* | *Q* [$m^3 s^{-1}$] | *Q** [$m^3 s^{-1}$] | *Definition* |
|---|---|---|---|---|---|
| $y_{max}$ | 4 | $Q_{max}$ | 455 | 750 | Maximum discharge within living memory |
| $y_y$ | 2.2 | $Q_y$ | 165 | 272 | Discharge from the average of the maximum yearly levels within living memory |
| $y_h$ | 1.8 | $Q_h$ | 118 | 194 | Discharge from the average level of a "high" level flow |
| $y_m$ | 1 | $Q_m$ | 44 | 72 | Discharge from the average level of a "medium" level flow |
| $y_l$ | 0.5 | $Q_l$ | 14 | 22 | Discharge from the average level of a "low" level flow |

rainfall while 2011 and 2012 showed an average production, even if high labour input was requested for diversion maintenance. 2013 was the driest year, with lower crop production. In particular, the following localised tendencies were observed:

- In 2011 and 2012, farmers far from the river received less water and obtained lower yields.
- During 2013, farmers who own land near to the river, managed to divert some water and obtained good crop production, ranging from 4 to 5.
- Farmers whose land is located on the bank are losing cultivable terrain due to bank collapses.

### *3.5 Problem analysis and ranking*

The problems of the systems were discussed and ranked during the final meetings. Farmers defined the following priority rank:

(1) *Weakness of diversion structures:* Diversion structures break down too often, leading to heavy maintenance works to assure system operation.

(2) *Lateral erosion:* Cultivable land loss due to bank collapses.

(3) *Flood risk for villages:* High discharges often causes floods in Harele *tabia* villages.

(4) *Flood risk for fields:* Fields are often flooded by Harosha river, leading to water logging problems, especially for Teff cultivation.

(5) *Size of the diversion structures:* The present size of diversion structure is too small and the diversion efficiency is low.

(6) *Sedimentation:* Due to upward slope diversion, sedimentation can be efficiently managed in the system and does not represent a significant problem.

(7) *Lack of manpower:* It was reported that some farmers do not participate to maintenance works because they are discouraged by system performances. However, this represents a management problem for the system as a whole.

(8) *Lack of materials:* Lack of building materials is not perceived as a problem.

(9) *Presence of weeds:* The presence of pests and parasite plants was reported, but is it not considered a relevant problem. Weeds, which are especially causing problems in case of early rains, are removed by farmers through ploughing.

During the discussions, farmers explained that they want to focus on technical solutions for the first two recognised problems. During the analysis of possible technical solutions, farmers strongly supported the use of gabions for new diversion structures, explaining that the gabions installed in D3 and D4 showed good resistance and that the community would be able to install and maintain the structures by itself. According to that, farmers also suggested to protect the river banks with gabions to reduce the damage of floods on crops and villages.

## 4 Discussion

Spate irrigation represents a vital source of livelihood for rural population of Ethiopia, and modernisation of spate systems, where correctly implemented, has been proven to significantly reduce the poverty level of farmers by providing stable access to water resources (Hagos *et al.*, 2014). However, the low consideration of local farmers knowledge of the technique resulted in a poor design and led to the failure of most of the modernised systems implemented over the last twenty years (Erkossa *et al.*, 2014; Yazew *et al.*, 2014).

The use of PRA for the Diagnostic Analysis of the traditional spate irrigation system proved to be a feasible method to learn about farmers' views and knowledge on the management of spate irrigation. The DA carried out for Harosha spate irrigation system showed that the main constraint of the system is represented by the structural weakness of traditional diversion structures, built with local materials. PRA analysis also provided a solid ground for the development of technical solutions.

There is no need for a heavy engineering intervention on the canals system, and a modification of irrigation rules and management is not requested. Proposed solutions should maintain farmers' own technical solutions, like upstream diversion canals and the use of fuse structures, which are working with good results in Harosha system.

In previous studies realised in the framework of spate irrigation modernisation, hydrological analysis has been identified as one of the critical factors for new systems development, in order to design structures that can be consistent to the typical nature of wadi floods (Erkossa *et al.*, 2014). In the present study, discharge calculation was made possible by farmers' experience. Results showed an extreme uncertainty related to hydraulic structures design, given by the typical characteristics of wadi contexts. Spate irrigation systems should be designed to be resilient to the impact of extreme flows, which can also modify the morphology of wadi bed. In this situation, the reconstruction of diversion structures can be considered an actual part of the management system (Mekdaschi Studer & Liniger, 2013; Van Steenbergen *et al.*, 2010). Nevertheless, like in other studies realised with the participation of local farmers in Ethiopia (Erkossa *et al.*, 2014), DA showed how the current diversion structures of Harosha system need to be improved to reach a sufficient stability in the medium term, reducing the heavy burden for maintenance and reconstruction to farmers.

While most of the literature on spate irrigation modernisation focuses on single-intake design solutions (Embaye *et al.*, 2012; Libsekal *et al.*, 2015), structure analysis and design discussion have shown that the current design of multiple spur diversion structures is suitable for a wadi context, where the strategy of multiple intake points could be more resilient as a single intake system. The latter could completely fail if the single diversion structure is damaged or destroyed.

Farmers emphasised that gabions could represent a suitable solution for building new structures, as they can be installed and maintained by farmers themselves. Gabion structures also represent a flexible solution that can resist to deformations imposed by large floods and river morphological modifications, typical of wadis, better than rigid concrete structures. In addition to this, a spur deflector made in gabions would maintain the use of traditional technical solutions adopted by farmers, like fuse structures and upstream slope canals for sediment management. Solutions for reducing the high sediment load, typical of wadi flows, have been tested by other authors by using a design approach based on computational hydraulic modelling, like diversion canals with changes in section (Embaye *et al.*, 2012). The PRA analysis showed that upstream canals have a good performance for sediment control and can be considered for further technical design.

Spate irrigation development strategies in Ethiopia have focused mainly on the modernisation of the irrigation structures (Mehari *et al.*, 2011; Embaye *et al.*, 2012; Erkossa *et al.*, 2014; Yazew *et al.*, 2014; Libsekal *et al.*, 2015). PRA results showed that lateral erosion of the river bank, flood damages to villages and flood damages to cultivations were indicated by the farmers as the second, third and fourth relevant problem of the system. Structures for flood risk mitigation are considered necessary by local population, in order to alleviate flood damages on rural systems, considering both crops and human settlements. To frame effective spate development strategies, an integrated approach is then necessary, considering both irrigation development and flood risk mitigation for rural communities.

#### Acknowledgements

This study was possible thanks to the funding of the International Fund for Agrarian Development IFAD, the coordination of UNESCO-IHE and the support of Mekelle University, Ethiopia. The authors are very grateful to the Harele people and their hospitality, kindness and collaboration. We want to thank Zinabu Gobezie and Kiflom Tesfay for their help in community involvement, translation and their sincere friendship born wondering around Alamata. We also thank the two anonymous reviewers and Dr. Eric Tielkes as editor for their comments, which considerably contributed to the improvement of this manuscript.

## References

Bresci, E. & Letterio, T. (2007). Diagnostic analysis for the rehabilitation of an irrigation system in the Stolac municipality (Bosnia Herzegovina). *In:* Lamaddalena, N., Bogliotti, C., Todorovic, M. & Scardigno, A. (eds.), *Water Saving in Mediterranean Agriculture and Future Research Needs. Vol.1 - Options Méditerranéennes: Série B. Etudes et Recherches; N. 56 Vol.I*. pp. 319–327, CIHEAM, Bari, Italy.

Bruscoli, P., Bresci, E. & Preti, F. (2001). Diagnostic Analysis of an Irrigation System in the Andes. *Agricultural Engineering International: the CIGR Journal of Scientific Research and Development*, III, 1–14.

Chambers, R. (1994). Participatory rural appraisal (PRA): Analysis of experience. *World Development*, 22, 1253–1268.

Clyma, W., Lowdermilk, M. K. & Corey, G. L. (1977). *A research development process for improvement of on-farm water management*. Water Management Research Project, Colorado State University. Fort Collins, Colorado, USA.

Dedrick, A. R., Bautista, E., Clyma, W., Levine, D. B. & Rish, S. A. (2000). The management improvement program (MIP): A process for improving the performance of irrigated agriculture. *Irrigation and Drainage Systems*, 14, 5–39.

Embaye, T. G., Beevers, L. & Mehari Haile, A. (2012). Dealing with sedimentation issues in spate irrigation systems. *Irrigation and Drainage*, 61 (2), 220–230.

Erkossa, T., Langan, S. J. & Hagos, F. (2014). Constraints to the Development, Operation and Maintenance of Spate Irrigation Schemes in Ethiopia. *In:* Erkossa, T., Hagos, F. & Lefere, N. (eds.), *Proceedings of the Workshop on Flood-Based Farming for Food Security and Adaption to Climate Change in Ethiopia: Potential and Challenges, Adama, Ethiopia*. pp. 5–22, International Water Management Institute (IWMI), Colombo, Sri Lanka.

Falciai, M. (1996). *Diagnostic Analysis of Farmer-Managed Irrigation*. University of Florence, Italy.

Hagos, F., Erkossa, T., Lefore, N. & Langan, S. (2014). Spate Irrigation and Poverty in Ethiopia. *In:* Erkossa, T., Hagos, F. & Lefere, N. (eds.), *Proceedings of the Workshop on Flood-Based Farming for Food Security and Adaption to Climate Change in Ethiopia: Potential and Challenges, Adama, Ethiopia*. pp. 43–52, International Water Management Institute (IWMI), Colombo, Sri Lanka.

Hagos, M. A. (2010). *Groundwater Flow Modelling assisted by GIS and RS Techniques (Raya Valley - Ethiopia)*. Master's thesis, ITC - International Institute for Geo-information Science and Earth Observation, Enschede, The Netherlands.

Kidane, H. (2009). *Community Spate Irrigation in Raya Valley: The Case of Three Spate Irrigation Systems*. Master's thesis, Addis Ababa University, Addis Ababa, Ethiopia.

Libsekal, H., Gebre-Egziabher, T., Mezgebu, A. & Yazew, E. (2015). Developments in the Design and Construction of Modern Spate Systems in Tigray, Ethiopia. *Civil Environmental Research*, 7, 108–113.

Mehari, A., Van Steenbergen, F. & Schultz, B. (2011). Modernization of spate irrigated agriculture: A new approach. *Irrigation and Drainage*, 60, 163–173.

Mekdaschi Studer, R. & Liniger, H. (2013). *Water Harvesting: Guidelines to a Good Practice*. Centre for Development and Environment (CDE), Bern; Rainwater Harvesting Implementation Network (RAIN), Amsterdam; MetaMeta, Wageningen; The International Fund for Agricultural Development (IFAD), Rome.

Nichols, G. (2009). *Sedimentology and Stratigraphy*. (3rd ed.). WILEY, Oxford.

Tesfai, M. & de Graaff, J. (2000). Participatory rural appraisal of spate irrigation systems in eastern Eritrea. *Agriculture and Human Values*, 17 (4), 359–370.

Van Steenbergen, F., Lawrence, P., Mehari, A., Salman, M., Anderson, I. M., Nawaz, K. & Ratsey, J. (2010). *Guidelines on spate irrigation*. FAO, Rome.

Van Steenbergen, F., Mehari, A., Alemehayu, T., Alamirew, T. & Geleta, Y. (2011). Status and Potential of Spate Irrigation in Ethiopia. *Water Resources Management*, 25, 1899–1913.

Yazew, E., Gebru, T., G/Egziabher, T., Mezgebu, A., Fessehaye, D., Habtu, S., Haile, A., Van Steenbergen, F. & Wester, F. (2014). *Spate Irrigation Systems in Raya Valley (Ethiopia)*. Overview Paper #13, Spate Irrigation Network.

# 6

# An analysis of household farm investment decisions under varying land tenure arrangements in Ghana

Michael Ayamga , Richard W. N. Yeboah *, Sylvester Nsobire Ayambila

*Faculty of Agribusiness and Communication Sciences, University for Development Studies, Tamale Ghana*

**Abstract**

Land tenure insecurity is widely perceived as a disincentive for long-term land improvement investment hence the objective of this paper is to evaluate how tenure (in)security associated with different land use arrangements in Ghana influenced households' plot level investment decisions and choices. The paper uses data from the Farmer-Based Organisations (FBO) survey. The FBO survey collected information from 2,928 households across three ecological zones of Ghana using multi-staged cluster sampling. Probit and Tobit models tested the effects of land tenancy and ownership arrangements on households' investment behaviour while controlling other factors. It was found that marginal farm size was inversely related to tenure insecurity while tenure insecurity correlate positively with value of farm land and not farm size. Individual ownership and documentation of land significantly reduced the probability of households losing uncultivated lands. Individual land ownership increased both the probability of investing and level of investments made in land improvement and irrigation probably due to increasing importance households place on land ownership. Two possible explanations for this finding are: First, that land markets and land relations have changed significantly over the last two decades with increasing money transaction and fixed agreements propelled by population growth and increasing value of land. Secondly, inclusion of irrigation investment as a long term investment in land raises the value of household investment and the time period required to reap the returns on the investments. Households take land ownership and duration of tenancy into consideration if the resource implications of land investments are relatively huge and the time dimension for harvesting returns to investments is relatively long.

*Keywords:* Ghana, Land tenure, soil improvement, investment decision, probit

## 1 Introduction

Tenure security enables farmers make long-term investments to improve soils in the expectation that they will hold their land rights long enough to reap the associated benefits. Secure land rights provide incentives for investing in soil conservation, land improvements and other productivity-enhancing operations since farmers are assured of reaping the stream of benefits associated with their investments (Platteau, 1995). In contrast, the effect of lack of land tenure security is uncertainty in a farmer's mind about the value of improvements made to the land. This uncertainty tends to increase as farming becomes more commercialised and expected to have a bearing on how farmers manage their land and thus influence agricultural productivity (Alemu, 1999; Sjaastad & Bromley, 1997).

Low agricultural investment and productivity in Ghana have among other things, been attributed to problems with security of land rights and gender disparity with respect to access to and control of productive resources including land (Agbosu, 2000; Tsikata, 2003).

* Corresponding author
Richard W. N. Yeboah
UDS, P. O. Box TL 1882, Tamale Ghana
Email: narteyrwn@hotmail.com

With public agricultural investment at nine percent of GDP, and trailing the Comprehensive African Agricultural Development Program (CAADP) - Maputo Declaration's target of 10 percent, it is worrying that Ghana's quest to modernise agriculture and maintain growth levels consistent with middle income economies would be severely compromised if the levels of private and public investments in agriculture remain low. Though a multiplicity of factors may account for the low agricultural investment in Ghana, the land tenure and administration system has often been cited as a disincentive for land investment. Aryeetey & Udry (2009) reports that land and its management in Ghana has often been poorly understood and is generally perceived to induce tenure insecurity.

Land tenure and administration system in Ghana is a combination of customary and statutory processes. Customary land tenure systems are usually managed by traditional rulers, land earth priests, council of elders, and family or lineage heads. Legislations and interventions by colonial and early post-independence governments led to the establishment of the pluralist legal system in which English common law was grafted onto Ghanaian communal societies, largely neglecting differences between the early 19th century capitalist economic structures and the egalitarian communal institutions of Ghana (Agbosu *et al.*, 2007). The resulting legal pluralistic land administration has laid the foundations for conflicts and is widely perceived to induce tenure insecurity which has hindered farm investments and agricultural growth (Agbosu *et al.*, 2007).

This paper examines smallholder farmers' land investment decisions under varying land tenure arrangements using plot level data from 2,928 households across Ghana. The paper is motivated by the increasing need for knowledge on how indigenous land tenure arrangements influence the willingness of plot holders to undertake long-term soil and farm improvement investments. Exploring how land relations in Ghana affect farm households' investment behaviour supports the achievement of the dual goals of identifying entry points for government to influence smallholder farm investment and also providing feedback for ongoing land tenure and administration reform.

## 2 Land tenure and smallholder farm investment: the theory and empirical evidence

Resource allocation under different land use arrangements has been a subject of debate in theory especially on the subject of enforceability of land and labour contracts (Otsuka *et al.*, 1992). Marshall (1890) and Mill (1848) conclude that share tenancy results in inefficient resource allocation, arguing that since the share tenant receives as marginal revenue only a fraction of the value of his/her marginal product of labour, the tenant's incentive to supply labour or other inputs at the optimum level is limited[1]. On the other hand, Cheung (1969) and Johnson (1950) have contended that, if effort is costlessly enforceable, sharecropping arrangements can be as efficient as owner-cultivated and fixed-rent tenancy. Ahmed *et al.* (2002) reports that relative to land cultivated by owners and fixed-fee rental land, alternative land tenure systems such as sharecropped and gifted land prevailing in Ethiopia were less efficiently cultivated and resulted in considerable economic losses to farmers due to the restrictions imposed by landowners on the latter. They suggested that policies that facilitate individualised land transactions would result in considerable gains to farmers in particular and the national economy at large.

The growing list of studies examining the relationship between land tenure security and long-term farm investment or productivity have done little to quell uncertainty and controversy over the much hypothesised nexus especially in Africa. Most of the studies in Africa, including those mentioned above have produced mixed results with the greater majority failing to establish the hypothesised strong links between tenure security, investments and productive efficiency.

Considering the relationship between land title and investment, most empirical studies have produced inconclusive results. In Ghana, Rwanda and Kenya for example, Migot-Adholla *et al.* (1991) found that increasingly individualised land rights do not appear to have any effect on agricultural investment and yields. In areas of Kenya with land registration, no link was found between land titling and long-term investments to improve land (Barrows & Roth, 1989). In Zimbabwe, Harrison (1987) found little variation in the productive performance between small holder farmers with no land title and large scale commercial farmers with land titles.

Evidence linking individualised land rights with higher productive efficiency is weak and inconsistent. Though Laffont & Matoussi (1995) found significant evidence of inefficiency under sharecrop arrangements in a study in Tunisia, the study failed to provide evidence

[1] The tenant receives a predetermined share of the output (e.g. 40 %) and also pays or decides on variable input use. The exogenously determined share of the output may under value his labour and variable input contribution or over compensate the landlord for use of the land.

of higher efficiency under individualised land holdings. Ahmed *et al.* (2002) found significant inefficiency on sharecropped land but not so on land under fixed-rent contracts. Unlike in Africa, the evidence on the relationship between titled land rights and productivity has been more consistent in Asia and Latin America, where some link is established between yields, farm investments and tenure security Feder *et al.* (1987); Salas *et al.* (1970); Shaban (1987); Villamizar (1984).

A cursory look at the approaches used in investigating the tenure security-productivity hypotheses in Ghana reveals a certain degree of proclivity to notions of superiority of individualised land rights and land titling as ultimate determinants of tenure security. As a result, the research into the land tenure security-productivity hypotheses appears lopsided with emphasis on identifying analytical and modelling deficiencies as opposed to interrogating issues that border on conceptualisation and operationalisation of land tenure security, the most significant parameter of the hypothesis.

The common response to the failure to observe expected relationships between tenure security and productivity is the attempt to argue that tenure security is endogenous and that earlier studies lacked the econometric rigour to adequately account for the perceived endogeneity of tenure security (Besley, 1995; Hayes *et al.*, 1997; Twerefou *et al.*, 2011). Many of the more recent investigations of the tenure security-productivity hypotheses have therefore focused on resolving the issues of endogeneity in tenure security mostly through multistage econometric modelling (Besley, 1995; Hayes *et al.*, 1997; Twerefou *et al.*, 2011). The findings of these studies however have not been radically different in terms of resolution of the ambiguity surrounding the relationship between tenure security and expected improvements in investments.

Using the same data set as Migot-Adholla *et al.* (1991), Besley (1995) assumed that land rights were endogenous with farmer investment aimed at improving their rights over land. He concluded that better land rights facilitated investment in Wassa but not in Anloga, a direct opposite of the findings made by Migot-Adholla *et al.* (1991). Twerefou *et al.* (2011) in their study of tenure security, investments and the environment in Ghana, set tenure security as endogenous. They found that investments in farmlands in Ghana were low, appeared not to enhance tenure security, and argued that the reverse causation assumption of tenure security enhancing investment seemed non-existent. They further concluded that tenure security appeared to be an incentive for investment and had a positive and significant effect on investment though the authors suggested the results were not robust because the model did not control for endogeneity. Twerefou *et al.* (2011) are contradicted by Dzanku (2007) who failed to establish an overwhelming link between land rights and investment in irrigation and soil improvements.

The farmer bases his investment decisions on his level of tenure security and chooses between investments in capital equipment, which is not lost in the event that he/she loses his rights to land and long-term soil improvement and irrigation-related investments, which are completely lost in an eviction (Feder *et al.*, 1987). Several studies on land tenure security and farm investment and productivity, including Place & Hazell (1993) in Ghana, Rwanda and Kenya, and Hayes *et al.* (1997) in Gambia have applied the Feder *et al.* (1987) framework.

The model assumes that the household's utility is increasing in present value of future income stream ($\pi$), and household characteristics and asset wealth ($H^c$):

$$max_{I_i} U[E(\pi_t); H_t^c] \quad \text{Subject to}$$
$$E(\pi_t) = \sum_{t=1}^{T} \delta^t (p_t q_t A_t E[\omega_t] - cI_i(\epsilon_e) I_{it}; H_t^c)$$
$$q_t = y(s_t, K_t)$$
$$s_t = s_0 \left(1 - er(\varphi^t, \sum_{\omega=1}^{t} I_{i\omega})\right) \tag{1}$$

Where $\pi_t$ the value of future income stream ($\pi_t$) at the end of the household's planning ($T$), $cI_i$ is the unit cost of conservation investment ($I_i$) discounted by $\delta^t$. $H_t^c$ denotes household asset wealth characteristics. The unit cost of conservation investment is assumed to be decreasing in level of farmer experience ($cI_i(\epsilon_e) < 0$). The expected crop revenues are given by the product of crop price ($p_t$) and yield ($q_t$), and land area ($A_t$), and dichotomous expectation of land tenure in the period $t(E[\omega_t])$. Yield in season t is assumed to be concavely increasing with soil depth[2] ($q'(s_t) > 0$) and also a function of other conditioning factors ($K_t$) such as weather, pests and soil fertility.

Unlike Place & Hazell (1993) and Hayes *et al.* (1997), this study adds the dimension of investment in the development of irrigation structures.

[2] Soil depth is a proxy of good soil conditions. Deep and friable soil is presumed good conditions for crop productivity and yields are assumed to increase with improving conditions. Soil depth is defined as depth in soil to which the roots of a plant can readily penetrate.

## 3 Methodology

### 3.1 Empirical model

Based on the theory and the large body of empirical research undertaken on the subject of land tenure and farm investment across a number of African countries by Clay *et al.* (1998), Feder & Feeny (1993) and Hagos & Holden (2006) the study sets the household as a utility maximizing entity that chooses between short-term complementary inputs use and long-term land investments based on the household's rate of time preference and specifies an estimable empirical model as:

$$I_i = f(tenure, wealth_{t-1}, H^c_{T-1}, plot, market, crop, Zone) \quad (2)$$

where: $I_i$ measures household conservation and irrigation investment in plot $i$. The survey solicited household responses on the investments made in soil and water conservation and irrigation expressed in Ghana Cedi (GHS). The tenure variable represents factors that influence the farmer's expectation of retaining tenure or land rights such as whether the plot is owner operated, rented, temporally transferred (loaned). The duration of tenure is also included and is expected to improve farmers perceived tenure security. The $wealth_{T-1}$ variable denotes household wealth and asset holdings including relative farm size, livestock holdings, labour and other resource endowments. $H^c_{T-1}$ represents household demographic characteristics such as age and education of household head. The variable *plot* represent farm characteristics such as soil type, drainage, degree of fragmentation (ratio of total number of parcels to total farm size), and access to irrigation. The variable *market* measures market access variables such as borrowing and access to agricultural extension information. The variable *crop*[3] denotes the type of crop cultivated, either short duration crops or annuals or perennials that require land for several seasons. The *zone* variable controls for location fixed effects such as distance to markets, population density and rainfall.

The household soil conservation and irrigation investment decision making is assumed to happen at two levels. The household first decides whether to invest or not to invest and upon deciding to do the former, make decision on the level of investment. Both decisions are influenced by factors including those outlined in the empirical model (equation 3.). The level of conservation is given by

$$I^L = X_1\beta_1 + \varepsilon_1 \quad (3)$$

where $I^L$ is the level of household conservation investment which depends on the vector of $X_1$ explanatory variables outlined in (2).

The decision to invest or not is given by

$$I^D = X_2\Omega_2 + v_2 \quad (4)$$

where ($X$ and $I^D$) are observed, whereas $I^L$ is observed only when $I^D = 1$.

The model assumes that $\varepsilon_1$and$v_2$ are independent of $X$ implying that $X$ is exogenous, and $v_2 \sim N(0, 1)$. Given such a model, if the error terms in Equations (3) and (4) are related, they must first be estimated jointly given the premise that the household chooses whether to invest and then, having decided positively chooses the level of conservation investment. This implies there could be problems of selection bias hence requiring that the two equations be estimated jointly. The estimation procedure therefore involved testing for the presence of selection bias using the Heckman selection model (Heckman, 1990), and examining the likelihood ratio test of independence.

Sample selection bias was tested using a Heckman two-step model (Deaton, 1997). Adjustment for the standard errors in the level of investment model (equation 3) for heteroscedasticity using Powell's Censored Least Absolute Deviations (CLAD) estimator was done. The CLAD estimator unlike the standard estimators of the censored regression model is robust to heteroscedasticity, consistent and asymptotically normal for a wide class of error distributions (Arabmazar & Schmidt, 1981).

The Heckman test for sample selection bias tests the null hypothesis of $H_0$: $\rho\varepsilon_1 v_2 = 0$ and the alternative hypothesis $H_A$: $\rho\varepsilon_1 v_2 \neq 0$. The measure of correlation between $\varepsilon_1$ and $v_2$ is the correlation coefficient $\rho$. If the study rejects the null hypothesis of $\rho\varepsilon_1 v_2 = 0$, then the decision to invest equation (the sample selection equation) and the level of investment (outcome equation) cannot be said to be independent and thus must be estimated jointly by the Heckman technique. The Wald's test of independence indicates that $\rho\varepsilon_1 v_2$ is not significantly different from zero hence failure to reject the null hypothesis $H_0$: $\rho\varepsilon_1 v_2 = 0$ i.e. the models revealed no significant selection bias. As indicated, the significance of this result is that the sample selection (decision to invest) and outcome equations (money value of investments) could be treated as two independent equations and estimated separately.

[3] Dummies for annual crops (including maize, millet, cassava, cowpea, yams, among others); perennial cash crops (comprising cocoa, oil palm and mango); non-perennial cash crops (pineapple, soybean, melons)

The binomial probit and censored regression (Tobit) models were used to estimate the decision to invest and level of conservation and irrigation investments, respectively. When the dependent variable is dichotomous (0, 1), the probit and the logit models are preferable but for continuous dependent variables that are censored at or below zero the Tobit model is preferable (Anley *et al.*, 2007). The Tobit model is used to allow for censoring of households that made zero investment in land improvement.

The standard Probit and Tobit models may be formulated as:

$$y_i^* = x_i'\beta + \epsilon_i$$
$$y_i = \begin{cases} 1 & \text{if } y_i^* = 1 \\ 0 & \text{otherwise} \end{cases}$$

$$y_i^* = x_i'\beta + \epsilon_i$$
$$y_i = \begin{cases} y_i^* & \text{if } y_i^* > 0 \\ 0 & \text{if } y_i^* \leq 0 \end{cases} \tag{5}$$

where $i = 1, 2, \ldots, N$, and $\epsilon_i$ is assumed to be NID $(0, \delta^2)$ and independent of $x_i$. This model is a censored regression model where observations may be censored from below.

### 3.2 *Study area and data*

The study used the Farmer Based Organisation (FBO) survey data collected by the Institute of Statistical, Social and Economic Research (ISSER) of the University of Ghana in 2008 (ISSER, 2008). The survey was intended to facilitate the monitoring and evaluation of the Millennium Challenge Compact signed between the Government of Ghana and the Millennium Challenge Corporation (MCC) of the United States of America.

The strength of the data used for this study lies in the geographical spread of the sample. Land tenure systems in Ghana have been largely influenced by ethnicity, population pressures and agriculture intensification, with the land tenure systems in the South differing from tenure systems prevalent in the Northern parts of Ghana. Geographically, the study covered six regions; Northern, Central, Eastern, Volta, Ashanti and the Greater Accra Regions located in three ecological zones.

A sample of 2,928 farm households drawn from 23 districts in six regions within the three distinct agro-ecological zones of Ghana, namely, the Northern Agriculture Zone (Northern Region of Ghana), the Afram Basin (Ashanti and Eastern Regions of Ghana), and the Southern Horticultural Belt (South-East Coastal Plains of Volta Region) was used (Figure 1).

The FBO survey collected information on the overall living circumstances and farming activities of members of FBOs and their respective households. In-depth household data was collected using two sets of questionnaires; a household questionnaire and a community questionnaire: The survey collected information on a wide range of household attributes including the demographic, education and health characteristics; migration; household transfers; information seeking behaviour of households; household assets and participation in financial markets (borrowing, savings and lending behaviour); household agriculture activities including land ownership and transactions and agriculture processing and, non-farm enterprises of households. Information was also collected on the location of households, community facilities and farm sizes using geographic position system units (GPS). The community questionnaire was essentially a market price survey.

The data was collected using multi-stage probability sampling, clustered, and stratified with probability proportional to the size of target population to sample households. The approach used a two-stage sampling technique. During the first stage, 27 enumeration areas (EAs) were selected using systematic sampling with probability proportional to size method (PPS) for each district. The EAs were the same as those used during the 2000 Population and Housing Census by the Ghana Statistical Service (GSS, 2000). The selected enumeration areas were listed fully to know the total number of households that served as sampling frame from which an appropriate sample size was selected.

## 4 Results

### 4.1 *Descriptive statistics*

Table 1 presents description and summary statistics of variables used in equations 3 & 4.

The standard errors were adjusted to correct for clustering effects. Only 20 % of sampled households are female-headed. Although about 70 % of household heads reported to have attended basic school, only about 20 % are able to read and write a simple sentence in English.

Only 14 % of sampled households perceive their tenure status as insecure. The distribution appears to be even across observations in terms of the modes of land acquisition and forms of land rights. Though 30 % of sampled households have land titles, only 20 % are reported to exercise complete rights over their land. Close to 20 % of households purchased their land outright while 30 % received land as gifts. Only one percent

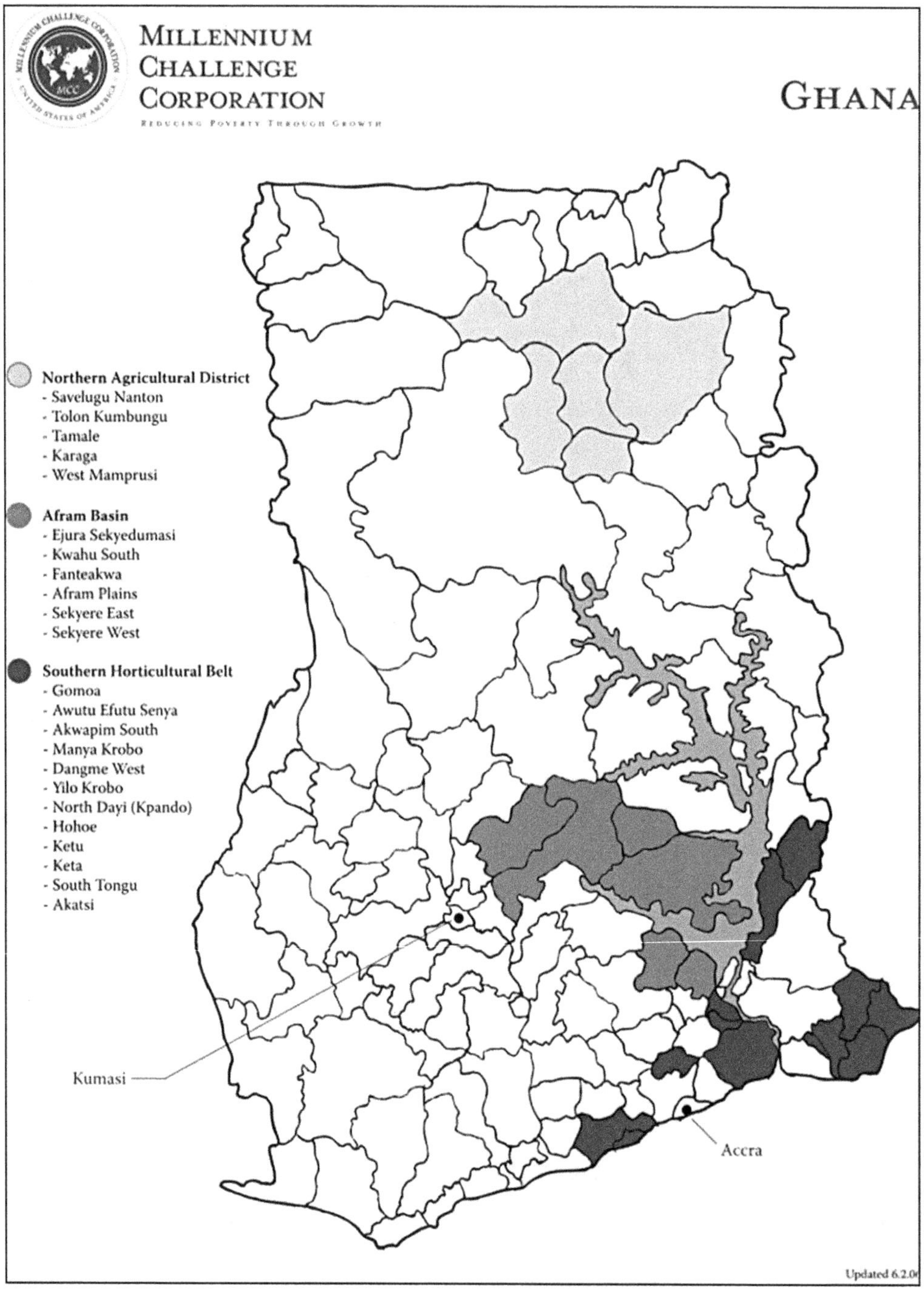


**Fig. 1:** *The study area in regions and districts*

of land is held in sharecropping arrangements. About 80 % of sampled households indicated they could vacate their lands and still maintain ownership but the period of time for which they could vacate the plots averaged at 6 months of the production season.

About a third of sampled households made land improvements and irrigation related investments. Households invested an average of about GHS 200[4] per acre on land improvement and irrigation. Compared with average value of farm produce and average non-farm income, households invest about 29 % of farm revenue (excluding livestock) or 60 % of non-farm revenue. Sampled households have owned and used land for over 9 years on the average with over 10 % of the farms enclosing pre-existing physical structures such as farm houses, fences, and storage yards among others.

[4] GHS 200 is equivalent to USD 95 at the then prevailing exchange rate of 2.1 Ghana Cedis (GHS)

**Table 1:** *Descriptive statistics*

| *Variable and Description* | *Estimate* | *Standard Error* |
|---|---|---|
| *Household Characteristics* | | |
| Sex (dummy 1= male 0= female) | 0.8 | – |
| Age (Number of years) | 47.1 | 12.7 |
| Dependency ratio (ages below 15 and above 64/ ages 15 to 64) | 1.1 | 0.9 |
| Basic education (Dummy 1=attended basic school 0=otherwise) | 0.7 | – |
| Literacy (dummy, 1= can read and write; 0=otherwise) | 0.2 | – |
| *Assets and Wealth variables* | | |
| Livestock Holding (in Tropical Livestock Units) | 2.5 | 7.9 |
| Land Holding (acre) | 12.2 | 11.7 |
| Value of output (GHS value of output per acre) | 695.1 | 3,888.7 |
| Family labour (man hours per season) | 270.8 | 573.5 |
| Non-farm income per household (GHS) | 336.5 | 2,001.1 |
| *Land Tenure and Security Variables* | | |
| Ownership with deed (Dummy, 1= registered land with deed; 0= otherwise) | 0.3 | – |
| Gift land (Dummy, 1= received land as gift; 0= otherwise) | 0.3 | – |
| Sharecropped land (Dummy, 1= land acquired under sharecropping; 0= otherwise) | 0.1 | – |
| Assurance of tenure security (number of years) | 0.5 | 0.7 |
| Years of land ownership (Number of years) | 9.1 | 6.7 |
| Probability of investment in land (1= if household made investment and 0= otherwise) | 0.3 | – |
| Intensity of investment land (in GHS) | 202.5 | 846.7 |
| Number of physical structures (farm houses, storage houses) | 0.1 | 0.5 |
| *Crop and Location Variables* | | |
| South (Dummy, 1= southern horticultural belt; 0= otherwise) | 0.2 | – |
| Northern agricultural zone (Dummy, 1= northern agricultural zone; 0 =otherwise) | 0.4 | – |
| Affram Basin (Dummy, 1= Affram Basin; Otherwise=0) | 0.3 | – |
| Cash crops (Dummy, 1= cash crops; 0 = otherwise) | 0.1 | – |
| *Plot Characteristics* | | |
| Index of land fragmentation (Number of plots/ acre) | 1.6 | 1.8 |
| Access to irrigation (Dummy, 1= irrigated, 0= land rain-fed) | 0.1 | 0.3 |
| Soil water retention (Number of hours it takes farms to drain) | 21.4 | 70.1 |
| Ratio of zonal to household farm[5] | 0.9 | 0.0 |
| *Market Access and Participation* | | |
| Receive extension visits (1= farm received extension visit 0=otherwise) | 0.5 | |
| Distance to major market (Kilometers) | 2.3 | 16.5 |
| Adoption of technologies (Dummy, 1= yes; 0= no) | 0.2 | |

Source: Authors' computations from *MIDA FBO Survey, 2008*

[5] Refers to the farm size per capita of the household divided by the average farm size per capita in the zone.

Larger dispersions occur in farm revenue, non-farm income and household labour use. Since farm sizes and scales of production vary it is expected that larger dispersions would occur in farm revenue. Larger families tend to have more access to family labour. The large standard deviation of family labour use is in tandem with the wide variation in family sizes.

The result of the probit model of the decision to invest in soil and water conservation and irrigation is presented in Table 2. Endogenous variable such as the number of physical structures established on the land by the household were excluded from the model because of lack of appropriate instruments that would enable us to predict the variable.

Land tenancy arrangements such as renting and sharecropping were found to have significant influence on households' decision to invest and had the expected negative signs. While land ownership (without deed) did not exert a significant influence on the probability of households deciding to invest, land documentation (ownership with deed) was found to exert a positive effect on the decision to invest and was significant at the 10 percent level. Households were also found to be less willing to invest in farm lands they received as gifts (gifted land) and which they had not formally documented as well as on sharecropped plots. The assurance of tenure security exerted a significant and positive effect on the probability of investing in soil and water conservation and irrigation.

The distance to major markets was found to exert a negative effect on the decision to invest and was significant at 1 %. Access to irrigation as well as the ratio of household to zonal land size also had the expected significant positive effects on the decision of households to invest.

The effect of household resource endowment on the decision to invest was found to be mixed. Household livestock expressed in Tropical Livestock Units (TLU)[6] had no significant influence on this decision. On the contrary, labour availability (family labour) and value of output per area (or crop income) did significantly increase the probability of households deciding to invest in land conservation and irrigation. Households with high labour would be in the position to implement complementary farm activities required to optimize the benefi ts from land improvement related investment.

Location of households and crop characteristics such as cultivation of cash crops were also found to have significant effects on the decision to invest. Compared with the southern horticultural belt (used as the reference category), households located in the Affram Basin and the northern agricultural zone had higher propensities to invest contrary to *a priori* expectations that the relatively high level of commercialisation of agricultural products and the higher population density in the South would make these households more willing to invest in land improvements.

Cultivation of vegetables and spices was found to positively influence the decision to invest in soil and water conservation and irrigation. This was expected since the production of vegetables and spices in many instances is done in the dry or minor seasons and would usually involve significant investment in irrigation. The cultivation of perennial cash crops and non-perennial cash crops had the expected positive and negative coefficient signs, respectively, but only the coefficient for non-perennial cash crops was statistically significant.

Land fragmentation[7] had a positive effect on the probability of investing and was significant at 1 %. The ratio of zonal to household farm size was significant at 10 % and had a positive coefficient. Soil depth was significant at 10 % and positively correlated with the probability of investing. Drainage (an indicator of water retention) had a negative coefficient but was not significant. Access to agricultural extension as well as the adoption of improved technology did not significantly affect households' probability to invest.

The level of soil conservation and irrigation investment was measured by the amount (GHS) spent by the household to improve and conserve soil water or fertility and to facilitate irrigation. Households' level of investment in soil conservation and irrigation was hypothesised to be influenced by a number of household socioeconomic and farm characteristics as well as the provenance of land (Table 3). Compared to the results of the probability of investment, differences in the amounts spent by households in soil conservation and irrigation were to a large extent explained by differences in types of land use and ownership arrangements as well as the duration of land ownership.

[6] Tropical Livestock Units are livestock numbers converted to a common unit. The Conversion factors are: cattle = 0.7, sheep = 0.1, goats = 0.1, pigs = 0.2, chicken = 0.01. Factors taken from HarvestChoice (2011)

[7] Land fragmentation ratio measures the ratio of total household farm size (in acres) to the number of parcels or plots. It is an indicator of whether household land holdings are continuous large tracts or small scattered plots.

**Table 2:** *Maximum Likelihood Estimates of the Determinants of the Decision to Invest*

| *Variable* | *Coefficient* | *Robust Std. Errors* | *Z* |
|---|---|---|---|
| *Household Characteristics* | | | |
| Sex of household head | 0.4065 | 0.0971 | 4.19*** |
| Age of household head | –0.0240 | 0.0091 | –2.62** |
| Age square of household head | 0.1754 | 0.0862 | 2.03* |
| Dependency ratio | 0.0101 | 0.0309 | 0.33 |
| Basic education | 0.2133 | 0.0382 | 5.58*** |
| No education | 0.2884 | 0.0856 | 3.37** |
| *Household Assets and Wealth variables* | | | |
| Livestock Holding (in TLU) | –0.0343 | 0.0386 | –0.89 |
| Household land holding | –0.0156 | 0.0075 | –2.08* |
| Value of output per area | 0.4721 | 0.0886 | 5.33*** |
| Family labour | 0.0733 | 0.0342 | 2.14* |
| *Land Tenure and Security Variables* | | | |
| Gifted land | –0.3087 | 0.06015 | –5.13*** |
| Sharecropped land | –0.2323 | 0.0676 | –3.44** |
| Rented land | –0.1173 | 0.0654 | –1.79* |
| Ownership with deed | 0.1633 | 0.0676 | 2.42* |
| Ownership without deed | –0.0136 | 0.0944 | –0.14 |
| Assurance of tenure security | 0.1389 | 0.0554 | 2.50* |
| Years of land ownership | –4.3728 | 3.9281 | –1.11 |
| *Crop and Location Variables* | | | |
| Afram Basin | 0.4404 | 0.0796 | 5.53*** |
| Northern agricultural zone | 0.7883 | 0.1149 | 6.86*** |
| Distance to major market | –0.0140 | 0.0038 | –3.65*** |
| Perennial cash crops | 0.0198 | 0.1002 | 0.20 |
| Non-perennial cash crops | –0.1275 | 0.0625 | –2.04* |
| Vegetables and spices | 0.2350 | 0.0624 | 3.77*** |
| *Plot Characteristics* | | | |
| Ratio of zonal to household farm size | 0.1805 | 0.0877 | 2.06* |
| Index of land fragmentation | 0.1829 | 0.0510 | 3.58*** |
| Access to irrigation | 0.9210 | 0.0710 | 12.96*** |
| Drainage | –0.1371 | 0.3398 | –0.40 |
| Soil depth | 25.8536 | 14.5829 | 1.77* |
| *Access to Innovation and Technology* | | | |
| Receive extension visits | –0.0130 | 0.0539 | –0.24 |
| Adoption of improved technology | 0.0771 | 0.0498 | 1.55 |
| Constant | –1.5085 | 0.2811 | –5.37*** |
| Log likelihood | –1866.34 | | |
| Wald $\chi^2$ (30) | 365.43 | | |
| Prob > $\chi^2$ | 0.0000 | | |
| Pseudo $R^2$ | 0.1009 | | |

Dependent Variable is the decision to invest into soil conservation and irrigation (1 if the household invested and 0 if household did not invest)
Source: Authors' computations from *MIDA FBO Survey, 2008*
*, **, and *** are levels of significance at 10 %, 5 %, and 1 %, respectively.

**Table 3:** *Determinants of Soil and Water Conservation and Irrigation Investment expenditure*

| *Variable* | *Coefficient* | *Robust Std. Err.* | *t-statistic* |
|---|---|---|---|
| *Household Characteristics* | | | |
| Sex of household head | −102.1851 | 36.4605 | −2.80** |
| Age of household head | 11.6229 | 6.0322 | 1.93* |
| Age of household head square | −146.8364 | 58.7225 | −2.50* |
| Household size | 28.5339 | 3.9584 | 7.21*** |
| Dependency ratio | −90.2414 | 17.4334 | −5.18*** |
| Attained basic education | −32.1379 | 19.5568 | −1.64 |
| No formal education | −121.8411 | 40.1405 | −3.04** |
| *Household Assets and Wealth variables* | | | |
| Livestock holding (in TLU) | −23.8377 | 21.5921 | −1.10 |
| Land holding | −7.8167 | 4.12557 | −1.89* |
| Non-farm income (per capita) | 17.5791 | 10.2533 | 1.71* |
| Value of output per area | 0.1187 | .018828 | 6.31*** |
| Family labour | 19.6628 | 21.6363 | 0.91 |
| *Land Tenure and Security Variables* | | | |
| Family land | −39.4991 | 44.0785 | −0.90 |
| Gifted land | −47.0821 | 45.5502 | −1.03 |
| Sharecropped land | 6.1550 | 39.7833 | 0.15 |
| Rented land | 93.5805 | 43.5905 | 2.15* |
| Ownership with deed | 150.3289 | 44.5321 | 3.38** |
| Ownership without deed | 67.4012 | 37.4213 | 1.80* |
| Duration of land contract [8] | 17.4348 | 4.15372 | 4.20*** |
| Years owning land [9] | −7.1505 | 4.26031 | −1.68* |
| Duration of tenure security [10] | 281.6785 | 26.7354 | 10.54*** |
| *Crop and Location [11] Variables* | | | |
| Affram Basin | 138.6426 | 39.2577 | 3.53*** |
| Northern agricultural zone | 326.5296 | 63.9384 | 5.11*** |
| Perennial cash crops | 61.9845 | 58.6894 | 1.06 |
| Non-perennial cash crops | −83.3227 | 29.2898 | −2.84** |
| Vegetables and spices | 39.7842 | 34.7841 | 1.14 |
| Distance to major market | −1.9830 | 1.54175 | −1.29 |
| *Plot Characteristics* | | | |
| Ratio of zonal to household farm size | 51.8104 | 50.2338 | 1.03 |
| Index of land fragmentation | −36.7666 | 16.1320 | −2.28* |
| Access to irrigation | 258.0953 | 33.1152 | 7.79*** |
| Drainage | 213.7851 | 108.114 | 1.98* |
| *Market Access and Participation* | | | |
| Receive extension visits | 88.5725 | 27.2893 | 3.25** |
| Adopted improved technology | 35.7479 | 24.5838 | 1.45 |
| Constant | −1062.186 | 206.2269 | −5.15*** |
| Log pseudo likelihood | −4034.3159 | | |
| F(38,1352) | 11.30 | | |
| Prob > F | 0.0000 | | |
| Pseudo $R^2$ | 0.0596 | | |
| Left-censored observations at INVSWCON_IRR ≤1 | 893 | | |
| Uncensored observations | 459 | | |
| Observations | 1352 | | |

Dependent variable: GHS (₵) invested in soil and water conservation and irrigation.
Source: Authors' computations from *MIDA FBO Survey, 2008*
*, **, and *** are levels of significance at 10 %, 5 %, and 1 %, respectively.

Land ownership (both with deed and without deed) had significant positive effects on households' investment expenditure. Consistent with *a priori* expectations, family owned, rented, gifted and sharecropped lands had negative coefficient but were however not statistically significant.

The signs of household wealth variables were mixed in terms of their consistency with *a priori* expectations. Livestock holding (in TLU) did not have the expected positive sign even though the coefficient was not significant. Non-farm income and value of output had the expected signs and was significant at 10 % and 1 % respectively. Location dummies which were included to capture site specific effects in investment indicated that autonomous investment in soil conservation and irrigation was relatively higher in the Afframm basin and the northern agricultural zone. The dummies for these two locations were significant at 1 %.

With respect to plot characteristics, the degree of land fragmentation and drainage were found to be significant in explaining the amounts households invested in soil and water conservation and irrigation. Households invested less on highly fragmented plots and more on irrigable plots and well drained plots. With regards to market and information access, farmers who received extension visits were found to invest higher amounts than those who did not. Adoption of improved technology had the expected positive sign but did not significantly influence household investment expenditure.

Some household and demographic factors seemed to significantly influence the levels of soil conservation and irrigation investments. Both age and age-squared of household head had significant effects on the amounts invested exerting positive effects on the amounts households invested in soil conservation and irrigation.

Household size was found to have a significant positive effect on the amounts invested and was significant at 1 %. Although not significant, formal education was inversely related to the amounts households invested in soil conservation and irrigation but the formal education variable was not significant. As expected however, not having formal education significantly reduced the amounts households invested in soil and irrigation.

## 5 Discussion

Land documentation and duration of tenure security positively influenced households' decision to invest and is an indication that land titling programmes and policies can facilitate households' farm investment by enhancing tenure security. This finding is an indication that formalisation of land rights though not a comprehensive indicator of land tenure security, contributes significantly to enhance land holders' perception of their tenure security and their decisions to invest in lands.

Household land relative to zonal farm size was included to test for farm size effects. Contrary to the finding by Holden & Yohannes (2002), the study found that marginal farm size was inversely related with tenure insecurity. Tenure insecurity was rather found to correlate positively with value of farm land and not farm size as Alemu (1999) posited. The reason for this observation is that owning large tracts of land may not be a source of insecurity if there are no competing claims. However, as land values rise, interests in land both as a factor of production and commodity increases. Local elites who probably were not interested in land re-entered the market and may appropriate land belonging to less powerful holders.

Contrary to expectations that households that owned more livestock would invest more in land because they were regarded as wealthy (Holden & Yohannes, 2002), the reverse held. One possible explanation is the use of livestock as security against crop failure. Households may not sell their livestock to invest in land conservation and irrigation but may do so in the event of crop failure.

The expectation on the relationship between location dummies was that farmers in the southern horticultural zone would invest more because land was scarce and there was the need to maximise output. This expectation was based on the Boserup (1965) population pressure hypothesis which posits that high population density (land scarcity) would stimulate investment and use of improved technology. It was expected that the high population density of the southern horticultural belt will lead to higher propensities to invest. The absence of this population pressure effect could however be due to abundant non-farm employment opportunities in the South which significantly increases the opportunity cost

[8] Duration of land contract refer to the number of years a land owner has agreed to allow a tenant use land in contract (formal or informal) arrangement.

[9] Refers to the number of years the farmer has owned the land. Households who have owned land for longer periods are perceived more secure.

[10] Duration of tenure security is the length of time that a given right is legally valid. The economic dimension requires, in addition, that the time horizon be sufficiently long to enable the holder to recoup (see Roth & Haase, 1998)

[11] Southern Horticultural belt is used as reference category

of farm labour and other resources and also offer opportunities that do not require ownership of land.

The study expected that highly fragmented land (measured by the land size in acres per number of plots) would discourage investment as the prospects of commercial scale production would be limited. The findings however suggest that households' decision to invest was positively correlated with land fragmentation. This is contrary to *a priori* expectation that highly fragmented plots, usually common with communal or family held land, would have a negative effect on the probability of investing. One possible explanation for this observation is that households may be using land fragmentation as some sort of risk spreading activity. Further, land ownership overrides fragmentation in investment decision.

The results on the relationship between the monetary value of investments and tenure security variables were mostly consistent with *a priori* expectations. Farmers who owned land invested more relative to those who had acquired land through tenancy. The negative relationship between the amounts invested and the cultivation of family owned, rented, gifted and sharecropped lands could be indicative of some degree of Marshallian inefficiency (Marshall, 1890) transmitting to households' willingness to invest and levels of investment expenditure on long-term soil improvement and irrigation.

In relation to the socio-demographic factors included in the models as control variables, the results showed that age and age-squared exerted significant positive effects on the amounts invested. The expectation was that age and age-squared would have positive and negative effects respectively on the amounts invested. That would have implied younger farmers invested more in soil and water conservation and in irrigation because they constituted the most productive group and most likely to be engaged in farming for commercial reasons. The results however suggest that investment did not vary much along age lines.

Household size was positively related to the amounts invested. Large households usually cultivate larger farms and hence are more likely to invest more. Households with lower number of dependents as opposed to workers were expected to invest more because relative to households with higher dependency ratios, they would have surplus production, and the proceeds could be reinvested in farming. The negative sign of the dependency ratio variable is thus consistent with this expectation. Higher dependency ratio reduces the amount of investment in soil conservation and irrigation.

## 6 Conclusion

This paper explored the relationship between households' perception of their tenure security and how these perceptions influenced plot level farm investment decisions in soil conservation and irrigation. The paper measured land tenure security both in terms of right entitlements and the duration within which holders perceived the rights to be secure. The household was modelled as a utility maximizing entity that choose between short-term complementary inputs use and long-term land investments based on the household's evaluation of its land tenure security.

Findings of the paper support the long-held view that better land rights facilitate long-term farmland improvement investments. Tenure security, often posited to influence household farm investment decisions appears to hold in the case of Ghana. Individual ownership of land increased the probability of investing and the level of investments made in land improvement and irrigation. This evidence probably points to the increasing importance households place on land ownership and also underscores the importance of tenure security in facilitating land titling programmes. Several studies, especially those conducted in the middle to late twentieth century, failed to plausibly establish strong links between land ownership and investment as this study finds. Two possible explanations for this particular finding of the study are: First, there is the possibility that land markets and land relations have changed significantly over the last two decades with increasing money transaction and fixed agreements propelled by population growth and increasing value of land. Second, the inclusion of irrigation investment as a long term investment in land raises the value of household investment and also the time period required to reap the returns on the investments.

The finding of this paper seems to reinforce Boserup (1965) and Ault & Rutman (1979) predictions that land tenure in Africa evolved in tandem with principles of economic efficiency and that households would acquire full rights (i.e. take full rights into consideration) when the need for such rights arose, propelled either by population growth and land scarcity or by emerging market opportunities. It could be deduced that households take land ownership and tenancy arrangements into consideration if the resource implications of land investments are relatively huge and the time dimensions for harvesting returns to investments is relatively short.

It is recommended that for productivity to increase due to intensification, land owners should allow for long-term lease to provide the needed security that will encourage huge investment into development of land for long-term benefits.

## References


Agbosu, L., Awumbila, M., Dowuona-Hammond, C. & Tsikata, D. (2007). Customary and Statutory Land Tenure and Land Policy in Ghana. Technical Publication, No. 70, Institute of Statistical Social and Economic Research (ISSER), University of Ghana, Accra.

Agbosu, L. K. (2000). Ghana Land Law: Contradiction between Anglo-American and Customary Conceptions of Tenure Practices. Working Paper, No. 33, Land Tenure Center, An Institute for Research and Education on Social Structure, Rural Institution, Resource Use and Development, University of Wisconsin-Madison, USA. URL `http://digital.library.wisc.edu/1793/21905`.

Ahmed, M. M., Gebremedhin, B., Benin, S. & Ehui, S. (2002). Measurement and sources of technical efficiency of land tenure contracts in Ethiopia. *Environment and Development Economics*, 7, 507—527.

Alemu, T. (1999). *Land Tenure and Soil Conservation: Evidence from Ethiopia*. Ekonomiska Studier, Goteborgs Universitet.

Anley, Y., Bogale, A. & Haile-Gabriel, A. (2007). Adoption decision and use intensity of soil and water conservation measures by smallholder subsistence farmers in Dedo district, Western Ethiopia. *Land Degradation & Development*, 18 (3), 289—302.

Arabmazar, A. & Schmidt, P. (1981). Further Evidence on the Robustness of the Tobit Estimator to Heteroskedasticity. *Journal of Econometrics*, 17, 253–258.

Aryeetey, E. & Udry, C. (2009). Towards Efficient and Equitable Land Tenure Reforms in Ghana-ISSER. November 2009.

Ault, D. E. & Rutman, G. L. (1979). The Development of Individual Rights to Property in Tribal Africa. *Journal of Law & Economics*, 22 (1), 163–182.

Barrows, R. & Roth, M. (1989). Land Tenure and Investment in African Agriculture: Theory and Evidence. Land Tenure Center, LTC Paper No. 136, University of Wisconsin-Madison, USA.

Besley, T. (1995). Property Rights and Investment Incentives: Theory and Evidence from Ghana. *The Journal of Political Economy*, 103 (5), 903–937.

Boserup, E. (1965). *Conditions of Agricultural Growth: The Economics of Agrarian Change Under Population Pressure*. New York, Aldine Publishing Co.

Cheung, S. N. S. (1969). *The Theory of Share Tenancy*. Chicago, University of Chicago Press.

Clay, D., Reardon, T. & Kanganiemi, J. (1998). Sustainable intensification in the highland tropics: Rwandan farmers' in land conservation and soil fertility. *Economic Development and Cultural Change*, 46 (2), 351—377.

Deaton, A. (1997). *The Analysis of Household Surveys. A Microeconometric Approach to Development Policy*. Baltimore, The Johns Hopkins University Press (for the World Bank).

Dzanku, F. M. (2007). *Land Rights, Sustainable Natural Resource Use and Agricultural Productivity in Ghana*. Institute of Statistical, Social and Economic Research (ISSER) University of Ghana.

Feder, G. & Feeny, D. (1993). The theory of land tenure and property rights. *In:* Hoff, K., Braverman, A. & Stiglitz, J. E. (eds.), *The Economics of Rural Organization: Theory, Practice, and Policy*. pp. 240–258, Oxford University Press, Oxford, UK.

Feder, G., Onchan, T., Chlamwong, Y. & Hongladorom, C. (1987). *Land Policies and Farm Productivity in Thailand*. John Hopkins University Press, Baltimore MD, 1988.

GSS (2000). Ghana – Population and Housing Census 2000. Ghana Statistical Service (GSS).

Hagos, F. & Holden, S. (2006). Tenure security, resource poverty, public programs, and household plot-level conservation investments in the highlands of northern Ethiopia. *Agricultural Economics*, 34, 183–196.

Harrison, P. (1987). *The greening of Africa: breaking through in the battle for land and food*. Paladin Grafton Books, London.

HarvestChoice (2011). Total livestock population (TLU) (2005). International Food Policy Research Institute, Washington, DC. and and University of Minnesota, St. Paul, MN. URL `http://harvestchoice.org/node/4788`.

Hayes, J., M. Roth, M. & Zepeda, L. (1997). Tenure Security, Investment and Productivity in Gambian Agriculture: A Generalized Probit Analysis. *American Journal of Agricultural Economics*, 79 (2), 369–382.

Heckman, J. (1990). Varieties of Selection Bias. *American Economic Review*, 80 (2), 313–318.

Holden, S. & Yohannes, H. (2002). Land Redistribution, Tenure Insecurity, and Intensity of Production: A Study of Farm Households in Southern Ethiopia. *Land Economics*, 78 (4), 573–590.

ISSER (2008). GLSS5+ and FBO Surveys and Impact Evaluation of the MiDA Program. Institute of Statistical, Social and Economic Research (ISSER).

Johnson, D. G. (1950). Resource allocation under share contracts. *Journal of Political Economy*, 58, 111–123.

Laffont, J. J. & Matoussi, M. S. (1995). Moral hazard, Financial Constraints and Sharecropping in El Oulja. *Review of Economic Studies*, 62, 381–399.

Marshall, A. (1890). *Principles of Economics*. Macmillan, London.

Migot-Adholla, S. E., Hazell, P. B. & Place, F. (1991). Indigenous Land Rights Systems in Sub-Saharan Africa: A Constraint on Productivity? *World Bank Economic Review*, 5, 155–175.

Mill, J. S. (1848). *Principles of Political Economy*. Longmans, London.

Otsuka, K., Chuma, H. & Hayami, Y. (1992). Land and labor contracts in agrarian economies: theories and facts. *Journal of Economic Literature*, 30 (4), 1965–2018.

Place, F. & Hazell, P. (1993). Productivity Effects of Indigenous Land Tenure Systems in Sub Saharan Africa. *American Journal of Agricultural Economics*, 75, 10–19.

Platteau, J.-P. (1995). Reforming Land Rights in Sub-Saharan Africa: Issues of Efficiency and Equity. UNRISD Discussion Paper No. 60. Geneva.

Roth, M. & Haase, D. (1998). Land Tenure Security and Agricultural Performance in Southern Africa. BASIS-CRSP Program, Land Tenure Center, Madison.

Salas, O. A., Knight, F. & Sáenz, C. (1970). Land Titling in Costa Rica: A Legal and Economic Survey. USAID and University of Costa Rica Law School.

Shaban, R. A. (1987). Testing between competing models of share cropping. *Journal of Political Economy*, 95 (5), 893–920.

Sjaastad, E. & Bromley, W. D. (1997). Indigenous Land Rights in Sub-Saharan Africa: Appropriation, Security and Investment Demand. *World Development*, 25 (4), 549–562.

Tsikata, D. (2003). Securing Women's Interests within Land Tenure Reforms: Recent Debates in Tanzania. *Journal of Agrarian Change*, 3 (1 and 2), 149–183.

Twerefou, D. K., Osei-Assibey, E. & Agyire-Tettey, F. (2011). Land tenure security, investments and the environment in Ghana. *Journal of Development and Agricultural Economics*, 3 (6), 261–273.

Villamizar, F. (1984). IBD Financing in Land Administration Programs. Paper presented at the International Workshop on Land Tenure Administration, Salvador, Brazil, August 20–24.

# 7

# Optimising contract design in modern food supply chains: The case of paprika sector in Central Malawi

Lana A. Repar[a,*], Stephen Onakuse[a], Joe Bogue[a], Ana Afonso[b]

[a]*University College Cork, Ireland, College Road, Cork; Food Business and Development Department, Cork University Business School*

[b]*Technical University of Madrid, Spain, s/n. 28040 Madrid; Technical College of Agriculture, Project Planning for Rural Development and Sustainable Management - GESPLAN research group*

**Abstract**

Contract farming remains one of the most efficient tools for integrating small-scale farmers into modern food markets. However, the literature lacks evidence on best practices in designing contracts for food supply chains in developing countries. The purpose of this study is to analyse the design of an existing contract in the emerging Malawian paprika sector using a qualitative analytical approach. The study compared a contract design in Malawi's paprika supply chain with four similar contracts from the horticultural sector in Malawi, India, Zambia and Cape Verde. This study employed a thematic qualitative analysis and developed 17 categories for analysing contracts. The findings showed that the Malawian contract contained in total eleven defined clauses. There were missing clauses that influenced the risk and power distribution between parties in the contract. The comparison of the Malawian contract with other contracts revealed that none of the analysed contracts included all clauses necessary for a sustainable and fair relationship. The study proposed improved contract design and future actions to contribute to decreasing inequalities between parties engaged in the Malawi's paprika supply chain. The implications of the study include initiation of changes in public policies related to contract farming strategies as the findings suggest that vulnerable small-scale farmers might be left unprotected and generate low gains through supply contracts due to poorly formulated contract clauses.

*Keywords:* contractual arrangements, developing country, small-scale farmers, thematic analysis

## 1 Introduction

Contract farming is an institutional arrangement that integrates farmers into modern food supply chains through an agreed set of formal rules and informal constraints (Minot & Roy, 2007; Narayanan, 2012). The primary role of contracts is to overcome existing market imperfections by providing inputs, credit and extension services to farmers (Jia & Bijman, 2013; Cai *et al.*, 2014). Also, contracts might serve to distribute the production and marketing risks between the contractor and farmers (Barrett *et al.*, 2012). Contracts oblige farmers to supply buyers with a particular product of certain specifications (Will, 2013). Contracts often establish terms for product purchase in advance, such as expected quantities, quality and price (Vermeulen & Cotula, 2010; Barrett *et al.*, 2012).

Contracting can be organised through five different models depending on the parties involved: centralised (processor/packer and farmers); nucleus estate (contractor, estate and farmers); multipartite (contractor, public entity, cooperative and farmers); informal (contractors and farmers) or intermediary (contractor, intermediary and farmers) (Eaton & Shepherd, 2001). Furthermore, Bijman (2008) and Will (2013) described three types of contract farming arrangements. First, in the market specification contracts, the farmer and contractor agree on the terms of delivery, such as quantities,

* Corresponding author
Email: l.repar@umail.ucc.ie, lana.repar@gmail.com

qualities and timing. The farmer bears most of the production risk but retains control over the majority of the production-related decisions. The marketing risk is distributed between the parties. Second, the production-management contracts involve transferring a substantial amount of decisions regarding production processes from the farmer to the contractor. The farmer agrees to follow the production and input regime suggested by the contractor as the contractor assumes the most of the marketing risk. Third, in the resource-providing contracts, the contractor secures the market and the key inputs for the production. The inputs are offered as in-kind credit and their value is recovered upon product delivery through deduction from the final price. The degree of control can vary widely in this type of contract.

Recent studies show that contracting increases farmers' incomes, encourages technology adoption and improves the quality of produce (Jones & Gibbon, 2011; Fréguin-Gresh & Anseeuw, 2013; Girma & Gardebroek, 2015). In contrast, being under contract can increase gender imbalance in the household, result in small farmers' indebtedness or enable contractors to turn small-scale farmers into dependent labourers (Birthal *et al.*, 2008; Prowse, 2012). In addition, farmers in developing countries often lack access to suitable land and related facilities, such as irrigation and roads needed for efficient contract farming (Reardon *et al.*, 2009; Valipour, 2015a,b). Depending on the contract design, the parties could receive benefits or carry the risks. For example, Warning & Hoo (2000) and Narayanan (2012) found food supply contracts in developing countries ambiguous and one-sided, lacking fairness and farmers' participation in the design stage. Contracts are often written in an incomprehensible technical language and can be designed to favour buyers (Maertens, 2006; Cotula, 2011; Pultrone *et al.*, 2012). Thus, even if the farmer increases productivity and the quality levels of the product, the contractor might capture the entire premium, depriving the farmer of any meaningful benefits.

More specifically related to the formulation of contract clauses, Pultrone (2012) stated that benefits from contract farming could diminish if a contract offered unclear, incomplete and misguiding contract clauses. Narayanan (2012) questioned whether farmers even comprehend commitments contained in the clauses. Barrett *et al.* (2012) argued that contracting companies designed and offered contracts to more profitable farmers to reach expected profit levels. In the empirical literature, Schipmann & Qaim (2011) assessed contracts for pepper in Thailand and concluded that clauses providing inputs and credits increased the attractiveness of the contract for farmers. Abebe *et al.* (2013) explored preferences of potato farmers in Ethiopia and found that farmers opted for written contracts, which secured inputs, technical assistance and seeds, and had variable quality and price options.

Current research does not provide sufficient evidence on the best practices in designing contracts for small-scale farmers in developing countries. The empirical examples mentioned above refer only to one part of the contract (preferences for provisions) but do not consider the whole contract. Thus, there is a lack of research on how companies design contracts (Brousseau, 2008). According to Smalley (2013), exit options that are given to small-scale farmers in contracts should be further explored. Sykuta & Parcell (2002) argued that access to contract documentation and the privacy surrounding contracts hindered a better understanding of contracting issues. Notwithstanding the comprehensive studies on contract clauses by Prowse (2012) and Pultrone (2012), and the Legal Guide by UNIDROIT/FAO/IFAD (2015) that serves as a reference point, the up-to-date, in-depth empirical research on contract design challenges in developing countries is scarce. To bridge the existing knowledge gap, this current study addresses the problem of contract design in food supply chains using an authentic case from the field and synthesises the available materials to make inferences. The purpose of this study is to analyse, through comparison, contract design in the Malawian sector of paprika (*Capsicum annuum* L.) using a qualitative analytical approach. The study answers the following research questions: (i) What is the structure of the Malawian contract? (ii) What is the content of the contract under study? and (iii) How does the Malawian contract differ from similar food supply contracts and how can it be improved? In this study, contract design is defined as the structure and content of the contract (Furlotti, 2007).

This work contributes to the existing knowledge in three aspects. First, the study identifies weaknesses of the contract design and discusses how to improve it to better serve the needs of small-scale farmers. Second, the study uses a qualitative research approach, which is underutilised in developmental and agricultural studies (Locke & Lloyd-Sherlock, 2011). Third, the study adds needed empirical evidence to the field by analysing the design of the entire contract in Malawi and enhances the overall understanding of contract formulation.

The remaining part of this introductory section focuses on the economic profile of Malawi and positioning of contract farming in the context of the Malawian paprika sector. The Republic of Malawi is a low-income country with a population of around 16.7 million and US$ 255 GDP *per capita* (World Bank, 2015a,b). About 84 % of the population lives in rural areas, while 49 % of the country's area is devoted to agricultural land (FAO, 2013; UN, 2016). The most important food crops grown in Malawi between 2011 and 2013 were: cassava, maize and potatoes (CountrySTAT Malawi, 2016). The crops with the highest export value in Malawi in 2014 were: tobacco, sugar, tea and coffee (NSO, 2015). Large landowners produced significant export commodities and small-scale farmers in Malawi cultivated mainly food crops and rarely engaged in high-value cash crop production. The research has more recently focused on the significance of the paprika sector, which reached US$ 1.9 million in export value in 2014 and has the potential to further contribute to Malawi's trade (Agar & Chiligo, 2008; NSO, 2015). Rising international demand for paprika made it an attractive export commodity for Malawian small-scale farmers due to their comparative advantage in the use of low-input technology (Kumwenda & Madola, 2005; Makoka *et al.*, 2010). However, Makoka *et al.* (2010) found that Malawian small-scale farmers faced many challenges that constrain them from linking with high-value markets and these included: lack of finance, low productivity, poor quality and little value addition.

Contract farming was introduced to Malawian small-scale farmers in the 1980s (UNIDROIT, 2014). Despite multiple efforts, Malawi does not have an official strategy for contract farming. Evidence from Malawi on contract farming's impact is so far limited and mixed. Chirwa & Kydd (2009) reported that small-scale farmers in the tea sector had insufficient voice and power to enforce contracts, while Kumwenda & Madola (2005) found that side-selling represented a significant issue reported by companies. Hence, one of the main reasons why contract farming has not been successfully established in Malawi is due to small-scale farmers' opportunistic behaviour in side-selling (CYE Consult, 2009). Agar & Chiligo (2008) explored contract farming in Malawi's cotton, sugar, tea, tobacco and paprika sector and concluded that contracting increased small-scale farmers' income and improved their productivity. Nevertheless, the authors identified two issues that severely hindered contract farming operationalisation: small-scale farmers perceived the contract price as intentionally too low and contractors complained about a breach of contract terms (Agar & Chiligo, 2008).

The general paprika supply chain in Malawi involved contractors, farmers as suppliers, input providers, vendors, processors, retailers and further customers. The enabling environment consisted of institutional elements (Government) and supporting services (Universities, consultancy sector, NGOs, civil society, aid organisations and farmers' associations). In Central Malawi, fresh paprika was sold on the green market by local farmers and in city supermarkets, which sourced paprika from South African suppliers or local small-scale farmers on a seasonal basis and through oral agreement. This present study focused on dried paprika for export regulated via contract farming. Small-scale farmers entered the contract mainly as individuals, while to a lesser extent, others joined the contract as members of a farmers' association since the contractor wanted to obtain higher volumes of paprika from established farmers' associations. The product flow of dried paprika chain consisted of three phases. First, small-scale farmers delivered whole dried paprika pods to the contractor. Second, the contractor exported whole dried paprika pods to the processor in South Africa. In the final phase, the processor extracted oleoresin from paprika and supplied customers in South Africa and Germany.

The structure of the paprika supply chain in Central Malawi was similar to paprika chains in neighbouring Zambia. Stevens (2004) explored the paprika supply chain in Zambia and found that small-scale farmers were selling their paprika to the contractor while the enabling environment involved donor groups and the Government. The paprika was processed into oleoresin and further exported to South Africa and Spain (Stevens, 2004). Kabungo & Jenkins (2016) reported that the contractor in Zambia had a dual role. The contractor was buying paprika from outgrower small-scale farmers and further selling it on the international export markets. The contractor was also providing extension services to small-scale farmers. A similar pattern can be observed in both export paprika supply chains in Malawi and Zambia. It is likely that the described pattern occurs in other south-eastern African countries, which provides a foundation for the generalisability of this study's findings in the form of learned lessons that can be applied in neighbouring countries.

## 2 Materials and methods

This study analysed the design of a contract used in Malawi's paprika supply chain through comparison with four other contracts. The study uses the terms *studied contract* and *Malawian contract* interchangeably to denote the contract that is the subject of the paper and to distinguish among different contracts. Due to confidentiality, the study refers to the contractor using the term Company.

The total of five contracts were analysed for the study. The emphasis of this paper is on the design of the paprika contract physically obtained during the interview with the Company's senior extension officer in Malawi in November 2014 [1]. The remaining four contracts were retrieved as secondary data from the FAO contract farming online database (FAO, 2016a,b,c,d). Since food supply contracts are unique in their context, the sample of contracts used for comparison had to resemble similar conditions as the studied contract. The selection criteria for the contracts obtained from the FAO database included: (i) the contract had to be administered in a developing country; (ii) the contract had to involve farmers and buyers; and (iii) contract had to relate to a horticultural crop. The number of contracts from the FAO database was restricted to four to allow for an in-depth analysis and comparison. The FAO online database already contained a sample of the Company's contract for paprika in Malawi used in 2010/11 season, but in practice, it was replaced by a new contract received during fieldwork. The remaining three contracts originated from Zambia (between a group of farmers and a purchaser of paprika for the 2003/04 season), India (between an individual farmer and a processing and exporting company for fruits and vegetables) and Cape Verde (between an individual farmer and a buyer of fruits and vegetables).

### 2.1 Contract setting

The studied contract outlines the relationship between the Company and the small-scale paprika farmers. The Company was established in 2008 and started commercial paprika farming in 2010 after the acquisition of a Malawi-based company that previously dealt with paprika. The Company offered its contracts for cultivating paprika to small-scale farmers (outgrowers) in 2010 and building on the existing outgrower programme from the acquired company. The main motivating factors to start contracting with paprika involved: suitable climatic conditions for growing paprika; market demand for paprika; suitable infrastructure in place from the previous company; and the suitability of paprika farming for small-scale farmers unlike other cash crops (tea, coffee and sugar) that require plantations. Apart from its own estates, the Company sourced paprika from 10,000–15,000 small-scale farmers across Malawi. Small-scale farmers produced 65 % more volume of paprika compared to the amount grown on the Company's estates. Paprika is the Company's biggest export cash crop and it accounts for 15–20 % of Company's net sale. The importance of paprika stems from its potential to generate high income from small land area. The Company has a long-term trading relationship with the processing company in South Africa where it sells the majority of dry paprika from Malawi. The Company sells paprika to other traders in South Africa in smaller volumes. During the course of the study, the Company was the only official contractor for dried paprika in Malawi's Central Region. The Company offered a written, nucleus estate model, production-management outgrower contract type. The purpose of the contract was to ensure a reliable supply of quality dry paprika from the small-scale farmers who followed precise production guidelines. The two-page contract in English was distributed on behalf of the Company through extension officers.

### 2.2 Analytical framework

The design of the five contracts was analysed using a template style of thematic qualitative analysis. A qualitative approach was applied due to characteristics of the material: contracts represent rich textual data, which reflect economic relationships through words, patterns and subtle meanings. Thematic analysis was used to identify, analyse and report themes emerging from selected contracts (see Braun & Clarke (2006) for the six phases followed in thematic analysis). A particular type of thematic analysis employed in this study was the template analysis, which combines a relatively high degree of structure in analysing the textual data and the possibility to adapt to a specific study (King, 2012). The coding template was developed using the Company's contract and it was then applied, revised and re-applied to the remaining four contracts, providing the consistency and direction in the overall analysis. The study used both inductive and deductive approaches to coding (see Fereday & Muir-Cochrane, 2006). NVivo™ 10 software facilitated the organisation and codification of the data. Contracts were first coded openly and focused on categories identified in the raw data and afterwards categories were refined to align with categories found in the literature.

The template analysis resulted in identification of contract categories (clauses), which were interpreted and developed into themes. The themes were divided in three main groups: (i) contract structure, (ii) contract content and (iii) comparison of designs. The theme labelled as contract structure was further classified in three sub-themes: defined, partially defined and missing clauses. The Company's contract was analysed in-depth while the other four contracts served primarily for comparison purposes.

[1] The fieldwork addressed numerous issues beside the contract design but the topics are beyond the scope of this paper.

## 3 Results

### *3.1 Contract structure*

Central features of the contract were the Company's and the small-scale farmer's obligations and general conditions. The Company's obligations included: organising purchasing points; purchasing the crop; determining grades; paying for the crop; and training the small-scale farmers. The small-scale farmer's obligations included: cultivating the crop; adhering to chemical usage instructions; grading; and selling the crop. General conditions were defined around quality failure, breach of contract and liability clauses. Table 1 displays the main contract clauses.

#### *3.1.1 Defined clauses*

The first part of the studied contract (2013/14 season) identified parties who had entered into an agreement. The contract continued with an outline of the nature of the relationship between parties and the exact duration of the contract. Under the general conditions, the contract defined the following clauses: consequences in the case of quality failure, clarification of breach of contract and the Company's disclaimer of liability regarding the price guarantee, input provisions and any incidents that might occur. The amount of paprika expected from small-scale farmers was emphasised several times throughout the contract. The contract specified the payment conditions and mandatory instructions on crop cultivation for small-scale farmers. The contract concluded with signatures of parties and their representatives.

#### *3.1.2 Partially defined clause*

The contract examined in this study partially defined the price. The clauses related to the price were mentioned twice in the contract: under the Company's obligations and within general conditions. In both cases, the amount to be reimbursed to small-scale farmers was not specified. However, clauses indicated the procedure for price determination. The delimitation of the Company's responsibility for price guarantee was explicitly stated in the contract.

#### *3.1.3 Missing clauses*

The clause that defined different grades for delivered paprika was not included in the studied contract. In addition, the contract did not specify conditions for termination of the agreement. Reference to the applicable law and options for settling possible disputes between parties were not contained in the contract. The contract also did not involve the clause that clarified responsibilities in the case of *force majeure* events.

### *3.2 Contract content*

The preamble of the contract defined that the outgrower entered the contract with an aim of growing contracted paprika for, and on behalf of, the Company. The agreed contract lasted for one season, which started on the day of purchase of the seeds and continued until the last part of the crop was sold. The seeds were the only input supplied through the contract and the Company sold packets of seeds to outgrowers on a cash basis. The crop in the contract was defined as all paprika that was produced from the purchased seeds packets. The contract stipulated that the outgrower was obliged to sell the entire crop exclusively to the Company. Grading was determined on the marketing day, where a complete description for each grade was provided (but not contained in the written contract), and parties agreed on the identified grades. The Company retained the right to determine the price. The price was influenced by the world market price, fluctuation in exchange rates and demand for paprika. The Company was obliged to make cash payments to the outgrower on the day of purchase or within two weeks. In addition, the Company organised central purchasing points not more than 30 kilometres from the original growing area if outgrowers delivered at least three tonnes of paprika.

The contract included training on cultivation and grading practices provided by the Company and the Ministry of Agriculture. The outgrower agreed to follow the Company's instructions, especially on recommended chemical usage. The failure to deliver the agreed quality of the crop was regulated through the contract by the Company's right to reduce the price of the crop, downgrade the crop or reject the crop. Moreover, the contract defined two cases for breach of contract. First, the outgrower breached the contract if paprika was not sold to the Company. Second, the Company breached the contract if the contracted paprika was not purchased.

The liability clauses were defined in the form of a disclaimer. The Company claimed no responsibility for an outgrower's injury, health issues or death during the growing and transporting processes. The Company was not liable for the supply of inputs, such as chemicals, fertilisers or finances. The Company did not guarantee the price at which the paprika would be purchased. The signatory clause required the contract to be signed under the condition that each party had read and fully understood the content. The signing parties included the chairperson and secretary of the farmers' association, one of the Company's representatives and a witness.

**Table 1:** *A comparison of the design of selected supply contracts*

| *Clause* | *Malawi (2013/14)* | *Malawi (2010/11)* | *Zambia* | *India* | *Cape Verde* |
|---|---|---|---|---|---|
| Crop | Paprika | Paprika | Paprika | Fruits and vegetables | Fruits and vegetables |
| Type | Production-management | Production-management | Resource-providing | Resource-providing | Production-management |
| Parties | Outgrower (individual or group), the Company | Outgrower (individual or group), company | Small-scale growers (group), purchaser | Farmer (individual), company | Seller (individual), buyer |
| Preamble | Outgrower grows paprika for, and on behalf of, the Company | Outgrower grows paprika for, and on behalf of, company | Purchaser contracts the grower to produce the crop | Farmer desires to enter into agreement with company | Seller offers to supply buyer, and buyer agrees to purchase the goods |
| Applicable law | X | X | X | Defined | Defined |
| Duration | One season | One season | One season | One year; extendable by company | Indicated number of years |
| Termination | X | X | X | Inputs misuse, quality failure in two consecutive seasons | Failure to supply or purchase, *force majeure*, breach |
| Input provision | Seeds only; cash in advance | Seeds only; cash in advance | Optional: seeds, chemicals and credit | Seeds, fertilisers and plant protection chemicals; written proof | X |
| Quantity | All crop grown using seeds sold by the Company | All crop produced by the outgrower | All crop, including surpluses above agreed quantity | All crop produced on agreed acreage | Amount/kg defined |
| Description of grades | X* | Four grades described | Four grades described | X* | Seven grades described |
| Payment method | Cash; immediately or within two weeks | Cash; immediately or within two weeks | Deduction + interest; within 15 work days | X | Cash, cheque, or bank transfer |

**Table 1:** *A comparison of the design of selected supply contracts* (continued)

| *Clause* | *Malawi (2013/14)* | *Malawi (2010/11)* | *Zambia* | *India* | *Cape Verde* |
|---|---|---|---|---|---|
| Price | Company's decision; based on the prevailing market prices, currency fluctuation, and demand; split by grade; not explicitly defined; no bargaining | Defined; fixed minimum amount/kg; split by grade; no bargaining | Defined; fixed amount/kg; no bargaining | Company's decision; amount/kg; split by grade; not explicitly stated; no bargaining | Defined; fixed amount/kg, but dependent on grade; mutual negotiation in case market price changes ± 10 % compared to contract price |
| Training | Instructions by the Company and third party, especially on grading and cultivation | Instructions by company and third party, especially on grading and cultivation | Instructions, newsletters, and educational visits by purchaser | Instructions by company | X |
| Quality failure | Downgrading; price reduction; rejection | Downgrading; price reduction; rejection | Downgrading; price adjustment; rejection; charging | Contract termination and return of provided inputs | X |
| Breach | Side-selling; purchase refusal | Side-selling | Side-selling (monetary penalty) | Side-selling (implicitly) | X |
| Liabilities | Outgrower: cultivation, chemical usage, grading, selling; Company: purchase | Outgrower: cultivation, chemical usage, grading, selling; company: purchase | Grower: cultivation, chemical usage, delivery, selling; purchaser: purchase | Farmer: cultivation, grading, selling; company: input provision, purchase | Seller: cultivation, delivery, selling; buyer: purchase |
| Disputes | X | X | X | Amicably; arbitration | Amicably; mediation |
| *Force majeure* | X | X | X | No party is liable; farmer must repay the loan | No party is liable; contract can be terminated |
| Signatory | Association, Company, witness | Company, outgrower, witness | Purchaser, grower, witness | Farmer, company, witness | Seller, buyer, witness |

* The price in the contracts was spilt by grades but grades' characteristics were not described in the contract.

### 3.3 Comparison of contracts

In this section, the clauses of the studied contract are compared with clauses in selected contracts from Malawi, Zambia, India and Cape Verde. The first part of this section emphasises the differences in two Malawian contracts.

#### 3.3.1 Differences in the contract structure of two Malawian contracts

The structure of the compared contracts varied considerably in four key aspects. First, the contract from the 2010/11 season outlined the minimum price for each of four grades listed in the contract. As noted earlier, the studied contract partially defined the clause on price and did not include grades. Second, regarding breach of contract, the studied contract expanded on the contract from 2010/11 and included the statement that the Company breaches the contract if it does not purchase the contracted crop. Third, the studied contract contained the clause specifying English as the working language. Fourth, while the contract in 2010/11 was signed between the Company and the outgrower as an individual, the 2013/14 contract required the signatures of representatives of the farmers' association. The rest of the clauses appeared in both Malawian contracts and remained similar in their content, thus the recent contract (2013/14 season) is the subject of further analysis.

#### 3.3.2 Comparison of the studied Malawian contract with three similar contracts

The contracts from Malawi and Cape Verde were classified as production-management contracts while the Zambian and Indian contracts were resource-providing contracts. Table 1 shows that the Malawian contract did not differ from similar food supply contracts regarding the parties involved - all contracts were made between companies and individuals or groups. The preamble of both the Malawian and Zambian contracts emphasised the farmers' obligations to cultivate the crop for the contractor, while contracts from India and Cape Verde defined the contractual relationship through an offer and acceptance form, i.e. the farmer offered to cultivate the crop and the company accepted the offer.

All contracts explicitly defined the duration of the agreement. Only the Indian contract included the possibility to extend the contract. The contracts showed specific variations in the provision of inputs as a result of different types of contracts and their purposes. The Malawian contract included the provision of seeds exclusively and the contract from Cape Verde did not involve any inputs. The contracts from Zambia and India involved seeds, fertilisers, chemicals and cash credit. The required crop quantities differed across contracts. The Malawian and Zambian contracts demanded the whole crop to be delivered to the contractor. In the Zambian case, even surpluses above the determined quantity had to be sold to the contractor. In the Indian contract, farmers were obliged to deliver the entire crop produced on an agreed acreage, while the Cape Verde contract defined the expected quantity in kilogrammes.

Price and grade clauses also varied among contracts. The Malawian and Indian contract did not describe grades. In contrast, the contracts from Zambia and Cape Verde provided detailed descriptions of four and seven grades respectively. The price in the contracts was either depended on market conditions or defined as fixed amount per kilogrammes. In the cases of Malawi and India, the contracting companies decided on the price, and although not explicitly defined in the contract, the price was split by grades. Only the Cape Verdean contract included a clause on mutual renegotiation if the market price declined or increased 10 % compared to the contract price.

Different payment methods were found in the compared contracts. The contracts from Malawi and Cape Verde provided cash payments to the farmers. While the Indian contract did not define any payment method, the Zambian contract involved the principle where the final price was subject to deductions based on input costs and interest rate. The Malawian and Indian contracts provided training on good agricultural practices. The contract in Zambia included newsletters and educational visits for outgrowers. If the delivered crop did not satisfy quality standards, contractors in Malawi, Zambia and India had the option to downgrade the crop, adjust the price, reject the crop or automatically terminate the contract. The Cape Verdean contract did not define actions in the case of quality failure.

Side-selling was considered as a breach of contract in the Malawian, Zambian and Indian contracts. In Zambia, side-selling incurred monetary repercussions for the grower. Within all contracts, farmers had obligations to cultivate and sell the crop. All contracts obliged the contractor to purchase the contracted crop under agreed quality standards. In the Indian contract, the company was liable for input provision.

The Malawian and Zambian contracts did not define clauses related to the applicable law, contract termination, dispute settlements and *force majeure*. In contrast, the Indian and Cape Verdean contracts included those clauses. The company in India had the right to terminate the contract if the outgrower failed to deliver a quality crop for two periods or misused the provided inputs. In the contract from Cape Verde, each party could terminate the contract if one party failed to deliver or purchase the crop, in the case of breach of contract and if *force majeure* events occurred. In the case of India and Cape Verde, disputes between the parties were first resolved amicably, while the second stage of the settlement involved arbitration and mediation respectively. In the Indian and Cape Verdean contracts, neither party was considered liable for the *force majeure* events. In India, the outgrower had to repay a given loan, while in Cape Verde *force majeure* was a valid foundation for contract termination.

The signatory clauses in the contracts from Zambia, India and Cape Verde required signatures from the company, the outgrower and at least one witness. In contrast, the Malawian contract did not require the outgrower to directly sign the contract, as the signatory clause referred to the representatives of the farmers' association.

## 4 Discussion

### *4.1 Contractual completeness*

The Malawian contract did not include clauses covering grades, applicable law, dispute settlement, exit options and *force majeure*. Thus, the Malawian contract had a certain degree of incompleteness, which is inevitable in almost all the contracts. The incompleteness in many contracts occurs mostly due to *ex-ante* and *ex-post* transaction costs related to drafting and enforcing contracts, asymmetric information and parties' bounded rationality (Hart, 1988; Hart & Moore, 2008). According to Hart (2010), the reason for leaving a contract incomplete might be to take advantage of an agent's cognitive limitations. Nevertheless, omitting some clauses in the contract is justified when the excluded clauses do not introduce additional risks or inequalities for the parties involved.

Defining the price of paprika in the studied contract involved some challenges for the Company. As an export commodity, the price of paprika is influenced by international prices and currency fluctuations. The comparison of the two contracts from Malawi revealed that the contract from 2010/2011 season included the definition of the price clause, which was not done in the contract from 2013/2014 season. Apart from identifying the parties and the objectives of the agreement, general contract law does not impose any requirements regarding the contract form (UNIDROIT/FAO/IFAD, 2015). One may infer that contracts do not necessarily need to include the price clause. In the Malawian contract, the Company retained the right to decide solely on the price based on market conditions. By doing so, the Company reduced its marketing risk and increased small-scale farmer's vulnerability to fluctuations on the international market. The Malawian contract required the entire crop to be sold to the Company at the market price corrected for currency fluctuations. Thus, the contract provided a secure outlet for small-scale farmer's produce but did not offer a stable or premium price. In addition, the contract did not involve the possibility to bargain over the price. According to the existing price clause, the small-scale farmer signing the Malawian contract accepted the risk that volatile international prices would shape the final price paid by the Company.

The literature on contracts provides further explanations on the price clause. In some cases, the contractors may outline the pricing mechanism in the contract. However, the pricing formula might be too complex and the real extent of the price premium for the small-scale farmer can be disguised (Pultrone, 2012). Small-scale farmers rarely participate in price determination and their bargaining power is often reduced since they cannot assess whether the price paid represents an appropriate remuneration (Echánove & Steffen, 2005). In cases where the price is not stated and is left to be shaped by the market as in the Malawian contract, the companies might closely track market trends to set delivery dates to favour low prices (da Silva, 2005). The consequences of inadequately defined and paid price via the contract might result in side-selling or cessation of production of that commodity (Baumann, 2000).

Omitting grades impacted the fairness of the Malawian contract to the least extent. International traders in the paprika sector tend to follow existing American Spice Trade Association (ASTA) guidelines that specify industry standards for spice quality and are known for ASTA colour values (ASTA, 2016). Therefore, the grades for paprika do not change as often as the price. Even though the contractor is usually familiar with ASTA specifications, small-scale farmers may lack access to essential information concerning grade levels. In this regard, the Malawian contract included training on grading for small-scale farmers and reduced the information asymmetry between the parties.

Nevertheless, it is argued in the literature that clauses on quality and grades are open to manipulation, which often causes disputes between companies and farmers (Eaton & Shepherd, 2001; Echánove & Steffen, 2005). There are three main reasons why companies might want to control these clauses. First, by arbitrarily controlling classification of commodities, companies might downgrade some proportion of the high-quality commodity into a second grade to obtain more quality products at a lower price (Baumann, 2000). Second, the price of the commodity on the market might be lower and companies could be tempted to supply the produce from the open market or export; thus, companies may 'invent' rigorous quality standards and reject a small-scale farmers' commodity (Pultrone, 2012). Third, in some cases the market for the commodity might be saturated and to reduce the risk of buying a contracted commodity with an uncertain market, companies might raise the quality standards to create a barrier for farmers' produce with an intention to reject it (Imbruce, 2008).

The Malawian contract did not define applicable law governing the contract. Hence, it is unclear which jurisdiction applied when interpreting and enforcing the contract. Also, the contract did not describe procedures for dispute settlement, so it was not possible to determine each parties' rights and responsibilities if disagreements arose. The Company is likely to have an advantage in directing dispute settlements by influencing either arbitration or mediation due to higher financial and information capacities. Small-scale farmers and even farmers' associations may be reluctant to pursue legal claims because of the costs involved in taking legal actions against the Company. The Legal Guide by UNIDROIT/FAO/IFAD (2015) confirms that the dispute settlement clause is among the ones most prone to manipulation in agricultural contracts. Pultrone (2012) further highlighted the importance of a clear definition of both dispute settlement clause and the law guiding the contract by suggesting that the contract should be explicit about which national law will govern the dispute settlement if parties come from different countries. Since the Malawian contract remained silent on the applicable law and the disputes clauses, the more powerful party gained the opportunity to determine rules *ad hoc*.

The termination clause was not indicated in the Malawian contract. As shown in Table 1, in some food supply contracts, the relationship can be automatically terminated in the case of any or a particular type of breach. The Malawian contract described acts that classify as a breach, without implying immediate termination as the consequence of the breach. On the other side, the Malawian contract did not specify conditions for terminating the agreement by either party's free will. Therefore, if any party intended to exit the contract in an amicable way, the contract did not define the required procedures, such as prior notice and the remaining liabilities.

Since the studied contract omitted the termination clause, it was left to each party's interpretation whether the contract locked-in the parties or gave the right to exit the contract arbitrarily. A similar case applied to *force majeure* events. The contract did not foresee natural disasters, which can result in delays in the delivery, reduced quality or complete destruction of the crop. It was unclear how costs related to crop damage would be distributed between parties. If the *force majeure* clause is not defined, it can be implied that risks and responsibilities associated with unpredictable events are borne by the farmer (Echánove & Steffen, 2005). The Malawian contract as designed could introduce the risk of disputes if natural disasters occur since each party could try to claim indemnity rights. Possible disputes could not be easily settled due to the lack of dispute settlement and applicable law clauses in the Malawian contract.

One of the peculiarities of the Malawian contract was the signatory clause. In the introductory part, the contract addressed the 'outgrower', which could be interpreted as either the individual farmer or a member of the farmers' association. Nevertheless, the contract required the signature of the association's representatives to conclude the agreement. This suggests that the Company preferred to establish a relationship with the farmers' association due to numerous advantages of collective action, such as the joint price negotiation, new production practices, information sharing, better input provision options and collective marketing of paprika (Abebaw & Haile, 2013; Wanglin & Abdulai, 2016). However, the membership in the farmers' association was not a condition to join the contract since the Company also contracted individual small-scale farmers.

This study's findings corroborate to conclusions from two recent studies. Prowse (2012) argued that companies often fail to incorporate even basic information in the contract, leaving farmers without an accurate idea of the nature of the agreement they signed. The author noted that often farmers signed contracts that could give away their rights over the crop to the company (Prowse, 2012). Moreover, Pultrone (2012) analysed key elements of typical agricultural contracts and concluded that more complete contracts would help each party perform their obligations in a more efficient and effective manner and avert misunderstandings. In particular, the inclusion of clauses on price, quantity and quality, *force majeure*, termination and dispute settlement clauses add to contract's clarity and certainty (Pultrone, 2012).

### 4.2 *Learned lessons for improved contract design in Malawi*

None of the contracts analysed in Table 1 included all the clauses needed for fair and sustainable relationships. Some contracts reflected 'more equal' relationship than others and this study suggested how to improve the existing Malawian contract based on examples from the compared contracts. The four main learned lessons from the contract design are outlined below.

First, as the production-management type of contract, the Malawian contract included only paprika seeds as the provided input. If the inclusion of the entire input package (seeds, fertilisers, pesticides and chemicals) guaranteed an increase in quality and volumes of the crop, then it would be justified to provide more inputs via the Malawian contract. By providing more inputs, the Malawian contract would resemble resource-providing contract type such as in the example of the Zambian and Indian contracts. Nevertheless, the threat of small-scale farmers defaulting in repaying loans and the higher costs for the Company might exceed the perceived benefits of including the entire input package. In this sense, additional input provision in the current contract would require a thorough assessment of costs and benefits for both parties.

Second, the Malawian contract could include an appendix with simple guidelines for grades and therefore reduce uncertainties among small-scale farmers. Indeed, the final CYE report (2009) on the Malawian paprika sector recommended the development of a unique guide for grades. The study reported on three different grading systems operating in the same marketing area, which only increased small-scale farmers' mistrust towards the companies (CYE Consult, 2009).

Third, the pricing mechanism from the Cape Verdean contract could be adjusted to serve the Malawian contract. The Company could reduce small-scale farmers' marketing risk by indicating the expected average price for paprika in the contract. Furthermore, the price for paprika could be modified for different grades. The Company could include the clause where additional renegotiation of the price could take place if the market price fluctuates over or under a certain percent, which would reduce the Company's marketing risk. In addition, the Company could set the new expected average price every season, referring to the general market trends and previous experiences.

Fourth, the contract in Malawi should explicitly refer to the applicable law and determine rules for contract termination, disputes and *force majeure*. These clauses could be defined as part of the contract that could be negotiated when the need arises. Contracts from India and Cape Verde provide an example of all four clauses defined with no ambiguity. The proposed amendments would mean a lengthier Malawian contract and, as Cotula (2011) argued, longer contracts are more likely to efficiently tie down financial and social aspects.

The findings showed that the Malawian contract was designed as a production-management contract and had eleven defined, one partially defined and five missing clauses. The content of the Malawian contract was in part vague and incomplete. In particular, the clauses that were omitted from the contract influenced the risk and power distribution and potentially weakened the small-scale farmer's position in the relationship. Although the contract required the signatures of the representatives of the farmer's organisation, their presence did not guarantee better bargaining power for the small-scale farmers since the joint negotiation of the contract terms was not enabled by any contract clause. This study concludes that the Malawian and other similar food supply contracts should unambiguously include (but not be limited to) the following clauses: parties, duration, inputs provided, grades, price and payment, quantity, quality and delivery terms, breach and consequences, liabilities, termination, dispute settlement, *force majeure*, applicable law and signatory.

The findings of the study involve implications for supply chain management and policy-making.

First, the responsibility to improve the current contract design in Malawi's paprika supply chain belongs to both the Company and contracted small-scale farmers. To form complete and sustainable contracts, the Company and small-scale farmers should interact and negotiate terms that reasonably represent mutual and individual interests. Third parties, including non-governmental and aid organisations, could enhance the farmer's position through advocating for transparency in the contract design and by including neutral parties in assessing the fairness of the contract.

Second, the Malawian Government can play a part in directing policies towards more inclusive contract designs. Key recommendation for the policy change is to develop and implement criteria of minimum requirements for contract design regarding agricultural commodities, especially in cases where contracts involve vulnerable groups. The Government could incorporate the guidelines for contract design in emerging National Contract Farming Strategy to promote better-designed contracts.

Third, researchers and practitioners have a role in providing the evidence and learned lessons that enable adoption of advanced contract features. Based on this study's outputs, future research should move in several directions. First, to estimate how the combination of different attributes in contract design (e.g. more input provision and stricter enforcement mechanisms) impacts on the small-scale farmers from a financial and social perspective. Second, there is a need to explore contracting companies' perspectives on issues of contract formulation. Third, to analyse countries' policies and strategies regarding contract design and examine the level of alignment with the Legal Guide by UNIDROIT/FAO/IFAD (2015) and the Sustainable Development Goals. Finally, to pursue the UN Agenda 2030 that seeks to enable inclusive and sustainable economic growth through wealth sharing and decreasing inequalities (UN, 2015), contracts should be designed and carefully formulated for more balanced relations, particularly where small-scale farmers are involved in markets.

Due to context-specific conditions, caution is advised when generalising findings on contract farming. Nonetheless, this study compared contract samples from Zambia, Cape Verde and India, and suggested that the contract design in Malawi's paprika supply chain can be improved by learning the lessons from other countries. The findings from this study are relevant to agricultural policymakers and development practitioners for advancing contract design that will legally protect and ensure fair conditions for small-scale farmers.

**Acknowledgements**

The authors would like to thank the editor and two anonymous reviewers for their invaluable suggestions. The authors are also grateful to respondents who took part in the interviews in Lilongwe, Malawi for their collaboration and support during the fieldwork in the Central Region.

**Funding body**

This work has been conducted as part of a PhD thesis project supported by the Agricultural Transformation by Innovation (AGTRAIN) Erasmus Mundus Joint Doctorate Programme, funded by the EACEA (Education, Audiovisual and Culture Executive Agency) of the European Commission.

## References

Abebaw, D. & Haile, M. G. (2013). The impact of cooperatives on agricultural technology adoption: Empirical evidence from Ethiopia. *Food Policy*, 38, 82–91.

Abebe, G. K., Bijman, J., Kemp, R., Omta, O. & Tsegaye, A. (2013). Contract farming configuration: Smallholders' preferences for contract design attributes. *Food Policy*, 40, 14–24.

Agar, J. & Chiligo, P. (2008). Contract Farming in Malawi. Kadale Consultants, Final Report for the Ministry of Agriculture and Food Security, funded by World Bank, Malawi. Available at: http://api.ning.com/files/oUFdw6A7Cbau6mybPWEZWRYRqJ45mxzABCawHTYv*zYhfk-y6ctURAVL0in4WByoNpOg842zBqlJzc6tywC4Jw-hervcBR9h/ContractFarmingMalawil.pdf (last accessed: 26.01.2013).

ASTA (2016). Analytical Methods Manual. American Spice Trade Association (ASTA). Available at: http://www.astaspice.org/food-safety/astas-analytical-methods-manual/ (last accessed: 11.02.2016).

Barrett, C. B., Bachke, M. E., Bellemare, M. F., Michelson, H. C., Narayanan, S. & Walker, T. F. (2012). Small-scale farmer Participation in Contract Farming: Comparative Evidence from Five Countries. *World Development*, 40 (4), 715–730.

Baumann, P. (2000). Equity and Efficiency in Contract Farming Schemes: The Experience of Agricultural Tree Crops. Working Paper 139. Overseas Development Institute, London. Available at: https://www.odi.org/publications/2022-equity-efficiency-contract-farming-schemes-agricultural-tree-crops (last accessed: 17.12.2016).

Bijman, J. (2008). Contract farming in developing countries: an overview. Working Paper. Wageningen University, Department of Business Administration, Wageningen. Available at: http://library.wur.nl/WebQuery/wurpubs/fulltext/1763 (last accessed: 14.12.2016).

Birthal, P. S., Jha, A. K., Tiongco, M. M. & Narrod, C. (2008). Improving Farm-to-Market Linkages through Contract Farming. A Case Study of Small Producer Dairying in India. Discussion Paper 00814. Markets, Trade and Institutions Division, International Food Policy Research Institute, Washington, DC. Available at: http://www.environmentportal.in/files/IFPRIDP00814.pdf (last accessed: 17.12.2016).

Braun, V. & Clarke, V. (2006). Using thematic analysis in psychology. *Qualitative Research in Psychology*, 3 (2), 77–101.

Brousseau, E. (2008). From Bilateral Sets of Incentives to the Multi-level Governance of Relations. *In:* Brousseau, E. & Glachant, J.-M. (eds.), *New Institutional Economics: A Guidebook*. pp. 37–66, Cambridge University Press, Cambridge.

Cai, J., Setboonsarng, S. & Leung, P. S. (2014). Contract Farming and Poverty Reduction: The Case of Organic Rice Contract Farming in Thailand. *In:* Setboonsarng, S. & Leung, P. S. (eds.), *Making Globalization Work Better for the Poor through Contract Farming*. pp. 71–108, Asian Development Bank, Mandaluyong City, Philippines.

Chirwa, E. W. & Kydd, J. G. (2009). From Statutory to Private Contracts: Emerging Institutional Arrangements in the Small Producer Tea Sector in Malawi. *In:* Kirsten, J., Dorward, A., Poulton, C. & Vink, N. (eds.), *Institutional Economics Perspectives on African Agricultural Development*. pp. 213–226, International Food Policy Research Institute, Washington, DC, USA.

Cotula, L. (2011). Land deals in Africa: What is in the contracts? International Institute for Environment and Development (IIED). Available at: http://pubs.iied.org/pdfs/12568IIED.pdf (last accessed: 15.01.2014).

CountrySTAT Malawi (2016). Distribution of the Production Quantity for Primary Crops by Year, Crop (Ton). Available at: http://www.countrystat.org/home.aspx?c=MWI&ta=130CPD010&tr=7 (last accessed: 15.01.2016).

CYE Consult (2009). Value Chain Analysis of Selected Commodities: Institutional Development Across the Agri-Food Sector (IDAF) – 9 ACP MAI 19. Final Report, Request for services n°2008/159774/1. Available at: https://www.yumpu.com/en/document/view/28886315/value-chain-analysis-of-selected-commodities-institutional- (last accessed: 18.01.2013).

Eaton, C. & Shepherd, A. W. (2001). Contract Farming: Partnerships for growth. FAO Agricultural services bulletin 145. Food and Agriculture Organization (FAO), Rome. Available at: http://www.fao.org/docrep/014/y0937e/y0937e00.pdf (last accessed: 17.12.2016).

Echánove, F. & Steffen, C. (2005). Agribusiness and farmers in Mexico: the importance of contractual relations. *The Geographical Journal*, 171 (2), 166–176.

FAO (2013). Country profiles. Malawi. Food and Agriculture Organization (FAO), Rome. Available at: http://www.fao.org/countryprofiles/index/en/?iso3=MWI (last accessed: 15.01.2016).

FAO (2016a). Purchase agreement of fruits and vegetables in Cape Verde. Food and Agriculture Organization (FAO). Available at: http://www.fao.org/fileadmin/user_upload/contract_farming/samples/PURCHASE%20AGREEMENT%20OF%20FV_cape%20verde.pdf (last accessed: 03.01.2016).

FAO (2016b). Sample contract for fresh fruits and vegetables in India. Food and Agriculture Organization (FAO). Available at: http://www.fao.org/fileadmin/user_upload/contract_farming/samples/Sample%20contract%20for%20fresh%20fruits%20and%20vegetables%20in%20India.pdf (last accessed: 03.01.2016).

FAO (2016c). Sample contract for paprika in Malawi. Food and Agriculture Organization (FAO). Available at: http://www.fao.org/fileadmin/user_upload/contract_farming/samples/1_Sample%20contract%20for%20paprika%20in%20Malawi.pdf (last accessed: 03.01.2016).

FAO (2016d). Sample contract for paprika in Zambia. Food and Agriculture Organization (FAO). Available at: http://www.fao.org/fileadmin/user_upload/contract_farming/samples/16_Sample_contract_of_paprika_in_Zambia.pdf (last accessed: 03.01.2016).

Fereday, J. & Muir-Cochrane, E. (2006). Demonstrating Rigor Using Thematic Analysis: A hybrid Approach of Inductive and Deductive Coding and Theme Development. *International Journal of Qualitative Methods*, 5 (1), 80–92.

Fréguin-Gresh, S. & Anseeuw, W. (2013). Integrating small producers in the South African citrus sector. *In:* da Silva, C. A. B. & Rankin, M. (eds.), *Contract farming for inclusive market access*. pp. 79–104, Food and Agriculture Organization (FAO), Rome, Italy.

Furlotti, M. (2007). There is more to contracts than incompleteness: a review and assessment of empirical research on inter-firm contract design. *Journal of Management and Governance*, 11, 61–99.

Girma, J. & Gardebroek, C. (2015). The impact of contracts on organic honey producers' incomes in southwestern Ethiopia. *Forest Policy and Economics*, 50, 259–268.

Hart, O. (1988). Incomplete Contracts and the Theory of the Firm. *Journal of Law, Economics, and Organisation*, 4 (1), 119–139.

Hart, O. (2010). Making the Case for Contract Theory. American Economic Association, Ten Years and Beyond: Economists Answer NSF's Call for Long-Term Research Agendas. Available at: https://ssrn.com/abstract=1889201 (last accessed: 15.12.2016).

Hart, O. & Moore, J. (2008). Contracts as reference points. *The Quarterly Journal of Economics*, 123 (1), 1–48.

Imbruce, V. (2008). The production relations of contract farming in Honduras. *GeoJournal*, 73 (1), 67–82.

Jia, X. & Bijman, J. (2013). Contract farming: Synthetic themes for linking farmers to demanding markets. *In:* da Silva, C. A. B. & Rankin, M. (eds.), *Contract farming for inclusive market access*. pp. 21–38, Food and Agriculture Organization (FAO), Rome, Italy.

Jones, S. & Gibbon, P. (2011). Developing agricultural markets in sub-Saharan Africa: organic cocoa in rural Uganda. *Journal of Development Studies*, 47 (10), 1595–1618.

Kabungo, A. M. & Jenkins, G. P. (2016). Contract Farming Risks: A Quantitative Assessment. *South African Journal of Economic and Management Sciences*, 19 (1), 35–52.

King, N. (2012). Doing template analysis. *In:* Symon, G. & Cassell, C. (eds.), *Qualitative organisational research: Core methods and current challenges*. pp. 426–450, SAGE Publications, London/Thousand Oaks, UK/USA.

Kumwenda, I. & Madola, M. (2005). The status of contract farming in Malawi. Report for Food, Agriculture and Natural Resources Policy Analysis Network (FANRPAN). Available at: http://www.fanrpan.org/documents/d00691/Contract_Farming_Report.pdf (last accessed: 10.01.2013).

Locke, C. & Lloyd-Sherlock, P. (2011). Qualitative Life Course Methodologies: Critical Reflections from Development Studies. *Development and Change*, 42 (5), 1131–1152.

Maertens, M. (2006). Trade, Food Standards and Poverty: The Case of High-Value Vegetable Exports from Senegal. Poster paper prepared for presentation at the International Association of Agricultural Economists Conference, Gold Coast, Australia, August 12–18, 2006. Available at: http://ageconsearch.umn.edu/bitstream/25614/1/pp062657.pdf (last accessed: 22.01.2015).

Makoka, D., Chitika, R. & Simtowe, F. (2010). Value chain analysis of Paprika and Bird's Eye Chillies in Malawi. Final report. University of Malawi, Centre for Agricultural Research and Development. Available at: https://mpra.ub.uni-muenchen.de/27785/ (last accessed: 08.01.2013).

Minot, N. & Roy, D. (2007). Impact of high-value agriculture and modern marketing channels on poverty: An Analytical framework. IFPRI Food Policy Review. International Food Policy Research Institute, Washington, DC. Available at: http://citeseerx.ist.psu.edu/viewdoc/download?doi=10.1.1.121.7367&rep=rep1&type=pdf (last accessed: 17.10.2014).

Narayanan, S. (2012). Notional Contracts: The Moral Economy of Contract Farming Arrangements in India. Working Paper 2012-020. Indira Gandhi Institute of Development Research, Mumbai. Available at: http://www.igidr.ac.in/pdf/publication/WP-2012-020.pdf (last accessed: 19.03.2015).

NSO (2015). Trade Map - Malawi. National Statistical Office (NSO). Available at: http://www.nsomalawi.mw/index.php?option=com_wrapper&view=wrapper&Itemid=110 (last accessed: 15.01.2016).

Prowse, M. (2012). Contract Farming in Developing Countries - A Review. French Agency for Development. Imprimerie de Montligeon, Paris. Available at: http://www.afd.fr/webdav/shared/PUBLICATIONS/RECHERCHE/Scientifiques/A-savoir/12-VA-A-Savoir.pdf (last accessed: 10.01.2014).

Pultrone, C. (2012). An Overview of Contract Farming: Legal Issues and Challenges. *Uniform Law Review*, 17 (1-2), 263–289.

Pultrone, C., da Silva, C. A. B. & Shepherd, A. W. (2012). Guiding principles for responsible contract farming operations. Food and Agriculture Organization (FAO), Rome, Italy. Available at: http://www.fao.org/docrep/016/i2858e/i2858e.pdf (last accessed: 08.06.2015).

Reardon, T., Barrett, C. B., Berdegue, J. A. & Swinnen, J. F. M. (2009). Agrifood Industry Transformation and Small Farmers in Developing Countries. *World Development*, 37 (11), 1717–1727.

Schipmann, C. & Qaim, M. (2011). Supply chain differentiation, contract agriculture, and farmers' marketing preferences: The case of sweet pepper in Thailand. *Food Policy*, 36, 667–677.

da Silva, C. A. B. (2005). The growing role of contract farming in agri-food systems development: Drivers, theory and practice. Food and Agriculture Organization (FAO), Agricultural Management, Marketing and Finance Service, Rome. Available at: http://www.fao.org/fileadmin/user_upload/ags/publications/AGSF_WD_9.pdf (last accessed: 10.04.2016).

Smalley, R. (2013). Plantations, Contract Farming and Commercial Farming Areas in Africa: A Comparative Review. Land and Agricultural Commercialisation in Africa (LACA), Working Paper 055. Available at: http://bdsknowledge.org/dyn/bds/docs/854/FAC_Working_Paper_055.pdf (last accessed: 19.05.2015).

Stevens, A. (2004). Paprika Supply Chain in Zambia. Technical Report. Available at: http://pdf.usaid.gov/pdf_docs/Pnacy603.pdf (last accessed: 12.12.2016).

Sykuta, M. & Parcell, J. (2002). Contract Structure and Design in Identity-Preserved Soybean Production. *Review of Agricultural Economics*, 25 (2), 332–350.

UN (2015). Transforming our World: The 2030 Agenda for Sustainable Development. United Nations (UN) - Sustainable Developmentknowledge Platform. Available at: https://sustainabledevelopment.un.org/post2015/transformingourworld (last accessed: 23.12.2015).

UN (2016). Country Profile: Malawi. United Nations (UN). Available at: http://data.un.org/CountryProfile.aspx?crName=malawi (last accessed: 15.01.2016).

UNIDROIT (2014). The Legal Dimension of Contract Farming - Promoting Good Contract Practices between Producers and Buyers in Contract Farming Operations in the African Context. Report on the Consultation Workshop on the UNIDROIT / FAO Legal Guide on Contract Farming, Addis Ababa, 31 October 2014, International Institute for the Unification of Private Law (UNIDROIT). Available at: http://www.unidroit.org/english/documents/2014/study80a/s-80a-22-e.pdf (last accessed: 23.12.2015).

UNIDROIT/FAO/IFAD (2015). Legal Guide on Contract Farming. Available at: http://www.fao.org/3/a-i4756e.pdf (last accessed: 25.08.2015).

Valipour, M. (2015a). Agricultural water management in the world during past half century. *Archives of Agronomy and Soil Science*, 61 (5), 657–678.

Valipour, M. (2015b). Future of agricultural water management in Africa. *Archives of Agronomy and Soil Science*, 61 (7), 907–927.

Vermeulen, S. & Cotula, L. (2010). *Making the most of agricultural investment: A survey of business models that provide opportunities for small producers.* IIED/FAO/IFAD/SDC, London/Rome/Bern. Available at: https://www.ifad.org/documents/10180/a314bc61-46ff-4df0-be1a-a9cc9ad1b3e9 (last accessed: 20.04.2017).

Wanglin, M. & Abdulai, A. (2016). Does cooperative membership improve household welfare? Evidence from apple farmers in China. *Food Policy*, 58, 94–102.

Warning, M. & Hoo, W. S. (2000). The Impact of Contract Farming on Income Distribution: Theory and Evidence. Paper prepared for presentation at the Western Economics Association International Annual Meetings, Sydney (June 30). Available at: http://www.pugetsound.edu/files/resources/6926_00-6.pdf (last accessed: 10.04.2014).

Will, M. (2013). *Contract Farming Handbook. A practical guide for linking small-scale producers and buyers through business model innovation.* GIZ, Bonn/Eschborn, Germany. Available at: https://www.giz.de/fachexpertise/downloads/giz2013-en-contract-farming-manual.pdf (last accessed: 01.06.2013).

World Bank (2015a). Data: GDP per capita. Available at: http://data.worldbank.org/indicator/NY.GDP.PCAP.CD?order=wbapi_data_value_2013+wbapi_data_value&sort=asc (last accessed: 15.01.2016).

World Bank (2015b). Data: Malawi. Available at: http://data.worldbank.org/country/malawi#cp_surv (last accessed: 15.01.2016).

# 8

# Risks, resources and reason: understanding smallholder decisions around farming system interventions in Eastern Indonesia

Clemens M. Grünbühel [a,*], Liana J. Williams [b]

[a] *School of Environment, Resources and Development, Asian Institute of Technology (AIT), Pathumthani, Thailand*
[b] *CSIRO Ecosystem Services, Commonwealth Scientific and Industrial Research Organisation, Brisbane, Australia*

**Abstract**

Adoption of new cattle management practices by Indonesian smallholders occurs less as a 'technology transfer' in the classical sense but rather as a series of conscious decisions by farming households weighing risks and resources as well as matching innovations to livelihood strategies. This paper uncovers the context of decisions and communication of innovations by way of social networks. The research looks at two geographically distinct cases where new cattle management practices have been introduced. We apply the lens of a *common sense* framework initially introduced by Clifford Geertz. Smallholder decisions are analysed within a socio-cultural context and a particular set of resources, risks and livelihood objectives. We show that the respective value placed on land, cattle and food security is central to adoption of new cattle management techniques. Far from accepting everything novel, smallholders are selective and willing to make changes to their farming system if they do not conflict with livelihood strategies. Innovations are communicated through a range of existing social networks and are either matched to existing livelihood strategies or perceived as stepping-stones out of agriculture.

*Keywords:* adoption, common sense, decision analysis, Indonesia, smallholders, social network analysis

## 1 Introduction

The research for this paper stems from the need to learn more about how decisions are made and mediated when farm management innovations are introduced to a subsistence-oriented agricultural society. The research centres on a qualitative analysis of household decisions and decision influences through social networks in response to introduced information about improved cattle management in South Sulawesi and Central Lombok—two geographically distinct areas of Indonesia.

Modern economics acknowledge the issues attached with the *homo oeconomicus* assumption of people acting as 'selfish rational actors' (Janssen & Anderies, 2011, p.1569). Instead, researchers face the challenge of uncovering the context of decision-making, which "*requires the development of a framework that specifically captures context. Furthermore, this requires the development and application of methods to measure this context, such as social networks, mental models and communication patterns*" (*ibid.*, see also Anderies *et al.*, 2011). Understanding smallholding household decisions is imperfect - neither do they accept everything novel because they are destitute; nor do they reject innovations because they are conservative or illiterate. Rather, smallholders use culture to mediate social action, and institutions to filter their choices. Far from being 'irrational', they are rational in considering the social ramifications of change.

The research presented here subscribes to the call for a departure from the rational actor paradigm when designing development research or agricultural extension interventions. The farming household's decisions

* Corresponding author
Email: clemens@ait.asia

and actions occur in a cultural and social environment, for which the household has developed skills, techniques and knowledge to navigate through. The argument articulated here supports the notion that development interventions cannot assume to be designed on a white canvas, or within an economic balance sheet. Rather, household decisions are based on *common sense*, or encultured knowledge, which is, in essence, an adaptation strategy to the historic and current socio-cultural context (Geertz, 1975, 1983).

### *1.1 Farming systems and decisions*

Any farming activity occurs within a system constrained by social, physical and economic resources (Giampietro, 2004). Changes to one part of the system will most likely have an impact on other parts, i.e., require adjustments by several elements in the system. In largely subsistent smallholder systems any adoption of a new practice is most likely to have an effect on land and labour demands, division of labour, and input/output ratios. However, we cannot assume that farmer decisions are based solely on rational choice as other factors, such as social pressure, cultural norms, aspirations, and risk perception feed into the decision (Douglas, 1985).

The *homo oeconomicus* model of classical economic theory assumes an independent agent who acts rationally with the objective to maximise self-interest. It usually assumes full access to relevant information (*cf.* Reeson & Dunstall, 2009; Mzoughi, 2011). Newer theory building in behavioural economics, however, has identified several 'anomalies' under which rationality is discarded for the sake of other goals, such as avoiding risk, minimising losses and the unwillingness to change between different activities within the potential livelihood portfolio. In addition, the relevance of institutions, social preferences and norms has been acknowledged. For example, conditional cooperation and equity as guiding values when making decisions (*ibid*.).[1]

In fact, recent research relates adoption decisions mainly to social and moral concerns, rather than economic ones. Mzoughi (2011) differentiates between intrinsic and extrinsic motivations; intrinsic being the satisfaction of personal standards and extrinsic being social acceptance and reward. The field of economic anthropology (e.g., Rössler, 1999) tends to focus on non-economic rationality, i.e. benefits (material and social) outside of the monetary realm.

It seemed relevant for our study to look into the *cultural system* of decision making. Therefore, we took a heuristic approach following the seminal works of anthropologist Clifford Geertz (1975, 1983). The interpretative – rather than functionalist – approach of his framework lies on understanding the *emic* or inside perspective of the decision rather than focusing on its results. It captures the different steps involved in the process of decision-making. The analysis of common sense, according to Geertz (1983, 75f) requires focus on *perception* – rather than objectified reality – *assessment, judgement* and *conclusion* when dealing with everyday problems within a given cultural setting [2]. Naturally, such an approach is highly contextual and merits a number of case studies of decision-making processes: by looking at institutions, influences and resource constraints of each case, decision patterns emerge.

Geertz' notion of *common sense* emphasises the importance of understanding the knowledge *commonly* shared in the social system and how the system reacts to perturbations of new knowledge for which there are no indigenous (pre-existing) concepts. When exposed to everyday phenomena, people are able to perceive information and process it in a sensible, intelligent, comprehensive and reflexive manner and respond appropriately and efficiently. Common sense responses affirm the cultural system. They also provide guidance when making everyday decisions and evaluating risk. Common sense is based on knowledge of the immediate environment that ensures household survival through everyday challenges. It is specific to culture and environment and constantly (re-)negotiated among the culture bearers. It is democratic (i.e., everyone can/should have it) and serves as a guide when specialised knowledge is lacking (Geertz, 1975, 1983).

The fields of development studies (e.g., Ellis, 2000) and ecological anthropology (e.g., Ellen, 1982) similarly observe the systemic nature of farming and the multi-layered decision space under which the household operates. Within this framework, household decisions are influenced by the following factors (see also Figure 1):

[1] Conditional cooperation and equity are analogous terms to redistribution and reciprocity in economic anthropology (Polanyi, 2001; Rössler, 1999).

[2] We have used these categories to guide our interviews for understanding how respondents arrive at their decisions. In addition, we added another category of inquiry, which was *implementation*, assuming that any decision does not necessarily lead to the desired action. Geertz (1983) does not mention implementation, since he is interested in shared knowledge systems that lead to the notion of common sense.

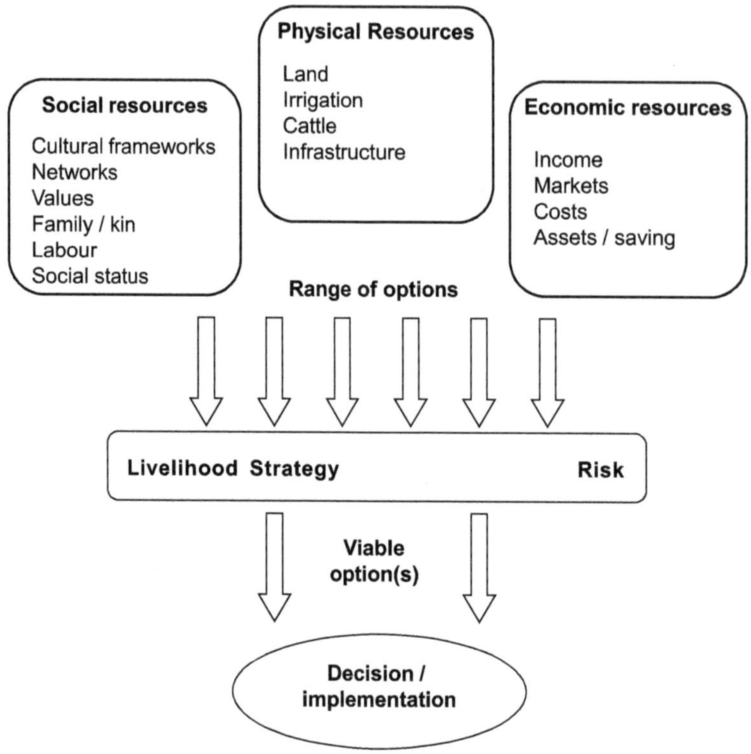


**Fig. 1:** *Interaction of resources, risk and livelihood strategy in the household decision-making process. (authors' own).*

- available resources which determine the viable options for the household's activity portfolio;
- household livelihood strategies;
- household perception and evaluation of risk.

*Resources* are defined here as material and immaterial assets enabling the household to select among different options depending on how the resources are applied or expended. Resources available to the household and expected economic effects determine the viability of options. When evaluating options, households select within the relevant institutional and cultural context. This includes social networks and institutions, which mediate how information is spread through the community and influence the interpretation of information (Preston, 1994; Wiersum, 1994).

Within the household, *livelihood strategy* as well as its evaluation of *risks* significantly influences the selection of new practices (Ellis, 2000; Sjah *et al.*, 2006; see Figure 1). Both livelihood strategy and risk evaluation are based on the *perceived* availability of resources which the household is endowed with. This is where the common sense notion comes to play. Only those resources, which pass the social filter and are deemed appropriate for application, become part of the decision process. Perceived resources are those commonly shared within the system (*cf.* Pastore *et al.*, 1999).

### *1.2 Supporting cattle management in Eastern Indonesia*

Holding large ruminants (buffalo and cattle) is a traditional ancillary activity of rice farmers in South East Asia (Devendra & Thomas, 2002). While the priority lies on the production of the staple crop, cattle are often held if there is additional land and labour available on the farm, and are particularly important for smallholder farmers (*ibid.*). Most farming households show a strong preference towards securing food stocks with rice, vegetables, and poultry (Pengelly *et al.*, 2003). Cattle, while rarely consumed within the household, are traditionally used as food source for larger festivities, such as weddings, and shared among relatives and neighbours. Additionally, cattle are bred, fattened and eventually sold off to traders or at local livestock markets where they mostly end up satisfying the growing beef demand in regional cities.

While cattle are generally not important for securing the farming households' food supply, they nevertheless fulfil an important socio-economic role in the farming system. Research has shown the multi-functionality of large ruminants in traditional Southeast Asian societies – as animal labour in cropping and transportation, as emergency protein source, as a symbol of status, as storage of wealth, and as cash savings for large investment needs (Devendra & Thomas, 2002; Paris, 2002; Mayrhofer-Grünbühel, 2004).

In eastern Indonesia, cattle are fed by either (a) tethered grazing on rice bunds, small patches of otherwise unused land and on the banks of waterways or (b) left to graze freely on harvested paddy fields or secondary forest. Stables or pens are used to protect cattle from natural hazards and cattle theft during the night. Supplementary manual feeding may occur if naturally available forage is insufficient. In this case, locally available forages, such as elephant grass (*Pennistum purpureum*) is used, which is either gathered from places where it grows wild or planted around the homestead.

The focus of our analysis is two livestock research projects in Lombok and South Sulawesi, Eastern Indonesia. The aim of these projects was to adapt techniques for improved cattle management and allow for a more rapid turnover and more secure income stream for smallholding famers (see Table 1; and, for more detail, van Wensveen *et al.*, 2010, 2011). Interventions included introduction of new forage varieties, strategic forage production, feed budgeting to improve cattle nutrition and health, as well as controlled mating and weaning to enable higher fertility rates. The newly introduced practices can be roughly separated into (1) practices requiring complementary resources and (2) practices requiring behaviour change only. Suggested practices support intensification by keeping the cattle in or near pens and planting forages. The emphasis, thus, shifts from predominantly free grazing to manual feeding.

The introduction of improved cattle management techniques in the two projects has been deemed widely successful (Martin, 2010) and a large number of households adopted one or more of the introduced techniques (455 known households in South Sulawesi, and 1,030 group members – out of a total 1,144 – in Lombok; for details see: van Wensveen *et al.*, 2010, 2011) [3]. The evaluation of uptake, however, revealed that the adoption of techniques by households was highly selective. Therefore, effort was put into understanding the social process of adoption, which is presented here.

We attempt to contribute to an understanding of household decisions under the conditions of constrained resource access. This is informed by the insight that culture and the social environment mediate the common understanding of how to manage the farming system. We seek to elucidate the smallholders' rationale – their *common sense* – of making changes to the farm system when exposed to new information and resources by providing a contextual description of the adoption process. This coincides with increased attention to the role of the smallholder in agricultural development (Wiggins *et al.*, 2010) and contributes to an understanding of the rural economy and decisions made under the uncertainty of system effects caused by introducing new practices in resource constrained environments.

## 2 Method

To understand household decision processes, which determine whether or not adoption takes place, *decision narratives* were developed. These provide insights into the livelihood strategies of households as well as the various steps in decision-making. Narratives were developed from in-depth interviews, which explored the various steps of deciding whether or not to accept new livestock management practices. To make them comparable across locations and contexts, the interviews were structured according to Geertz' (1975) categories of common sense [4] and, more specifically, along the analytical framework described above (*cf.* Figure 1). The categories enable outlining the perceptions, risks considered, and persons involved in the decision making process and identify relevant institutions and drivers.

The decision to make changes to the farming system does not occur in isolation of social relationships. Rather, there is a process of (formal or informal) exchange with other actors and institutions such as neighbours, village heads, religious leaders, government agencies – all of which feed into and possibly influence – the decision-making process. In fact, for development interventions to be successful, more attention needs to be paid to accessing and building links with existing networks (Mahanty, 2002).

Social Network Analysis (SNA) was used to analyse the spread of information and explore actors and institutions influencing household decisions. SNA allows for analysis of the structure and composition of social connections and subsequent implications on access to information or resources as well as the articulation of norms, and values and behaviour (Wasserman & Faust, 1994). The structure and context of social networks can highlight access to (or lack of) information and resources as well as the factors supporting decision-making (Lai & Wong, 2002; Hoang *et al.*, 2006).

[3] The projects have been completed and approved by the commissioning agency. The research presented here represents a comparatively small activity within the project.

[4] See our interpretation of Geertz' (1983) categories under *Farming systems and decisions*, above, as well as footnote 2.

**Table 1:** *Table 1: Summary of introduced practices for each area. Slight differences in recommendations reflect adaptations to local conditions and priority of implementation in each region*

| *Lombok:* | *South Sulawesi:* |
|---|---|
| 1. *Controlled mating*<br>Bull service facilitated through the *kandang* group; aims at mating once yearly. | 1. *Better use of existing forage* †<br>Improved management of forage to ensure productivity. |
| 2. *Preferential feeding for pregnant and lactating cows*<br>Requires additional forage for cows. | 2. *Introduction of new forages* ‡<br>Planting and management of new varieties on farmland or backyard plots. |
| 3. *Early weaning and preferential feeding for calves*<br>Requires pen to separate cow and calf, and quality forage for calves. | 3. *Controlled mating*<br>Aims to time mating so that the calf is born during the wet season when more feed is available. No bull provided through project. |
| 4. *Forage management* *<br>Requires management of forage on communal land or on dedicated farmland. | 4. *Early weaning and preferential feeding*<br>Requires pen to separate cow and calf, and quality forage for calves |
| | 5. *Feed budgeting*<br>Meeting anticipated forage demand throughout the year |

* Forages newly introduced to Lombok included Mulato (*Brachiaria hybrid* cv Mulato) and Panicum (*Panicum maximum*), pre-existing forages included Leucaena (*Leucaena leucocephala*), Gliricidia (*Gliricidia sepium*), Elephant Grass (*Pennisetum pupureum*) and Sesbania (*Sesbania grandiflora*). For both, the focus was on forage management.

† Existing forages in South Sulawesi included Gliricidia and Elephant Grass.

‡ Introduced grasses in South Sulawesi included Paspalum (*Paspalum artatum* cv Higane), Mulato, Panicum and Setaria (*Setaria sphacelata* cv Narok). Introduced herbaceous legumes included Clitoria (*Clitoria ternatea* cv Milgarra), Centrosema (*Centrosema pubescens* cv Cardillo) and Stylo (*Stylosanthes guianensis* CIAT 184, *Stylosanthes scabra* cv Seca, *Stylosanthes hamata* cv Verano).

In this study, SNA was used to examine how knowledge about new practices spread among households and communities as well as the type of households, relationships or institutions critical for promoting adoption (Bandiera & Rasul, 2006; Crona & Bodin, 2006). The underlying assumption is that communication through cultural institutions would be more efficient than communication of knowledge introduced by external project staff. Data for SNA was collected as part of the decision narratives and focused on capturing:

- Interactions and influence between households and local institutions;
- The spread of information; and
- The spread of resources as concomitant to information.

Respondents were asked questions to determine if and where they heard about new cattle management practices; who they had discussed the practices with, and if/how they were able to obtain resources to support adoption, such as cuttings or seeds for forage. Different types of networks constitute inherently different ties or relationships between actors (Bodin & Crona, 2009).[5]

[5] For example, formal associations *versus* kinship ties.

Interviews sought to reveal the nature and effectiveness of different networks for the spread of information. Data collection included both volatile as well as institutionalised relations, i.e., both the exchange of information as coincidental part of people's every day interactions, and part of the regular flow of information by the project, with the aim of identifying patterns of information and resource flow in the target community [6].

Key challenges in SNA relate to definition of analytical boundaries (either geographic or relational) and sampling strategies. Scott (2000) highlights a range of strategies depending on the focus of analysis, each with its own limitations and requirements for justification. In our case, working at a village level (and in multiple villages) meant it was not feasible to include all households within a village in the analysis. Instead, we used snowball sampling, starting with a sample of households, which the project had been communicating

[6] The project, of which this research was a part, shared limited resources (forage seed and mating bulls) among target farmers. Resources were not "handed out", however, but were one part of the communication package containing information on suggested practices. Therefore, the possible cases measured distinguished between the transfer of "information only" or the combined "information + resources".

with. We asked them whom they had provided information and resources to. A sub-set of these "scale-out" households would then be invited to participate in interviews. An equivalent number of households were also selected randomly and invited for interview.

The method used creates inherent limitations. For example, in a network created from snowball sampling, it becomes irrelevant to analyse the connectedness of the network as it is, by virtue of the sampling technique, highly connected (Scott, 2000). At the same time, the networks in this study represent only a small fraction of the relations present in the village. Nevertheless, rather than representing *minutiae*, the objective of the method was to arrive at an *orientation hypotheses*, which would allow us to understand selected relations between actors and provide more insight in the structure of information dissemination throughout the community (Schweizer, 1988)[7].

The two methods applied, decision narratives and network analysis, allow for observation of the structural-institutional as well as the cognitive-emotive aspects of processing information within the community. Hence, the results not only contain a formal analysis of relations among actors but also elucidate the process of evaluating risks and adapting livelihood strategies by households as their environment changes through the project intervention.

The study observes two cases in Eastern Indonesia, which differ substantially in terms of their agricultural systems, socio-economic characteristics and cattle management. One is characterised by single cropping, extensive livestock husbandry and a relatively low human population density, while the other is a (relatively) intensive farming and husbandry system with high population density (more details in results section, below). A total of 216 interviews were conducted in South Sulawesi (Baru, Bone and Gowa Regencies), and 80 in Nusa Tenggara (Central Lombok Regency) between July and November 2009. For the main characteristics of each Regency see Table 2. Interviews included respondents directly involved with the project, as well as those with limited or no involvement.

[7] The »orientation hypothesis« does not make strictly valid statements of relations between network variables. Rather, it identifies important phenomena within the network, which need to be considered when developing stringent hypotheses (Schweizer, 1988; Kelle, n.d.)

## 3 Results

### 3.1 *CASE STUDY 1: Central Lombok, West Nusa Tenggara*

#### 3.1.1 *Resources*

Central Lombok shows high population density of 709 people per square kilometre (see Table 2). Landlessness is a common feature among rural households. On average, smallholders in Lombok who own or rent land have access to less than 0.3 ha (Table 2). As a result, spare land for grazing or forage production is limited.

Constraints on the available land area in Central Lombok limit the potential for adoption of practices that require additional land, such as forage cultivation. Available resources in the farming system do not allow for a significant increase in land used to grow forage without corresponding adjustments to other types of land use, such as replacement of other crops or land purchase.

**Table 2:** *Comparison of population density and household resources, Nusa Tenggara Barat and Sulawesi*

| *Province* | *Regency* | *Population Density* (people per km$^2$) | *Paddy land* (% of regency) | *Estimated average paddy land ownership per household* (ha)* | *Estimated average cattle ownership per household** | *Average paddy land of households interviewed* (ha)† | *Average number of cattle owned by households interviewed*† |
|---|---|---|---|---|---|---|---|
| Nusa Tenggara Barat | Central Lombok | 709 | 43 | 0.28 | 0.3 | 0.25 | 2.6 |
| Sulawesi Selatan | Baru | 135 | 13 | 0.70 | 1 | 0.74 | 5.5 |
| Sulawesi Selatan | Bone | 153 | 24 | 0.38 | 0.7 | 1.13 | 6.6 |
| Sulawesi Selatan | Gowa | 311 | 21 | 0.28 | 0.3 | 0.60 | 4.1 |

* Average calculated based on total number of households in regency – therefore includes non-farming households
† Based on household interviews conducted in 2009-10, includes landless households
Source: Lombok data is for the year 2009, taken from the BPS – Statistics Lombok Tengah Province and DINAS Peternakan. Sulawesi data is for 2006, taken from the BPS – Statistics Sulawesi Selatan Province.

Cattle theft has been a significant problem in Central Lombok and a system of communal pens (*kandang*) was established in the 1980s to address this. Large sheds with multiple individual pens are used by a group of households from one or several communities. Cattle are taken out to graze during the day, and are returned to the pen each night, where group members take turns guarding to protect against theft. Each member is required to pay a fee (rice or cash) and participate in night watch duties. In return, they receive pen space for their cattle. Space constraints in communal *kandang* limit the number of cattle a household can keep at any one time. Access to government support for inputs such as fertiliser is often tied to group membership.

A household in Lombok with access to land will typically sow one crop of rice in the wet season, followed by a dry season crop, such as maize, soybean or tobacco. In addition, maize, cassava, chilli or forages are grown on bunds between rice fields. Grass for cattle is collected from riverbanks, roadsides and unused land. Unless explicitly claimed by a farmer (through signs or symbols), forages and grass are considered common property resources and thus can be grazed or collected by anyone. This fact greatly inhibits adoption of forage plantations, as even claims on forages often lead to conflicts among farmers.

Most households engage in some degree of off-farm activities for cash income (e.g., brick making, transportation). While this provides additional income security to the household, it also means less time available for on-farm activities. Income from off-farm activities is most often used for paying daily expenses, such as commodities, taxes and additional food.

Cattle provide the farming household with an opportunity to pay for major items by selling animals whenever needed. This practice allows for cash liquidity when required. When questioned about the income function of cattle farming, respondents in Lombok distinguished between:

- calf production: provides annual income for large expenses;
- fattening of bulls: provides 6-monthly regular income;
- poultry and goats: provides income for daily expenses.

At certain times in the year cattle prices tend to rise, e.g., during religious celebrations when demand is greater. Where possible, farmers try to plan the sale of their cattle according to the fluctuating market price. This is because households generally do not find it difficult to sell cattle. Most prefer to sell to traders who frequently visit villages. For farmers, this removes risk and costs required to transport animals to the market.

Respondents perceive the lack of available mating bulls as the key resource constraint to cattle production (*cf.* van Wensveen *et al.*, 2011). Prior to the project intervention, farmers either left cows in communal areas in hope of serendipitous mating, moved cows to neighbouring villages for mating, or used commercial artificial insemination services. Even where households kept bulls for fattening, they were not perceived as suitable for mating, since it was believed that this would make the bull lose weight and result in loss of income. Thus, farmers responded well to the project's provision of bull mating services. It was by far the single most widely adopted practice out of the package of cattle management techniques offered by the project (*cf.* Table 1). Perceived requirement and options offered seemed to coincide well in this case (less so in others).

Households estimate rearing cattle to be one of their most labour-intensive activities, due to time investments in the cut and carry of forage, relocation of cattle to grazing sites and watering spots during the day and returning them to the pen each night. The tasks of looking after cattle are shared between men, women and children within the same household. When feed resources are scarce, men generally travel longer distances to search for forage. Most households see these activities to be a major burden on their time budgets and react cautiously to any new investments into cattle management if these are perceived to put further strain on time resources.

Other than keeping mating bulls, households were reluctant to adopt practices relating to forage production and feed management. As productive land is limited, forage production would require redistribution of land to grow feed instead of food crops. Reduction of paddy land under land-scarce conditions is not considered a viable option among farming households in Lombok. Smaller forage plantations along bunds and in home gardens can be an alternative, but amounts are usually not sufficient to feed the herd. Thus, households are required to devote significant amounts of labour to "cut and carry" (i.e. retrieving forage from areas distant to the homestead). Significantly, however, small forage plantations are seen as a safety net for times when farmers are unable to travel in search of fodder, (e.g., due to illness, or during peak labour times).

Other than the limited availability of mating bulls most households did not perceive major problems with their cattle management systems. They were eager to use the group's bull for mating; thereby saving cash resources from being invested into expensive and often unreliable artificial insemination services. Satisfaction with other aspects of their cattle production, such as feed availability, meant little incentive to try new varieties of forage introduced by the project, even where there would have been nutrition benefits for cattle and production benefits for the household.

Respondents mentioned that, to be a successful farmer, a Sasak from Central Lombok must have both cattle and rice [8]. This is typified by the Sasak language expression *'ngaro ngarit'* which conveys the sentiment that cattle and cropping are complementary and together provide for a good income. Decisions on where to allocate resources into the future depend on perceptions of security. Investment in land is seen as providing concrete and tangible results as food and marketable crops. Cattle, on the other hand, cannot provide for a predictable income due to price fluctuations and risk of animal theft. Rather, households tend to breed or fatten cattle to provide for additional income with little (physical or financial) investment [9].

### 3.1.2 *Networks*

The project staff deliberately focused their efforts on *kandang* groups for information provision and support. Key members within each group, often the *kandang* leader or the bull keeper, were used to demonstrate the results and impacts of adopting different practices.

Use of the mating service, was readily accepted within the group as well as non-member households [10]. This provided a pathway for information to reach households outside the *kandang* groups and beyond usual familial or neighbourly connections. When non-*kandang* group members obtained mating services they were provided with information regarding other cattle management practices by group members who have been trained in these activities. However, actual uptake of these practices by non-*kandang* group members has been limited.

While non-member households were provided with information when they take their cows for mating, weak ties to the *kandang* group (i.e., relationship between members and non-members based on the provision of a service only) result in limited follow-up support and lack of consistency in the information provided. In Figure 2(a) and 2(b) the majority of non-group members interviewed had only adopted one practice – controlled mating through the group's bull service. These farmers felt unable to properly implement any of the other practices suggested.

In cases where non-group members have been able to successfully adopt practices in addition to the bull service, they are part of the social network of *kandang* group members (through family, neighbour or friendship ties). Of the six non-member farmers in Figure 2(b) who have adopted more than one practice, all have close relationships to group members, either by family, friendships or as previous members of this group.

In Figure 2(c) more than half of non-members who used the bull mating service adopted more than one practice. These respondents refer to family and neighbourly relations or, more generally, to repeated discussions with group members and being able to observe the development of forage plots and quality calves.

## 3.2 *CASE STUDY 2: Baru, Bone and Gowa Regencies, South Sulawesi*

### 3.2.1 *Resources*

In comparison to Lombok, Sulawesi has a lower population density (between 135–311 people per square kilometre across three study Regencies; see Table 2). More land availability allows for diverse land use distributed between paddy, other crops and grazing areas.

Due to the absence of *kandang* groups, the project worked with individual farmers in South Sulawesi. While the project staff would work with any farmer willing to learn about the introduced practices, five farmers in each village (n=60) volunteered as 'champions' to promote new forages and practices among their fellow community members It was anticipated that direct relationships with champion farmers would support demonstration of benefits, and farmer-to-farmer communication of practices.

In the South Sulawesi production systems, feed availability was a key constraint to cattle production, particularly during the dry season. While a shortage of bulls for mating was also seen as a constraint to cattle production in Sulawesi, establishing a mating service akin to that in Lombok was not a viable option for the lack of functional farmer groups and communal pens. Lack of infrastructure and resources (existing pens or time, money, materials to build one) were also a barrier to the separation of cows and calves for early weaning. Hence, the primary focus for the introduction of practices was on forage and feed management. The introduction of new forage varieties by the project was deemed viable due to land availability and the potential for land use substitution. This proved to be a successful project intervention.

[8] The Sasak are the majority ethnic group on the island of Lombok (`http://www.ethnologue.com/language/sas`).

[9] High labour costs do not seem to factor in these decisions.

[10] The *kandang* group subsequently began to commercialise the mating service to non-members within the community.

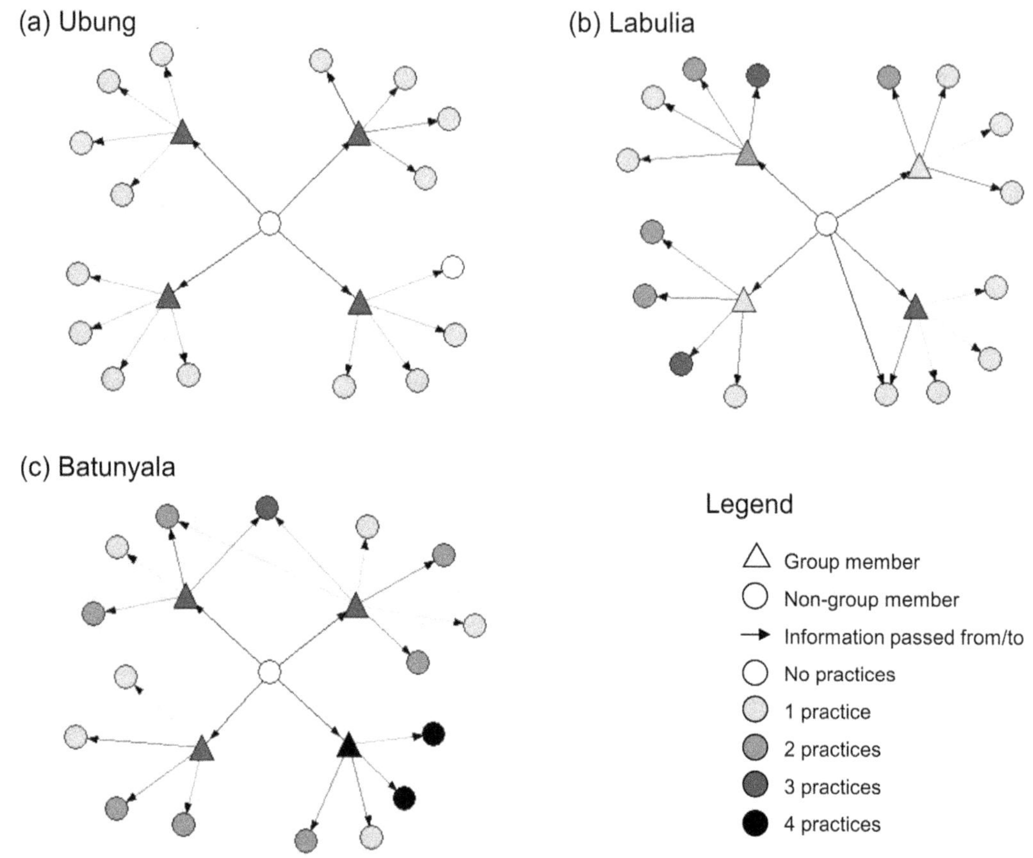


**Fig. 2:** *Information provision and adoption in a kandang farmer group, Lombok. Centre nodes represent project field staff; the nodes closest to the centre are members the group; nodes further out are non-group members (grey shading based on number of practices).*

Households typically produce a minimum of one rice crop, plus a secondary crop using residual soil moisture (maize or peanut). Most farmers leave their cattle to graze on harvested cropland, spare land, or dedicated grazing land. Forage production, largely elephant grass, was in common usage prior to the project, but used only to supplement grazing in the dry season. Farmers maintain that this practice is well established and adapted to the farming system as cattle provide for additional income without large investment requirements of cash or labour.

Some households engage in off-farm activities (e.g., brick making, transportation) to supplement farm income, and this is particularly pronounced in areas close to urban centres (i.e., Gowa Regency). The trade-off between labour spent on- and off-farm further restricts the ability to adopt new practices.

In Sulawesi, cattle are not typically part of the regular income stream of the household. Rather, cattle are a form of wealth saving and tend to be sold for particular events or purposes. Income from cattle production is used for three main purposes:

- major investments (agricultural inputs, vehicle, agricultural machinery);
- exceptional expenses (health costs, ceremonies, festivities);
- emergency expenses (in case of crop failure).

Most households rely on free grazing and homesteads rarely include cattle pens. Due to the distances between homes and grazing locations collection of forage is a task mainly carried out by men. Forage varieties introduced by the project proved popular as they were planted on spare land closer to the household, which reduced travel time to collect grass.[11] In addition, these varieties have higher nutritional value than traditional grasses harvested on roadsides and river banks (Pengelly *et al.*, 2003; Lisson *et al.*, 2010). Where households adopted new forage varieties, a redistribution of labour within the household became possible. Men spent less

[11] See Table 1 for a list of introduced and existing varieties.

travel time to collect forage and were able to reinvest spare time into cropping, rest, or other activities. Due to shorter distances and lighter carrying loads, however, it also led women to take on more responsibility to feed cattle, thereby adding to their time burden of cutting and carrying forage to sheds. On balance, the adoption of forage banks was seen to significantly decrease time spent in cattle feeding, albeit proportionally less so for women. These labour savings were stated by interviewees and had been quantified in previous studies of the same areas by Lisson *et al.* (2010).

While many farmers in South Sulawesi aspire to increasing cattle production as a proportion of their farming system, meeting household food requirements through rice production remains a priority. For many farmers cattle are a good complement to rice production, but are unlikely to replace rice as the central pillar of their livelihood strategy. Interviews suggest households only consider allocating land for the purpose of growing forage if they (1) have sufficient land for household rice production; and (2) already own enough paddy land to hand down to their children. Households with sufficient land area to fulfil these needs, or who had limited labour to manage additional paddy land would turn to cattle production. To the farmers interviewed, cattle are of secondary importance to acquiring sufficient land.

### *3.2.2 Networks*

Households with ties to champion farmers received an increased amount of information, resources and support. Thus, champion farmers were central to the spread of information, through relationships based around the mosque, neighbours and family ties. [12]

Champion farmers in Sulawesi were usually located close to main roads and their highly visible forage banks sparked interest both within and outside their existing networks. In addition, formal promotional activities were arranged by the project, such as farmer field days to demonstrate practices as well as cattle weighing events at the mosque. While these events help to spread basic knowledge, only those households living in the same village as the champion farmers tended to successfully implement a number of practices. Figure 3 illustrates the importance of proximity for adoption in a village in Bone Regency, South Sulawesi. The node with the highest number of connections in the diagram represents a champion farmer who hosted a field day on his farm. While many farmers visited from outside the

[12] Staff employed as extension officers by the project were clearly essential to the spread of information. However, since they were artificial to the existing networks, we decided not to focus on their role here.

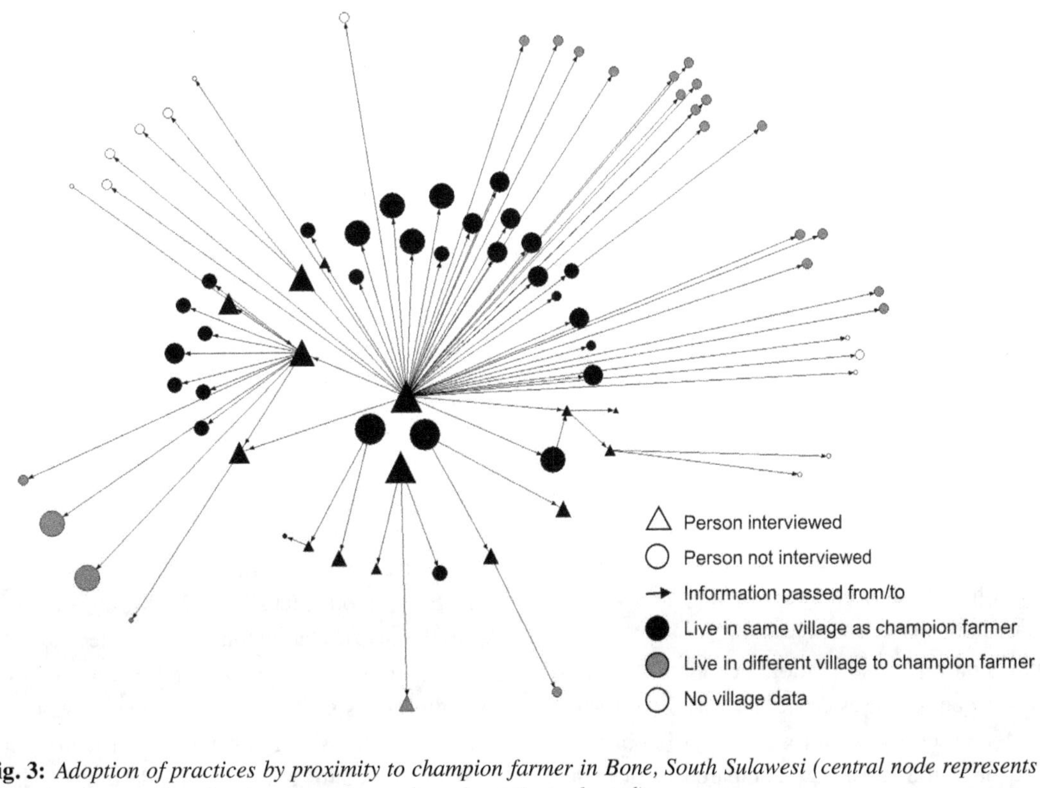


**Fig. 3:** *Adoption of practices by proximity to champion farmer in Bone, South Sulawesi (central node represents champion farmer, size of node represents number of practices adopted).*

village, these farmers adopted only one, if any, practices. In contrast, farmers residing in the same village as the champion farmer adopted more than one practice as they experienced higher exposure and access to information to support decisions on innovations.

Champion farmers in South Sulawesi were key to providing proof that the new practices (in particular, new forage varieties) yield positive results within the farm context. Their activities reassure potential adopters and provide feedback to those hesitant to take risks (see also Millar & Connell, 2009). Unsurprisingly, family and neighbours are often the first to receive information from the champion farmers, reflecting existing networks. Family receives higher preference in the distribution of resources concomitant to information, such as seeds or seedlings to start forage banks.

## 4 Discussion

Resource requirements for adoption of new cattle management techniques include, at a minimum:

(1) land for forage banks;

(2) pens to separate cows and calves;

(3) a bull for mating; and

(4) time to manage forages and implement the other practices.

If resources are tied up in other uses (e.g., land for other crops, or labour for other activities) then the household needs to make a decision whether substitution is worthwhile. This decision will depend on whether a problem with current production systems is perceived, and whether increasing cattle production is desirable. Culturally specific *common sense* determines this perception, not what is feasible from a rationalist perspective.

South Sulawesi and Central Lombok offer two distinctly different sets of resource conditions in which the household decisions are framed. In Central Lombok, with high population density, limited land, and high levels of landlessness, land is a key consideration. This impacts on inputs (e.g., land for forage production) as well as outputs (e.g., increased herd size) of cattle production. At the same time, limited land also encourages a diverse management strategy that places importance on cattle, as often the area of land owned by households is not sufficient to ensure subsistence.

In South Sulawesi, land is more abundant and substitutions between land uses do not necessarily threaten household food security. As a result, lower importance was placed on cattle rearing as compared to rice and dry-season crops. Instead, lack of bulls for mating and limited labour availability inhibited higher investments into cattle management.

Yet in both cases, even households with sufficient resources often decided not to change farming practices despite strong evidence for increased cattle production. Geertz' *common sense* understanding of household strategies for wellbeing and perceptions of risk provides a rationale for such decisions. While there is a complementary relationship between rice and cattle production, East Indonesian households tend to favour household food security through rice production. Many of these households selectively adopted new practices rather than the full package. They preferred investing into activities that added value to cattle production, while minimising risk in rice production. Many households saved time through the adoption of forage banks. The saved time, however, was not re-invested into cattle production, but into rice or other livelihood activities (*cf.* Lisson & Corfield, 2010).

Champion farmers and *kandang* group members freely shared information about new practices. However, multiple links in social networks (community, economic, family, proximate location) and/or redundant provision of information supported adoption. Farming households seem to rely on these redundant links to develop trust in new techniques and build evidence for their usefulness. Mating services and farmer field days provided an opportunity to introduce new practices, but do not necessarily lead to uptake if they do not resonate with shared cultural values.

Farming households tended to rely on a variety of sources and influences when considering new cattle management practices. Champion farmers in South Sulawesi and *kandang* group heads were mentioned as the most influential sources for decision-making. However, informal sources, such as conversations at Friday mosque meetings, with neighbours or consultation with family members can be vital when deciding for changes in farm management.

The research presented here presents complex household decisions under constrained resource conditions typical for modern-day smallholder societies. While the argument that resource constraints limit the capacity to invest in changes to the farming system is not new (*cf.* Ellis, 2000; McGregor, 2008), our data points toward the interpretation that adoption decisions by farming households occur within a rationality context, which is culturally specific. Just as green revolution farmers in the 1960s might have chosen to introduce new varieties for status reasons or modern-day French farmers adopt integrated crop protection and organic farming techniques for social and moral reasons (Mzoughi, 2011), current Indonesian smallholders make selective decisions based on (a) long-term resource availability, (b) form and source of communication, (c) risk and (d) livelihood strategies.

## 5 Conclusion

Subsistence-oriented smallholders are willing to experiment with changes to their farming systems where they do not conflict with existing livelihood strategies. It is easier for households in South Sulawesi to experiment with forage production, where land is relatively more abundant, than in Lombok, where land is scarce and prioritised for crop production. New practices are selectively applied and adapted as perceived through cultural rationality. While one could assume that increased cattle production would mark a gradual transition of the farming system it is more likely that subsidiary role of cattle to rice persists.

The continued focus on crop production and the low preference for expanding cattle has implications for Indonesia's policy on beef production, specifically, the Government's policy to establish Nusa Tenggara Barat as the 'Land of One Million Cattle' (*cf.* Jakarta Post, 2009). As long as smallholders conduct farming at a subsistence level and land is scarce they are reluctant to invest heavily in growth of herd size, increased calf turnover or fattening. Smallholders may see cattle as a supplementary income or even as a stepping-stone out of farming. Hence, the feasibility of current policy goals needs to be revisited.

While demonstration of the benefits and savings of new practices is important to inform a larger number of households about new practices, it is equally important to rely on existing networks for continued communication of ideas. This is paramount to providing legitimacy to the information and re-interpretation of the information along indigenous cultural categories. Decontextualised information is less valuable and useful than information, which can be accommodated by or integrated into *common sense* knowledge systems.

As research and development projects continue to provide innovations for smallholders to reduce poverty and transition into sources of global supply chains, "best practice" demonstrations seemingly provide rationalistic solutions for transforming farming systems through technology and practice change. Understanding the cultural context of household decisions requires a departure from economic decision-making models in both the design of and expectations for agricultural intervention. At stake is the potential for development agencies and practitioners to achieve improved rates and – more importantly – socially adapted adoption of innovations to traditional farming systems.

**Acknowledgements**

Funding for this research was provided by the Australian Centre for International Agricultural Research of the Australian Government (ACIAR-SMAR/2006/061 and ACIAR-SMAR/2006/096). The authors wish to thank Dr. Dahlanuddin (University of Mataram); Ketut Puspadi and Rachmat Rachman (Assessment Institute for Agricultural Technology, Indonesia); and Dr. Asmuddin Natsir (Hassanudin University); Monica van Wensveen, Bruce Pengelly, Cam McDonald and Jeff Corfield (CSIRO); Henky Widjaja; and Peter Horne (ACIAR) for sharing their knowledge and their support in this research. Data gathering was conducted with the enthusiastic effort of project field staff in Lombok and South Sulawesi. The authors wish to thank Zvi Hochman, James Butler (CSIRO) and two anonymous reviewers for providing us with valuable input to the various drafts of this paper.

## References

Anderies, J. M., Janssen, M. A., Bousquet, F., Cardenas, J. C., Castillo, D., Lopez, M. C., Tobias, R., Vollan, B. & Wutich, A. (2011). The challenge of understanding decisions in experimental studies of common pool resource governance. *Ecological Economics*, 70 (9), 1571–1579.

Bandiera, O. & Rasul, I. (2006). Social Networks and Technology Adoption in Northern Mozambique. *The Economic Journal*, 116 (514), 869–902.

Bodin, Ö. & Crona, B. I. (2009). The role of social networks in natural resource governance: What relational patterns make a difference. *Global Environmental Change*, 19 (3), 366–374.

Crona, B. I. & Bodin, Ö. (2006). What you know is who you know? Communication patterns among resource users as a prerequisite for co-management. *Ecology & Society*, 11 (2), Art. 7.

Devendra, C. & Thomas, D. (2002). Crop-animal systems in Asia: importance of livestock and characterisation of agro-ecological zones. *Agricultural Systems*, 71 (1–2), 5–15.

Douglas, M. (1985). *Risk acceptability according to the social sciences*. Russell Sage Foundation, New York.

Ellen, R. (1982). *Environment, Subsistence and System. The ecology of small-scale social formations*. Cambridge University Press, Cambridge.

Ellis, F. (2000). *Rural livelihoods and diversity in developing countries*. Oxford University Press Inc., New York.

Geertz, C. (1975). Common sense as a cultural system. *The Antioch Review*, 33 (4), 5–26.

Geertz, C. (1983). *Local knowledge: further essays in interpretative anthropology*. Basic Books, New York.

Giampietro, M. (2004). *Multi-scale integrated analysis of agroecosystems*. CRC Press, Florida.

Hoang, L. A., Castella, J. C. & Novosad, P. (2006). Social networks and information access: Implications for agricultural extension in a rice farming community in northern Vietnam. *Agriculture and Human Values*, 23 (4), 513–527.

Jakarta Post (2009). West Nusa Tenggara strives to meet 'One Million Cattle' target. The Jakarta Post, July 6, 2009. PT. Bina Media Tenggara, Jakarta.

Janssen, M. A. & Anderies, J. M. (2011). Governing the commons: Learning from field and laboratory experiments. *Ecological Economics*, 70 (9), 1569—1570.

Lai, G. & Wong, O. (2002). The tie effect on information dissemination: the spread of a commercial rumor in Hong Kong. *Social Networks*, 24 (1), 49—75.

Lisson, S. & Corfield, J. (2010). Impact assessment of project related activities from ACIAR project AS2/2004/005 Improving smallholder crop-livestock systems in Eastern Indonesia. CSIRO, Hobart.

Lisson, S., MacLeod, N., McDonald, C., Corfield, J., Pengelly, B., Wirajaswadi, L., Rahman, R., Bahar, S., Padjung, R., Razak, N., Puspadi, K., Dahlanuddin, Sutaryono, Y., Saenong, S., Panjaitan, T., Hadiawati, L., Ash, A. & Brennan, L. (2010). A participatory, farming systems approach to improving Bali cattle production in the smallholder crop-livestock systems of Eastern Indonesia. *Agricultural Systems*, 103 (7), 486–497.

Mahanty, S. (2002). Conservation and development interventions as networks: The case of the India Ecodevelopment Project, Karnataka. *World Development*, 30 (8), 1369–1386.

Martin, G. (2010). ACIAR investment in research on forages in Indonesia. ACIAR Impact Assessment Series Report No. 65. ACIAR, Canberra.

Mayrhofer-Grünbühel, C. (2004). *Resource use systems and rural smallholders. An analysis of two Lao communities*. Ph.D. thesis, University of Vienna, Vienna, Austria.

McGregor, A. (2008). *Southeast Asian Development*. Routledge, London.

Millar, J. & Connell, J. (2009). Strategies for scaling out impacts from agricultural systems change: the case of forages and livestock production in Laos. *Agriculture and Human Values*, 27 (2), 213–225.

Mzoughi, N. (2011). Farmer's adoption of integrated crop protection and organic farming: Do moral and social concerns matter? *Ecological Economics*, 70 (8), 1536–1545.

Paris, T. R. (2002). Crop-animal systems in Asia: socio-economic benefits and impacts on rural livelihoods. *Agricultural Systems*, 71 (1–2), 147–168.

Pastore, G., Giampietro, M. & Ji, L. (1999). Conventional and land-time budget analysis of rural villages in Hubei Province, China. *Critical Reviews in Plant Sciences*, 18 (3), 331–357. doi: 10.1080/07352689991309252.

Pengelly, B., Whitbread, A., Mazaiwana, P. R. & Mukombe, N. (2003). Tropical forage research for the future — better use of research resources to deliver adoption and benefits to farmers. *Tropical Grasslands*, 39, 198–209.

Polanyi, K. (2001). *The Great Transformation. The political and Economic Origins of our time*. Beacon Press, Boston. (orig. 1944)

Preston, D. (1994). Rapid household appraisal: a method for facilitating the analysis of household livelihood strategies. *Applied Geography*, 14 (3), 203–213.

Reeson, A. F. & Dunstall, S. (2009). Complex decision making. Implications for the Australian Tax and Transfer System. CSIRO Mathematical and Information Sciences Report 09/110. CSIRO, South Clayton.

Rössler, M. (1999). *Wirtschaftsethnologie. Eine Einführung*. Dietrich Reimer, Berlin.

Schweizer, T. (1988). Netzwerkanalyse als moderne Strukturanalyse. *In:* Schweizer, T. (ed.), *Netzwerkanalyse. Ethnologische Perspektiven*. pp. 1–32, Dietrich Reimer, Berlin.

Scott, J. P. (2000). *Social Network Analysis: A Handbook*. Sage Publications Ltd., London.

Sjah, T., Cameron, D. & Woodford, K. (2006). Extension service and farmer decision making on new copping lands in East Lombok Indonesia. *Journal of International Agricultural and Extension Education*, 13 (2), 39–55.

Wasserman, S. & Faust, K. (1994). *Social network analysis: methods and applications*. Cambridge University Press, Cambridge.

van Wensveen, M., Dahlanuddin, Corfield, J., Grunbuhel, C. M., Hasanuddin, Lisson, S., McDonald, C. & Puspadi, K. (2010). *Scaling up herd management strategies in crop-livestock systems in Lombok*. ACIAR, Canberra.

van Wensveen, M., Rahman, R., Bahar, S., Corfield, J., Grunbuhel, C. M., Lisson, S. & Marsetyo (2011). *Building capacity in the knowledge and adoption of Bali cattle improvement technology in South Sulawesi*. ACIAR, Canberra.

Wiersum, K. F. (1994). Farmer adoption of contour hedgerow intercropping, a case study from east Indonesia. *Agroforestry Systems*, 27 (2), 163–182.

Wiggins, S., Kirsten, L. & Llambí, L. (2010). The Future of Small Farms. *World Development*, 38 (10), 1341–1348.

9

# The determinants of the performance of dairy smallholders in Malawi

Ivo Baur[a,b,*], Christian Gazzarin[a], McLoyd Banda[c], Markus Lips[a]

[a]*Agroscope; Institute for Sustainability Sciences, Tänikon, Switzerland*
[b]*Ostrom Workshop, Indiana University Bloomington, USA*
[c]*Department of Agricultural Research Services, Mbawa Research Station, Embangweni, Mzimba, Malawi*

**Abstract**

Milk production in Malawi is stagnating although neighbouring countries steadily increase outputs. In this paper, we explored the determinants of the performance of dairy smallholders. As indicators of on-farm performance we used annual milk yield, calving intervals, and annual dairy income. Regression models revealed that milk yield was negatively related with farmer's age, female farmer, and household size, but positively influenced by farmers' experience. Calving intervals were strongly associated with labour costs and breeding method. Income from dairying was only associated with farmers' education but varied strongly with region. Regional effects existed for all performance indicators which may partly be rooted in land scarcity in the south and the efforts of development agencies to promote dairying in the northern and central regions. Results also revealed a tendency for pure breeds to produce higher outputs, but crossbreeds due to lower costs provided better income. Thus, we recommend that experienced farmers become involved in extension programs to provide comprehensive services that help farmers make more efficient use of their scarce assets, and thus realise more of the animals' genetic potential with regard to the three observed performance indicators.

*Keywords:* dairy production, performance, milk yield, calving intervals, dairy income

## 1 Introduction

Dairy production in developing countries improves the diet of consumers and the income and livelihood of smallholders (McDermott *et al.*, 2010; Banda *et al.*, 2012). Sedentary smallholders produce almost 80 % of the milk in southeastern Africa (Phiri *et al.*, 2010). According to the International Farm Comparison Network (IFCN), from 2010 to 2014 the annual milk production in Tanzania increased by 23 % to 2.0 million tons, in Kenya by 10 % to 4.3 million tons, and in Uganda by 52 % to 1.8 million tons, but in Malawi outputs remained constant at 0.05 million tons (Hemme, 2015). These developments show that dairying constitutes a growing segment in agricultural production in Sub-Saharan countries, but that development of the dairy sector involves major challenges as the figures for Malawi suggest.

Key on-farm constraints include land and fodder scarcity (Moffat & Austin, 2003; Kawonga *et al.*, 2012), animal health problems (Tebug *et al.*, 2011; Banda *et al.*, 2012; Tebug *et al.*, 2012b), poor infrastructural endowment, in particular cow housing systems (Banda *et al.*, 2012), breeding strategies (Chagunda *et al.*, 2004, 2015), and limited management skills (Chagunda *et al.*, 2006; Tebug *et al.*, 2012a). We therefore suggested that the development of the dairy value chain is best advanced by increasing on-farm performance (Baur *et al.*, 2016). However, despite the efforts of governmental and

* Corresponding author
Email: ibaur@indiana.edu

non-governmental organisations to improve farm performance (Duteurtre & Atteyeh, 2000; USAID, 2012), no study exists that identifies determinants of economic on-farm performance in Malawi. This paper aims at filling this gap by identifying the factors that determine on farm performance with regards to milk yield, calving intervals, and dairy income. Based on the observed determinants, we identify avenues for farmers and extension services to improve on farm-performance of smallholder dairy farmers in Malawi.

## 2 Methods

### 2.1 Data collection

Data were collected by survey during June and July 2013 (Banda *et al.*, 2016). The survey was performed in Malawi's three major milk-producing regions, Blantyre (South), Lilongwe (center), and Mzuzu (North). The study involved 12 research assistants responsible for data collection and entry. Data collection was jointly administered by the Ministry of Agriculture and Food Security, the Department of Agricultural Research Services, and Lilongwe University of Agriculture and Natural Resources. Before the data collection process started, the survey was pretested, and the interviewers received training on how to complete the English questionnaire.

The sample was drawn from a population consisting of farmers from 15 milk bulking groups (MBGs), equally distributed across the three study regions. Memberships in the MBGs differed substantially, with largest numbers appearing in the South. After collection of the questionnaires, the field supervisor checked each questionnaire for errors. Nevertheless, the dataset included inconsistencies, in particular with regard to the dependent variables milk yield and calving intervals. Implausible information led to exclusion in many cases and finally to the reduced sample size used in this study. As displayed in Table 1, 540 surveys were collected from farmers, with a majority of 54 % of the farmers located in the southern region of Blantyre, followed by 27 % of the farmers from the central region of Lilongwe, and 19 % of the farmers living in northern region of Mzuzu (Banda *et al.*, 2016).

Due to missing or implausible information we excluded more than half of the cases, and finally worked with a sample of 251 farmers. The reason for the exclusion of cases was missing information for the three lactation phases upon which we calculated milk outputs. We excluded cases when the calculation of average annual milk yield exceeded 7000 litres. Furthermore, we excluded cases with regards to calving intervals if the farmer couldn't clearly indicate when the cow last calved. Finally, we excluded outliers with regards to dairy income. When farmers realised, according to our calculations, more than 1,000,000 Malawi Kwacha (MWK) per cow, the figures for cost items were not deemed to be realistic and the case was excluded. As a consequence, the regional distribution changed, with the northern region being slightly overrepresented at the cost of the southern region (Table1).

**Table 1:** *Sample description: the actual regional distribution of farmers, the study sample and the considered cases.*

| | *Blantyre (South)* | *Lilongwe (centre)* | *Mzuzu (North)* | *Total* |
|---|---|---|---|---|
| Dairy cows in the region* | 22,870<br>76 % | 2,708<br>9 % | 4,514<br>15 % | 30,092<br>100 % |
| Farmers (1,000) | 1,167<br>63 % | 475<br>26 % | 215<br>12 % | 1,857<br>100 % |
| Study sample | 292<br>54 % | 146<br>27 % | 102<br>19 % | 540 |
| Cases considered | 131<br>52 % | 61<br>24 % | 59<br>24 % | 251 |

* Source: Chagunda *et al.* (2010)

### 2.2 Calculating the dependent variables

The study investigated three indicators of performance, namely, annual milk yield, calving intervals, and annual dairy income. Milk yield was measured as average output per day for three lactation phases. Because annual yields also depend on time in lactation, annual yields were calculated by splitting the total lactation phase into three periods, namely, first until 60$^{th}$ day, 61$^{st}$ until 150$^{th}$ day, and 151$^{st}$ day until the end of the lactation phase. For each period, the number of days is represented by $p_i$. In addition, for each period, the average yield ($y_i$) per day and cow together with calving interval ($c$) allowed us to calculate annual milk yield per cow ($Q$) as follows (Equation 1):

$$Q = 365 * \sum_{i=1}^{3} \frac{p_i * y_i}{c} \quad (1)$$

Calving interval was estimated based on rough indication of calving frequencies and detailed figures for lactation length. Calving frequency was indicated in years (every year, every second year, etc.) whereas lactation length was indicated in days. Hence, if lactation length was longer than 365 days and the farmer indicated that the cow calved every year, we estimated calving intervals based on lactation length plus an assumed dry period of 60 days.

The third dependent variable, dairy income ($DI_i$), was calculated for each farm separately from the total annual milk production ($Q_i$) multiplied by annual average price ($P_i$) minus direct costs ($DC_i$) such as costs of concentrates, by-products, veterinarian care, artificial insemination, salt, and depreciation costs (Equation 2).

$$DI_i = Q_i * P_i - DC_i \qquad (2)$$

An overview and description of dependent variables used in the regression models is given in Table 2. Mean annual milk yield per cow was almost 2,400 litres, with calving intervals of almost 500 days and a mean income from dairy production above MWK 140,000. As expected, the three variables were all significantly correlated. The strongest correlation existed between milk yield and dairy income with $r = 0.81$, $p < 0.001$, and there were negative correlations between calving interval and annual milk yield with $r = -0.47, p<0.001$ and between calving interval and annual dairy income with $r = -0.36, p<0.001$.

### 2.3 *Model specification*

Table 3 lists the independent variables included in the model. Considering demographic factors, the farmers in the sample had an average age of around 50 years, with the majority being female farmers (55 %). Farmers had on average 7 years of experience, and households consisted on average of 5.5 people. According to Tebug *et al.* (2011), farmers' milk yield in Malawi is positively associated with education and experience in dairying. In addition, the southern region Blantyre served as a reference category for a dummy variable for regional effects. We controlled for regional conditions such as temperature, rainfall, or the effect of development projects aiming at promoting the dairy sector regionally (USAID, 2012).

All farmers in the sample owned one dairy cow. Overall, two thirds of the cows were exotic breeds – predominantly Holstein Friesians, and one third of the animals were crosses with local zebus. The regional distribution of the breeds was unbalanced. In Blantyre the crossbreeds (52 %) and pure breeds (48 %) were about equal, while in Lilongwe and Mzuzu pure breeds dominated over crossbreeds with 90 % and 86 % respectively.

The milk price was considered as an explanatory variable as it is an incentive for dairy farmers to use more inputs (e.g. concentrates) which would result in an increase of milk yield per cow. Amiani (2011) shows a positive impact of milk price on milk yield for the Bungoma district in Kenya. Furthermore, we assumed that pure exotic breeds allow for higher yields than crossbreeds (Banda *et al.*, 2012) and included breed type as a dummy. For all the various cost items, we expected positive effects on yield except for veterinary costs. Among the cost items, by-products accounted for the highest costs followed by concentrates (mostly soya) and costs for veterinary services.

Concerning determinants of calving intervals, the literature suggests that costs that directly relate to reproduction, such as artificial insemination, increase with calving intervals, and that artificial insemination results in longer calving intervals than natural mating (Banda *et al.*, 2012). Accordingly, we controlled for the effects of breeding methods when specifying the ordinary least squares (OLS) model for calving intervals. With regard to the models explaining dairy income, the literature does not offer any farm-level analysis; we therefore specified models using similar variables as for calving intervals, except for the various cost positions that could not be included as they were used to calculate the dependent variable.

For the statistical analysis, we distinguished three models for each indicator: all breeds, only crossbreeds, and only exotic breeds. This approach allowed us to consider the breed types as different breeding methods. In addition, for the models explaining milk yield, we present a version with disaggregated cost items. For the models explaining calving intervals for subsamples distinguished by breeds, we could not consider the independent variables housing type and health programs because the cases were reduced below acceptable numbers. For the models explaining dairy income, we could not include the direct costs and the respective items as the inclusion would have violated endogeneity assumptions.

We estimated OLS models with robust standard errors to deal with heteroscedasticity (Verardi & Croux, 2008). To test for the quality of the model specification, we performed a link test (Pregibon, 1980) and the Ramsey regression specification-error test (Ramsey, 1969) for omitted variables in order to check for goodness of fit. Furthermore, we applied the Chow-test (Chow, 1960) to models with the subsamples defined by breed type to check if parameters differed. The analysis was performed using the software package StataSE 12.

**Table 2:** *Descriptive Statistics for the three dependent variables indicating performance in the study.*

| *Description* | *Number* | *Mean* | *SD* | *Min.* | *Max.* |
|---|---|---|---|---|---|
| Annual milk yield per cow (in litres) | 251 | 2,394 | 1,256 | 283 | 6,622 |
| Calving intervals (days) | 230 | 493 | 182 | 365 | 1,095 |
| Annual income from dairy farming (MWK)† | 251 | 141,429† | 141,089 | –151,286 | 778,378 |

† Malawi Kwacha (MWK) 141,429 ~ USD 410 at time the study was conducted.

**Table 3:** *Descriptive statistics of independent variables.*

| *Variable* | *Description* | *N* | *Mean* | *SD* | *Min.* | *Max.* |
|---|---|---|---|---|---|---|
| Age | Age of respondent in years | 243 | 50.5 | 13.2 | 20 | 83 |
| Gender | Binary variable: Value 1 for male farmer. Otherwise 0 | 251 | 0.45 | 0.50 | | |
| Education (primary) | Dummy variable: Value 1 if farmer attained primary education. Otherwise 0 | 166 | 0.64 | 0.45 | | |
| Education (secondary and higher) | Dummy variable: Value 1 if farmer attained secondary or higher education. Otherwise 0 | 56 | 0.21 | 0.41 | | |
| Experience | Years of experience in dairy farming | 250 | 6.83 | 5.07 | 1 | 40 |
| Household size | Number of people belonging to the same household | 251 | 5.57 | 2.04 | 1 | 10 |
| Grassland | Share of area used for pasture cultivation | 251 | 0.16 | 0.29 | 0 | 1 |
| Dummy Lilongwe | Dummy Variable: Value 1 if region is Lilongwe. Otherwise 0 | 251 | 0.24 | 0.43 | | |
| Dummy Mzuzu | Dummy Variable: Value 1 if region is Mzuzu. Otherwise 0 | 251 | 0.24 | 0.42 | | |
| Milk price | Milk price (MWK†/litre) | 251 | 101 | 18.3 | 56.5 | 200 |
| Dummy breed | Dummy Variable: Value 1 if the breed is a crossbreed. Otherwise 0 | 251 | 0.33 | 0.51 | | |
| Labour costs | Annual labour costs for dairying (MWK†) | 251 | 87,702 | 25,939 | 12,684 | 116,800 |
| Direct costs | Expenditures total (MWK†) | 251 | 97,481 | 70,909 | 0 | 404,900 |
| Concentrates | Expenditures for feed concentrates (MWK†) | 251 | 25,547 | 42,683 | 0 | 167,000 |
| By-products | Expenditures for by-products (MWK†) | 251 | 52,196 | 45,498 | 0 | 255,000 |
| Veterinarian care | Expenditures for veterinarian service (MWK†) | 251 | 9,217 | 11,399 | 0 | 62,000 |
| Artificial insemination | Expenditures for artificial insemination (MWK†) | 251 | 3,012 | 3,512 | 0 | 35,000 |
| Salt | Expenditures for salt (MWK†) | 251 | 4,937 | 7,907 | 0 | 72,000 |
| Other | Other expenditures (MWK†) | 251 | 2,571 | 12,246 | 0 | 120,00 |
| Dummy breeding: Artificial insemination | The cow is fertilised with artificial insemination | 251 | 0.41 | 0.49 | | |
| Dummy breeding: Natural mating | The cow is fertilised by natural mating | 251 | 0.23 | 0.42 | | |
| Dummy breeding: Artificial insemination & natural mating | The cow is fertilised using both artificial insemination and natural mating | 251 | 0.09 | 0.29 | | |
| Dummy housing type 1: Open kraal | The cow is kept in a very basic barn | 251 | 0.14 | 0.34 | | |
| Dummy housing type 2: Closed kraal | The cow is kept in a barn with a roof | 251 | 0.36 | 0.48 | | |
| Dummy housing type 3: Modern kraal | The cow is kept in a barn with a roof and solid ground floor | 251 | 0.24 | 0.43 | | |

† Malawi Kwacha (MWK) 1000 ~ USD 2.90. min and max. not reported for dummies.

## 3 Results

### 3.1 Determinants of milk yield

Results from OLS regression for the full sample as well as for the two breed types are presented in Table 4. Between 25 % and 40 % of the variance could be explained. Accordingly, age of the farmer had a negative effect on yields, and male farmers achieved significantly higher outputs than female farmers. Larger household size negatively affected milk yield, potentially caused by reduced resource availability in terms of capital and land that could be allocated to dairying. Region also had an influence as farmers in Lilongwe achieved significantly higher outputs than farmers in Blantyre, who served as the reference category. Surprisingly, the milk price had a negative effect on milk output. This might be related to differences within the three main regions.

Considering the subsamples for breeds, regional effects were more pronounced for exotic breeds than crossbreeds. Exotic breeds achieved much higher yields in Lilongwe and Mzuzu than in Blantyre. For crossbreeds, labour costs had a positive effect on milk yield, whereas the yield of exotic breeds increased only marginally with labour investments, suggesting that output from crossbreeds was much more sensitive to labour inputs than output from exotic breeds.

### 3.2 Determinants of calving interval

Calving intervals in the full sample were slightly negatively associated with grassland, suggesting that the farmers who dedicated more of their land to grassland and roughage production provided better feeding rations resulting in better reproductive performance. Elderly farmers also had longer calving intervals, when considering the complete sample. In Mzuzu, calving intervals were significantly longer than in the reference category Blantyre. Furthermore, crossbreeds had shorter calving intervals than exotic breeds. Also natural mating resulted in significantly longer calving intervals compared against artificial insemination which served as the reference category.

As displayed in Table 5, the analysis also revealed that artificial insemination was a more efficient reproduction method than natural mating. In the sample for crossbreeds, experience of the farmer also helped to reduce calving intervals.

### 3.3 Determinants of dairy income

Considering determinants of dairy income (Table 6), demographic effects were small whereas regional effects persisted. Accordingly, the farmers in Lilongwe and Mzuzu achieved higher incomes than the farmers in Blantyre.

Considering the sample for holders of crossbreeds, we found that education had a weakly positive effect upon dairy income, suggesting that higher education improved efficiency in the use of inputs but only for the smaller sample of crossbreeds and not for the larger sample of exotic breeds.

## 4 Discussion

We analysed three indicators and their respective determinates of the performance of smallholders engaging in dairying for three regions in Malawi. The performance indicators were milk yield, calving intervals, and dairy income. Results revealed an annual average milk yield of 2,394 litres per cow, which equals 6.5 litres per day. The yields observed in this study are thus similar to the yields reported for Malawi by Tebug *et al.* (2011) (5 to 8 litres) but lower than those stated by USAID (2012) (9 to 11 litres). Calving intervals were 493 days, with crossbreeds (377 days) having shorter calving intervals than exotic breeds (499 days). These figures deviate from figures reported elsewhere, with crossbreeds having calving intervals of 382 days (Banda, 1996) and pure exotic breeds having calving intervals of 396 days (USAID, 2012). The annual income from dairy products was estimated at MWK 141,400 which equals USD 210 at the time the study was conducted.

The regression analysis revealed demographic factors such as age, gender, experience, and household size to be major determinants of milk yield. Age and gender effects may result from low adoption rates of innovations for elderly farmers and women (Tebug *et al.*, 2012a). In addition, the lower performance of female than male farmers could be related to other aspects such as lower assets. Furthermore, education had a positive, but non-significant effect upon milk yield as observed also by Tebug *et al.* (2012b). The finding that education and experience were positively associated with milk yield shows that management skills of farmers play an important role for the improvement in milk yields.

Regarding calving intervals, we found the share of land allocated to grassland and roughage production to positively impact on fertility as it reduced calving intervals. Furthermore, we found that crossbreeds have shorter calving intervals then pure exotics which points to the better adaptation traits of crossbreds to climatic conditions. Results also suggest that natural mating is less efficient than artificial insemination which may be rooted in farmers' decisions to first apply artificial in-

**Table 4:** *Ordinary least square estimates for explaining milk yield.*

| *Specification* | *(1) All breeds, direct costs aggregated* | *(2) All breeds, direct costs disaggregated* | *(3) Crossbreeds, direct costs disaggregated* | *(4) Exotic breeds, direct costs disaggregated* |
|---|---|---|---|---|
| Age | −11.95 ** | −12.24 ** | −6.43 | −8.88 |
| | (−5.52) | (−5.72) | (−11.03) | (−6.84) |
| Gender | 344.97 ** | 354.96 ** | 630.92 *** | 266.91 |
| | (−154.71) | (−155.8) | (−295.08) | (−191.33) |
| Education (primary) | 276.99 | 265.15 | 519.33 | 117.52 |
| | (−222.87) | (−224.77) | (−455.34) | (−279.7) |
| Education (secondary & higher) | 303.00 | 278.10 | 93.44 | 411.96 |
| | (−258.67) | (−262.83) | (−510.03) | (−349.22) |
| Experience | 27.62 * | 27.20 | 26.18 | 32.41 |
| | (−14.16) | (−14.5) | (−24.9) | (−15.84) |
| Household size | −75.03 ** | −76.12 ** | −135.01 *** | −32.13 |
| | (−33.35) | (−34.19) | (−61.07) | (−40.15) |
| Grassland | 218.07 | 231.96 | 441.19 | 361.16 |
| | (−323.17) | (−323.49) | (−496.18) | (−393.83) |
| Dummy Lilongwe | 1479.18 *** | 1491.38 *** | 108.30 | 1637.78 *** |
| | (−244.48) | (−253.15) | (−620.33) | (−268.29) |
| Dummy Mzuzu | 255.21 | 245.06 | −244.87 | 498.47 *** |
| | (−175.34) | (−178.04) | (−424.4) | (−207.05) |
| Milk price | −7.66 ** | −8.16 ** | −5.14 | −11.88 |
| | (−3.67) | (−3.83) | (−4.80) | (−5.32) |
| Dummy breed | −186.35 | −178.62 | | |
| | (−157.90) | (−164.10) | | |
| Labour cost (000) | 4.74 | 4.62 | 10.25 | 1.75 |
| | (2.84) | (2.88) | (4.61) | (3.82) |
| Feed (000) | | 0.18 | | |
| | | (0.98) | | |
| Vetcare (000) | | 0.74 | | |
| | | (0.98) | | |
| AI (000) | | −9.52 | | |
| | | (16.19) | | |
| Salt (000) | | 8.54 | | |
| | | (7.28) | | |
| Other (000) | | 0.67 | | |
| | | (2.62) | | |
| Direct costs (000) | 0.44 | | −0.94 | 0.55 |
| | (0.83) | | (2.17) | (0.99) |
| Constant | 2703.71 *** | 2799.26 *** | 1930.22 *** | 2879.65 *** |
| | −494.18 | −516.28 | −848.40 | −653.81 |
| *N* | 233 | 233 | 77 | 156 |
| $R^2$ | 0.34 | 0.35 | 0.25 | 0.41 |

Robust standard errors in parentheses
*** $p<0.01$, ** $p<0.05$, * $p<0.1$

**Table 5:** *Ordinary least square estimates for explaining calving intervals.*

| *Specification* | *(1) All breeds, direct costs aggregated* | *(2) All breeds, direct costs disaggregated* | *(3) Crossbreeds, direct costs disaggregated* | *(4) Exotic breeds, direct costs disaggregated* |
|---|---|---|---|---|
| Age | 1.75 * | 1.83 | 1.79 | 1.21 |
| | (–0.97) | (–1.24) | (–1.91) | (–1.17) |
| Gender | –10.17 | –53.51 | 10.19 | –16.41 |
| | (–25.53) | (–33.47) | (–41.21) | (–33.31) |
| Education (primary) | –68.29 | –19.46 | –129.27 * | –45.28 |
| | (–52.51) | (–59.69) | (–139.56) | (–60.21) |
| Education (secondary and higher) | –28.80 | 42.78 | –76.35 | –26.83 |
| | (–57.85) | (–67.7) | (–154.71) | (–67.25) |
| Experience | –0.76 | 2.23 | –6.20 | 0.68 |
| | (–2.64) | (–2.98) | (–5.23) | (–3.32) |
| Household size | 0.43 | –0.21 | 3.59 | –2.13 |
| | (–6.35) | (–7.74) | (–13.78) | (–7.26) |
| Grassland | –82.00 * | –98.92 | –129.72 | –66.70 |
| | (–42.62) | (–74.08) | (–75.31) | (–51.97) |
| Dummy Lilongwe | 26.77 | 52.97 | 144.90 | 15.01 |
| | (–38.16) | (–51.08) | (–89.76) | (–42.47) |
| Dummy Mzuzu | 63.64 * | 100.12 ** | 127.76 | 47.85 |
| | (–35.79) | (–47.42) | (–115.15) | (–38.66) |
| Milk price | –17.59 | 11.23 | | |
| | (–29.73) | (–40.58) | | |
| Dummy breed | –0.01 ** | –0.01 * | | |
| | (0.00) | (0.00) | | |
| Labour cost | –1.27 | –1.40 | –1.12 | –1.23 |
| | (0.57) | (0.73) | (1.36) | (0.62) |
| Dummy breed method (Natural mating) | | 126.11 *** | | |
| | | (–37.84) | | |
| Dummy breed method (AI & Natural mating) | | –6.29 | | |
| | | (–46.13) | | |
| Dummy Housing type (Closed Kraal) | | –34.19 | | |
| | | (–39.98) | | |
| Dummy Housing type (Modern Kral) | | 32.12 | | |
| | | (–41.47) | | |
| Health program | 30.72 | 30.00 | 67.63 | 30.01 |
| | (–30.33) | (–44.79) | (–61.74) | (–36.29) |
| Constant | 569.33 *** | 464.21 *** | 559.68 *** | 588.46 *** |
| | (–91.95) | (–126.29) | (–221.19) | –102.36 |
| *N* | 213 | 151 | 61 | 152 |
| $R^2$ | 0.129 | 0.230 | 0.295 | 0.082 |

Robust standard errors in parentheses
*** $p<0.01$, ** $p<0.05$, * $p<0.1$

**Table 6:** *Ordinary least square estimates for explaining dairy income in Malawi Kwacha (measured in thousands).*

| *Specification* | *(1)* *All breeds* | *(2)* *Crossbreeds* | *(3)* *Exotic breeds* |
|---|---|---|---|
| Age | –0.93 | –0.29 | –0.86 |
| | (–0.65) | (–1.17) | (–0.82) |
| Gender | 4.31 | 35.74 | –9.12 |
| | (–17.38) | (–33.05) | (–21.77) |
| Education (primary) | 22.3 | 90.02 * | –13.16 |
| | (–25.91) | (–48.59) | (–32.71) |
| Education (secondary and higher) | 28.02 | 101.15 * | –0.96 |
| | (–30.94) | (–57.18) | (–40.15) |
| Experience | –0.36 | –0.36 | –0.06 |
| | (–1.44) | (–3.13) | (–1.55) |
| Household size | –5.12 | –12.30 * | –2.18 |
| | (–3.85) | (–6.59) | (–4.7) |
| Grassland | –3.48 | 5.12 | 11.38 |
| | (–39.42) | (–60.86) | (–49.4) |
| Dummy Lilongwe | 159.88 | 34.87 | 179.47 |
| | (–26.27) | (–85.56) | (–27.83) |
| Dummy Mzuzu | 75.54 | 5.81 | 101.64 |
| | (–19.73) | (–37.05) | (–22.74) |
| Dummy breed | 11.23 | | |
| | (–18.1) | | |
| Constant | 128.51 | 84.97 | 125.80 |
| Experience | (–45.42) | (–77.26) | (–59.00) |
| *N* | 233 | 77 | 156 |
| $R^2$ | 0.22 | 0.115 | 0.287 |

Robust standard errors in parentheses
*** $p<0.01$, ** $p<0.05$, * $p<0.1$

semination and then switch to natural mating in case of repeated failures. Concerning dairy income, we found a slightly significant and positive effect between income and education for the subsample of crossbreeds. It is however unclear while this effect exists only in crossbreeds.

Furthermore, we found regional effects across all three performance indicators which finally result in significant variance in incomes in the three regions. In Lilongwe and Mzuzu, dairy income was significantly higher than in Blantyre, which served as the reference category. However, the regional effects may result from different causes: Whereas farmers in Lilongwe achieved significantly higher yields compared with the reference category (farmers in Blantyre), the farmers in Mzuzu had longer calving intervals pointing to problems with the fertility of their cows. Accordingly, regional effects are rooted in different causes: In the South it is mostly the unfavourable conditions for dairy production including demographic pressure, very limited access to pastureland, and the long distances to be covered to gather forage. Furthermore, there are more farmers in the South with fewer extension workers per farmer compared to the northern and central regions. But unfortunately, this study could not quantify the benefits streaming from interaction with extension workers. In addition, the central and northern regions potentially enjoyed more and longer support from aid agencies than the southern region. Projects such as Land O'Lakes targeted central and northern regions, explaining why exotic breeds dominate in these regions with 90 % in Lilongwe and 86 % in Mzuzu, while in Blantyre only 50 % of the animals were exotics. Furthermore, the presence of development agencies also explains why milk yields in central and northern regions with an average of 3,507 litres in Lilongwe and 2,140 litres for Mzuzu, are higher as in Blantyre with 1,989 litres.

## 5 Conclusion

Performance on smallholder dairy farms in Malawi is largely determined by demographic variables, such as age, gender, education, and experience, but also varies with region. We showed that education and experience contribute positively to milk yield while advanced age had a negative effect. The reversing effects of age and experience shows that farmers in the sample are rather new to the business, and do not necessarily engage for lifetime in dairying since older farmers are not necessarily more experienced. Accordingly, educated, experienced farmers should be leading in extension programs to pass on their management knowledge to less experienced farmers. Finally, the regional effects — farmers in Blantyre (southern region) are clearly less efficient than the farmers in Lilongwe (centre) and Mzuzu (northern region) — may result partly from land scarcity and associated lower feeding ratios offered in the South and to some degree from the longer presence of development agencies in the northern and central regions. As regards the stagnation of milk production in Malawi, we can conclude that the ongoing gain of experience will have a beneficial impact, and that support for smallholders is essential when they start milk production. But equally important avenues to improve on-farm performance are investments in education, self-administered cross breeding programs, and improved roughage production.

## References

Amiani, L. V. (2011). *Assessment of the socioeconomic effects of the Napier stunt disease on small scale dairy production in Bungoma district, Kenya.* Master's thesis, Egerton University.

Banda, J. (1996). Evaluation of the productive and reproductive performances of Friesian × zebu and Jersey × zebu crossbred cattle at Bunda College farm in Malawi. *Malawi Journal of Science and Technology*, 3, 25–36.

Banda, L. J., Kamwanja, L. A., Chagunda, M. G. G., Ashworth, C. J. & Roberts, D. J. (2012). Status of dairy cow management and fertility in smallholder farms in Malawi. *Tropical Animal Health and Production*, 44, 715–727.

Banda, M. C., Chigwa, F. & Chiumia, D. (2016). Developing and evaluating sustainable integrated farming systems for improvement of smallholder dairy production in milk-shed areas of Malawi–baseline survey report. Unpublished manuscript.

Baur, I., Tabin, L., Banda, M., Chiuma, D. & Lips, M. (2016). Improving dairy production in Malawi: a literature review. *Tropical Animal Health and Production*, 49 (2), 251–258.

Chagunda, M. G. G., Bruns, E., King, J. & Wollny, C. (2004). Evaluation of the breeding strategy for milk yield of Holstein Friesian cows on large-scale dairy farms in Malawi. *The Journal of Agricultural Science*, 142, 595–601.

Chagunda, M. G. G., Gondwe, T. N., Banda, L., Mayuni, P., Mtimuni, J. P., Chimbaza, T. & Nkwanda, A. (2010). Smallholder dairy production in Malawi: current status and future solutions. Optimising Smallholder Dairying Project: Scottish Government, Scottish Agricultural College, Government of Malawi, and University of Malawi.

Chagunda, M. G. G., Msiska, A., Wollny, C., Tchale, H. & Banda, J. (2006). An analysis of smallholder farmers' willingness to adopt dairy performance recording in Malawi. *Livestock Research for Rural Development*, 18 (5), article #66.

Chagunda, M. G. G., Mwangwela, A., Mumba, C., Dos Anjos, F., Kawonga, B. S., Hopkins, R. & Chiwona-Kartun, L. (2015). Assessing and managing intensification in smallholder dairy systems for food and nutrition security in Sub-Saharan Africa. *Regional Environmental Change*, 16 (8), 2257–2267.

Chow, G. C. (1960). Tests of equality between sets of coefficients in two linear regressions. *Econometrica*, 591–605.

Duteurtre, G. & Atteyeh, A. (2000). Le lait à Moundou, témoin de l'intégration marchande des systèmes pastoraux au sud du Tchad. *Revue d'élevage et de médecine vétérinaire des pays tropicaux*, 53, 299–306.

Hemme, T. (2015). *IFCN dairy report 2015–For a better understanding of the dairy world.* IFCN AG, Kiel, Germany.

Kawonga, B. S., Chagunda, M. G. G., Gondwe, T. N., Gondwe, S. R. & Banda, J. W. (2012). Characterisation of smallholder dairy production systems using animal welfare and milk quality. *Tropical Animal Health and Production*, 44, 1429–1435.

McDermott, J., Staal, S., Freeman, H., Herrero, M. & Van de Steeg, J. (2010). Sustaining intensification of smallholder livestock systems in the tropics. *Livestock Science*, 130, 95–109.

Moffat, K. & Austin, N. (2003). Forage demand and constraints to adoption of forage technologies by livestock keepers in Malawi. *Tropical Grasslands*, 37, 274–278.

Phiri, B., Benschop, J. & French, N. (2010). Systematic review of causes and factors associated with morbidity and mortality on smallholder dairy farms in eastern and southern Africa. *Preventive Veterinary Medicine*, 94, 1–8.

Pregibon, D. (1980). Goodness of link tests for generalized linear models. *Journal of the Royal Statistical Society. Series C (Applied Statistics)*, 29, 15–23.

Ramsey, J. B. (1969). Tests for specification errors in classical linear least-squares regression analysis. *Journal of the Royal Statistical Society. Series B (Methodological)*, 31, 350–371.

Tebug, S. F., Chikagwa-Malunga, S. & Wiedemann, S. (2012a). On-farm evaluation of dairy farming innovations uptake in northern Malawi. *Livestock Research for Rural Development*, 24 (5), article #83.

Tebug, S. F., Kasulo, V., Chikagwa-Malunga, S., Chagunda, M. G. G., Roberts, D. J. & Wiedemann, S. (2011). Smallholder dairy production in northern Malawi: production and health constraints. *Tropical Animal Health and Production*, 44, 55–62.

Tebug, S. F., Njunga, G. R., Chagunda, M. G. G. & Wiedemann, S. (2012b). Health constraints and farm management factors influencing udder health of dairy cows in Malawi. *Journal of Agricultural Science*, 4, 136–141.

USAID (2012). Final evaluation of the Malawi Dairy Development Alliance, Land O'Lakes Malawi. United States Agency for International Development (USAID).

Verardi, V. & Croux, C. (2008). Robust regression in Stata. KU Leuven, Leuven, Belgium. Available at: https://ssrn.com/abstract=1369144

# 10

# The role of State Forest Enterprises in the payments for Forest Environmental Services Programme in Vietnam

Florence Milan [a], Tran Thi Thu Huong [b,c,*], Chu Thai Hoanh [a], Diana Suhardiman [a], Nguyen Duy Phuong [d], Manfred Zeller [b]

[a] *International Water Management Institute Southeast Asia Regional Office, Vientiane, Lao People's Democratic Republic*
[b] *Institute of Agricultural Sciences in the Tropics (Hans-Ruthenberg-Institute), University of Hohenheim, 70599 Stuttgart, Germany*
[c] *Vietnam National University of Agriculture, Trau Quy, Gia Lam, Ha Noi, Vietnam*
[d] *Soils and Fertilizers Research Institute, Dong Ngac, Tu Liem, Ha Noi, Vietnam*

**Abstract**

To promote pro-poor payments for environmental services, it is necessary to identify institutional options that reduce transaction costs and organisational problems associated with establishing and maintaining contracts with small-scale environmental service providers. This study examined the dual functionality of state forest enterprises (SFEs) in the implementation of the Payments for Forest Environmental Services (PFES) Program in Vietnam. We considered whether SFEs' involvement in the programme could reduce transaction costs and organisational problems. Data were collected from Tu Ly SFE in Hoa Binh province, northern Vietnam and from implementing agencies at various institutional levels. A survey of households participating in the SFE loan programme, and two stakeholder workshops were executed in 2014. The results revealed that Tu Ly SFE plays an important role in the livelihood of many farmers. A SWOT analysis exhibited SFEs' advantage over other state agencies in implementing national forest management programmes as there are fewer parties involved with greater autonomy and outreach in the district. This study proposes the acknowledgment of SFEs as environmental service providers in their own forestlands and to use SFEs as intermediaries in the Payments for Forest Environmental Services Programme activities.

*Keywords:* environmental services, intermediaries, services providers, transaction costs

## 1 Introduction

Payments for environmental services (PES) schemes have been implemented in different forms to encourage watershed protection, forest protection, erosion control, climate regulation and biodiversity conservation worldwide. PES focuses on bringing together service providers and users where providers are paid to maintain or improve environmental outcomes. There is an increasing interest in private investments, especially in developed countries such as the United States, Australia and France, based on Coasean economics (Coase, 1960), where transaction costs are assumed to be low, property rights are clearly defined, enforcement agencies are well funded, and an external monitoring system is credible (Clements *et al.*, 2010). In developing countries, government-funded PES plays a major role (Ecosystem Marketplace, 2008; Scherr & Bennett, 2011; Qi, 2014). Unfortunately, these countries have often unclear land ownership, weak law enforcement, and government agencies have poor capacity and little political support.

With the inception of the Rewarding Upland Poor for Environmental Services (RUPES) (CIFOR, 2013) and the Reducing Emissions from Deforestation and

* Corresponding author
Email: tthuong79@gmail.com

Forest Degradation (REDD+) programmes (Zhu *et al.*, 2010), PES schemes are on the rise in developing countries, especially in Asia. Vietnam leads Southeast Asia in PES with a programme supported by the government under the decree issued in 2010 on the Payments for Forest Environmental Services Program (hereafter, PFES programme) (Government of Vietnam, 2010). Households, individuals, village communities, and organisations working in protection forests, special-use forests, and production forests (those that supply environmental services) are eligible for payments. While environmental benefits can be generated from production forests (Kile *et al.*, 1998; Nambiar, 1999), in practice, PES programmes are often mixed with the Government of Vietnam's effort to promote plantation of fast-growing tree species (e.g. *Acacia mangium*, *A. auriculiformis*) as came up from the Five Million Hectare Reforestation Program, known as Programme 661 (Government of Vietnam, 1998).

The PFES programme in Vietnam involves both high transaction costs (Liss, 2008; Thuy *et al.*, 2013) and operational costs due to the centralised management system (Phuc *et al.*, 2012) and many contracts with small-scale environmental service providers. Transaction costs of institutions are the costs incurred when targeting, negotiating, contracting, executing and implementing forest management contracts, and for activities such as monitoring and coordinating tasks related to the management and use of forest resources. Effective PFES implementation requires substantial coordination between several government agencies in Vietnam. To promote sustainable PFES, it is necessary to identify institutional options that reduce transaction costs and organisational problems.

To date, limited research is available regarding innovative ways in reducing transaction costs of PES programmes. Therefore, we examined the potential role of the re-vamped SFEs in managing some aspects of Vietnam's PFES programme. We explored the dual functionality of SFEs (1) as environmental service providers in their own forest lands and (2) as intermediaries in PFES programme activities outside their areas of administration. As providers, they can offer environmental services such as (1) watershed protection, (2) forest protection by off-setting pressure on primary or old growth forests, (3) carbon sequestration, (4) water quality control, (5) degraded land rehabilitation, and (6) landscape enhancement (Fuhrer, 2000; Shelton *et al.*, 2001; Lamb et. al., 2005; de Groot & van der Meer, 2010). Potential goods from plantation forests are (1) sustainable source of renewable energy and industrial raw materials, (2) non-timber forest products (NTFPs), and (3) local employment (Shelton *et al.*, 2001). As intermediaries, SFEs can serve as implementing agencies for contracting, directing, monitoring and evaluating government forest programmes. We reviewed existing policies, while considering the feasibility of the proposed arrangement, and acknowledging the perceptions of the stakeholders. We accounted for the shortcomings of SFEs without a lengthy repetition of the details which can be found in the many reports on the processes of forestry reforms (Nguyen *et al.*, 2010). From a scholarly perspective, we contributed to current research on transaction costs (e.g., Liss, 2008; Sikor & Tan, 2011; Sommerville *et al.*, 2011; Phuc *et al.*, 2012), which largely addressed implementation issues, particularly with regard to individual farmer contracts. While this paper focuses on the challenges it faces from the perspective of transaction cost and the overall institutional set up for PES programme implementation, it also acknowledges the fact that plantations as a form of land use do not always correspond with sustainable forest management, especially in relation to heavy loss of biodiversity it may cause (McElwee, 2009; Šálek & Sloup, 2012; Šálek & Výlupek, 2012).

### *State Forest Enterprises in Vietnam*

SFEs have played an important role in the forestry sector in Vietnam. After the country's independence from French colonial rule, 6.3 million hectares of forest were managed by SFEs from 1954 until 1986 (MARD, 2001 as cited by Sikor & Tan, 2011). SFEs were mandated to protect forests and manage silviculture. However, SFEs were criticised for ignoring their role in forest protection and for prioritising optimisation of timber production to meet the increasing demand for forest products (Sikor, 1998; de Jong *et al.*, 2006; Tan *et al.*, 2008; Nguyen *et al.*, 2014). In 1987, when the *Doi Moi* economic reform was launched with the goal of creating a socialist-oriented market economy, the budget for SFEs from the central government were gradually reduced with less centralised control of the forestry sector (Artemiev, 2003). The reforms did not provide sufficient incentives to develop sustainable and commercial forestry (ibid.). Consequently, Vietnam faced a continuing decline in area under forest cover until the mid-1990s (Nguyen *et al.*, 2010).

Since 1991, forest management policies and practices in Vietnam have substantially changed. A state-run system has evolved into a new system that included households and communities as actors in forest and land management (1991 Law on Forest Protection and Development). Over the following years, numerous decrees, decisions and guidelines were promulgated regarding the reallocation of land and the devolution of land use rights to private organisations and households (e.g., 1992 Decree No. 327-CT: National forest protection programme 327; 1994 Decree 02/CP: Allocation of forestry land to organisations, households and individuals for stable and long term use for forestry purposes; 1995 Decree No. 1-CP: Regulations on the allotment of land by State-owned businesses for agricultural production, forestry and aquaculture). SFEs' authority was limited over natural forests, which contract farmers for the management and protection (1993 Law on Land). In 1998, the government launched Programme 661 (Decision 661/QD-TTg: Afforestation of five million hectares of forest). SFEs participated in the programme under the category of large forest owners. The programme provided a continuing source of government cash flow to SFEs that own protection forest (EASRD, 2005). SFEs played the role of implementing government agencies (intermediaries) by contracting, directing, monitoring and evaluating contract fulfilments with household beneficiaries of forestland in their areas. At the same time, SFEs were assigned responsibility for achieving the programme's objective of planting production forests in their own land in two phases (1998–2000 and 2001–2005), via preferential loans. In 2004, through Decree No. 200/2004/ND-CP: Rearrangement and innovation of forest enterprises, SFEs became fully autonomous commercial enterprises, while SFEs managing the more protected forests were transformed into Protection Forest Management Boards (hereafter, PFMBs).

To date, SFEs manage 15 % of Vietnam's natural forest and 17 % of its production forest, a substantial portion of the reported 13.36 million ha of forest cover in the country (MARD, 2012). The policies allowed SFEs to manage government projects, such as Program 661, by entering into contracts with farmers to plant and protect new forests or to plant production forests.

## 2 Conceptual framework and methodology

### *2.1 Conceptual framework*

From literature we considered four criteria essential in ensuring long-term success of a PES program: (i) acceptability, (ii) impact, (iii) costs, and (iv) financial sustainability of PES schemes (Fig. 1).

*Acceptability* – PES programs can be formulated to account for the different motivations of service providers and service users. On the one hand, PES schemes

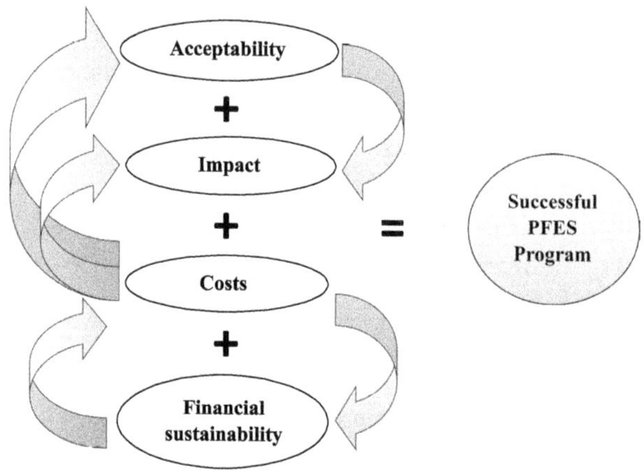


**Fig. 1:** *Key criteria for a successful Payments for Forest Environmental Services Programme (PFES).*
*Source: Own depiction.*

must generate revenue which is necessary for service providers to ensure they implement and maintain sustainable forest management or land use changes that will, in turn, produce environmental services (Nguyen *et al.*, 2013). Acceptability of the terms and transaction costs (monetary and non-monetary) of participating in PES schemes must be addressed (Falconer, 2000; Mettepenningen *et al.*, 2009). If incentives are not acceptable, potential service providers are likely to ignore them in their private decision making, leading to environmentally sub-optimal land use decisions. Among others, payments must account for the opportunity costs of the service providers. Pricing and other income generating opportunities are important in the design of PES programs, especially when service providers must modify their livelihood strategies or change their methods of production. On the other hand, payments should be within the capacity of the service users and set at a fair level (Kronenberg & Hubacek, 2013). High transaction costs can influence price setting (Vatn, 2010). The issue of transaction costs concerns to how costly it is to coordinate PES.

*Impact* – In developing countries, PES schemes are often designed to achieve both environmental and poverty reduction objectives (Tallis *et al.*, 2008; Gauvin *et al.*, 2010; Dunn, 2011) but this can be challenging (Zilberman *et al.*, 2008). Some authors have tried to link the benefits of PES to poor service providers (Bulte *et al.*, 2008; Wunder, 2008; Zilberman *et al.*, 2008; Milder *et al.*, 2010). Poor farmers can benefit from PES (Pagiola *et al.*, 2005) if they can provide the services at low cost and if the labour requirement is reasonable (Scherr *et al.*, 2006).

*Costs* – This criterion refers to the costs that government incurs due to the implementation of the PFES programme. These costs may need partially or fully be financed by taxpayer's money be it of a domestic or foreign source. It is therefore important to review whether a PFES programme is effective given its cost. The targeting, negotiating, contracting, and monitoring costs of PES schemes can be substantial in many contracts with small-scale service providers (FAO, 2007; Sommerville *et al.*, 2011). Strategic use of intermediaries can improve coordination, while reducing monitoring and transaction costs (Dunn, 2011). Agencies (such as SFEs) may have the capacity to reduce coordination costs when there are interdependencies (Vatn, 2010). Given the extensive geographic distribution of forests, a variety of organisations and persons may be involved in monitoring efforts. Consequently, the mandate given to these agencies, their capacity (funding, skills and experience of personnel, organisational design), and the way in which they interact (institutional structures and arrangements) will determine the success of the system (FAO, 2001).

*Financial sustainability* – Financial sustainability requires that revenues are sufficient to cover the ongoing costs of a PES programme (Mayrand & Paquin, 2004). Revenues can come from taxes, user fees, state subsidies, and grants from international organisations. If PES users accrue large benefits, such as in the case of hydropower operators benefitting from wise management of land and water resources in upstream areas, they will have an incentive to participate in a PES programme (Arias *et al.*, 2011). Pagiola *et al.* (2005) note that the financial sustainability of a PES scheme ensures the stability of income for environmental service providers. In some contexts, government financed PES may be the only option.

### 2.2 *Research methodology*

To understand the organisational strengths and weaknesses of SFEs, and how these correspond with their potential roles as environmental service providers, we look at how various actors, including key government agencies at different administrative levels, NGOs, international donor representatives, and farming households participating in the program, perceive and view the role of SFEs. Complementing this key stakeholders analysis, we also examined: (1) the policy and legal framework of SFEs in Vietnam; (2) pilot studies of PFES implementation in Lam Dong and Son La provinces; and (3) Tu Ly SFE's involvement in PFES pilot programmes in Hoa Binh province in northern Vietnam. We further examined the operational procedures of the Tu Ly SFE and its access to resources with a SWOT analysis. We reviewed the policy and legal frameworks of SFEs in Vietnam to determine if the new organisational and institutional frameworks are conducive for SFEs to participate in and mediate PFES projects. The pilot PFES in Lam Dong and Son La provided a basis for discussion. The PFES scheme in Lam Dong was implemented through SFEs, while the pilot study in Son La involved communities and households (Tan, 2011; Phuc *et al.*, 2012; Thuy *et al.*, 2013; Bac *et al.*, 2014).

Our empirical data are based on (i) interviews with Tu Ly SFE employees ($n = 4$), (ii) interviews with civil servants from implementing agencies at various institutional levels ($n = 16$), (iii) a survey of households participating in the SFE loan programme ($n = 14$), and (iv) the outcomes of two stakeholder workshops held in Hanoi and Ho Chi Minh in 2014. We also interviewed key stakeholders at the provincial and district levels ($n = 14$) to find out more about the implementation process and the stakeholders' roles in implementing the program. We used semi-structured questions in our quantitative household survey, with which we gathered information on the costs and benefits of joining the Tu Ly SFE loan programme. We also interviewed individuals paid by the enterprise to plant and manage forest parcels.

## 3 Results

### 3.1 *Tu Ly SFE in Da Bac district, Vietnam*

Tu Ly SFE was established in 1978 with operations on 4,612 hectares of sloping lands. Like all SFEs in Vietnam, Tu Ly SFE began as a provincial government programme. Its operations included the management of a private *Acacia* plantation forest and management of government projects, such as Program 661 in 1998. Tu Ly SFE offered two types of contracts under the government projects to the households in the area. Type 1 contracts were agreements for planting, tending and protection of new forest. In return, the households had access to the government subsidy (i.e., cash), firewood for consumption, NTFPs (e.g., herbs, bamboo shoots), and adult timber products (e.g., bamboo, *Acacia*). Type 2 contracts were for planting, tending and protection of production forest in the SFE own forest. After 7 years, households were allowed to harvest and sell adult timber. Via Type 2 contracts households managed 10 % of the SFE own forest. Tu Ly SFE provided non-collateral loans to households under the said contract, with annual interest rates, ranging from 5.4 % to 8 % in 2000.

After the restructuring of SFEs in 2004 (Decree No. 200/2004/ND-CP), Tu Ly SFE found itself operating as a subsidiary of Hoa Binh Forestry One Member Limited

**Table 1:** *Household characteristics of respondents under wage labour and Type 2 contracts with Tu Ly SFE.*

| *Household characteristics* | *Unit* | *(i) Wage labour contracts (n=6)* | *(ii) Type 2 contracts (n=5)* | *(iii) Type 2 contracts plus wage labour contracts (n=3)* |
|---|---|---|---|---|
| 1. Average size of household | People | 5 | 4 | 3 |
| 2. Average literacy rate | % | 88.0 | 86.0 | 83.0 |
| 3. Agriculture as primary source of income of household head | Household | 6 | 3 | 2 |
| 4. Farming as source of income | Household | 6 | 1 | 1 |
| 5. Livestock as source of income | Household | 3 | 1 | 1 |
| 6. Bamboo and NTFP collection as source of income | Household | 1 | 3 | 1 |
| 7. Non-agriculture wage or business as source of income | Household | 2 | 3 | 0 |
| 8. Recipients of government aid | Household | 3 | 1 | 2 |
| 9. Access to livelihood trainings | Household | 1 | 1 | 2 |
| 10. Owns red book for forest land | Household | 3 | 0 | 0 |
| 11. Average contracted forest area | ha | – | 2.06 | 1.77 |
| 12. Average loan taken with SFE | 1,000 VND | – | 9,000 | 23,000 |
| 13. Man-days per year hired | man-day | 32.5 | – | 83.3 |
| 14. Income from SFE employment | % | 26.0 | – | 40.0 |
| 15. Economic situation improved from SFE employment | % | 66.7 | – | 66.7 |
| 16. Received training from SFE | % | 16.7 | 60.0 | 33.3 |

Note: 1 USD = 20,800 VND; Source: Own data

Company, and at the same time as a district PFMB protecting 1,000 ha of forest for the government. As district PFMB, they continued to receive payment from the government for planting and protecting the forest through Type 1 contracts with farmers in the area. As a commercial enterprise, Tu Ly SFE also managed 1,000 hectares of production forest. The 90 % of this land was managed directly by the enterprise, with hired local farmers to plant and maintain the SFE forest. For the remaining 10 %, Tu Ly SFE continued offering Type 2 contracts and provided non-collateral loans to the households for the purchase of forest plantation inputs. The households can apply for a maximum of US$ 500 per hectare, per year, with end-of-term payment after seven years, when farmers are allowed to harvest the plantation forest. Between 2000 and 2010, Tu Ly SFE had Type 2 contracts with 73 households to manage 314 hectares of SFE's forest.

To date, they only have 15 Type 2 contracts with households in Da Bac. Since 2011, the contracts included loan interest rates as high as 16 %, twice the value of the loans they offered in 2000 under the government projects. One of the changes and challenges in the implementation of Decree No. 200/2004/ND-CP is the reduced availability of SFE funds and loans (EASRD, 2005). Tu Ly SFE was able to continue operating dependent on the Hoa Binh Forestry One Member Limited Company, which funded their operations through loans from the Bank for Investment Development of Vietnam (at 8.4 % interest), the Vietnam Forest Corporation (at 9.6 % interest), and from their own revenues. The main concerns of the enterprise are the high interest rates and the more stringent lending criteria recently imposed by the banks, making access to financing difficult. The burden of high interest rates has been passed on to the contracts with the farmers, making it difficult for the SFE to recruit more farmers to agree to the conditions of the Type 2 contracts.

### *3.2 Households participating in forest management in Da Bac*

We explored the perspectives and experiences of farmers participating in forest management through the Tu Ly SFE to shed light on the impact of PES. We summarised the household characteristics and forest activities of groups representing three models of SFE involvement with local residents: (i) households hired by Tu Ly SFE to establish and manage the forest (i.e., wage labour contracts), (ii) households under Type 2 contracts with Tu Ly SFE to plant forests (i.e., forest plantation contracts), and (iii) households having both Type 2 contract and wage labour (Table 1). In 2013, Tu Ly SFE hired 321 people on wage labour contracts (i) for 14,200 person-days for land preparation, digging holes, planting, and tending new forest plantations. Farmers were paid, on average, US$ 6 per day.

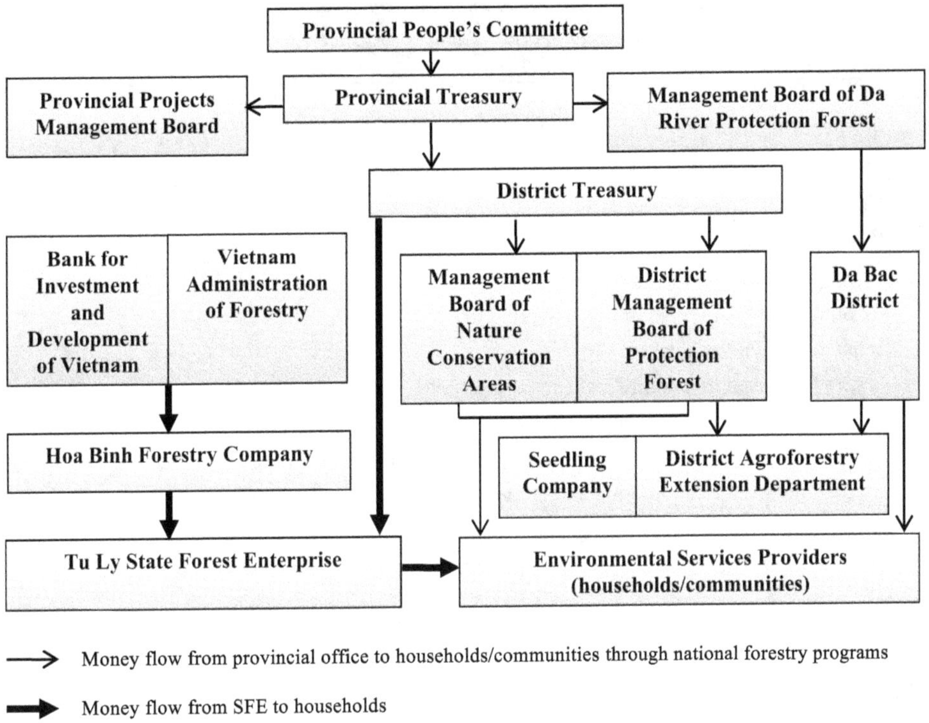


**Fig. 2:** *Money flow from the province to households and communities managing forest in Da Bac district, Hoa Binh province.*

Majority of the respondents rely on agriculture for their source of income. Most households with Type 2 contracts (ii) also rely on NFTPs for income. For those with Type 2 contracts (ii and iii), average contracted forestland is 2 ha. Those with combination of Type 2 contracts and wage labour (iii) have bigger loans with Ty Ly SFE. 67 % of households interviewed revealed that employment from Tu Ly SFE improved their economic situation (Table 1). Those hired (i and iii) also reported an increase in their income.

All of the respondents gained knowledge in forest plantation and protection during their employment. Although Tu Ly SFE does not provide formal training to their contract farmers, they provide technical guidance as needed. According to Dunn (2011), exposure to these kinds of activities can develop long-term behavioural change among households and individuals toward environmental issues such as slash and burn, illegal logging and land degradation. In a PES experiment in Thua Thien Hue province in Central Vietnam in 2003 (Bui & Hong, 2006), training and interaction with technical experts enhanced the environmental awareness of participating households.

### 3.3 *Management structure and money flow of forest programme in Da Bac*

The Vietnam Forest Protection and Development Fund (VNFF) was established in 2008 (Decision No. 114/2008/QD-BNN) to enable the forest sector to meet the demand for environmental services and to increase revenues through PFES. The Fund supports the PFES programme in Vietnam and is expected to raise US$ 2 billion by 2020 (FAO, 2009). According to the VNFF agency, hydropower contributes 98 % of total PFES payments.

The main threats to the PFES programme in Vietnam are the high transaction costs, at provincial- and district-levels, which reduce the net funds available for households and communities with Type 1 and Type 2 contracts to manage the forest, as revealed in the implementation of Program 661 and the money flow of the programme funds. Figure 2 shows the flow of funds received by farmers who adopt management schemes designed to protect the forest in Da Bac district. Funds of Program 661 were channelled through its provincial offices to three district management boards and community

**Table 2:** *A comparison of some variables affecting transaction cost of managing Program 661 by Tu Ly SFE and the Protection Forest Management Board.*

| *Independent variables affecting transaction costs* | *Tu Ly SFE* | *Management Board of Da river Protection Forest* |
|---|---|---|
| Number of communes served | 9 | 11 |
| Start-up variables | | |
| Staff involved in implementation planning | 4 | 10 |
| Staff involved in programme dissemination | 3 | 2 |
| Implementation variables | | |
| Staff involved in monthly and annual meetings | 7 | 10 |
| Aggregated man-day spent on meetings | 501 | 618 |
| Staff involved in contract signing and disbursement | 1 | 4 |
| Staff involved in monitoring and enforcement | 1 | 4 |

Source: Own data, key informant interviews in 2012.

levels. The authorities involved in the implementation and monitoring that constitute the web of commands include: the Agroforestry Planning Department, Forestry Department, Forest Protection Department, Department of Agriculture and Rural Development, Department of Natural Resources and Environment, and the People's Committee at different levels. Since the Management Board of Da River Protection Forest implemented the program across all districts in the Da River Reservoir catchment area, funds came directly from the provincial treasury, while Management Board of Nature Conservation Areas and Management Board of Protection Forest received funds from the district treasury. All three management boards were responsible for transferring government subsidies to households and communities as payment for their forest management activities. Funds were also used to hire technical experts and pay seedling suppliers. Needless to say, administrative and transaction costs escalate when more parties are involved. Reducing administrative and organisational costs, stemming from the heterogeneity and quantity of public authorities, will reduce the costs and enhance the financial sustainability of the PFES programme.

Funds provided to households under Tu Ly SFE contracts follow a different channel (shown in Fig. 2). For government programmes, Tu Ly SFE received funds directly from the District Treasury. For enterprise contracts, the funds are discharged via Hoa Binh Forestry One Member Limited Company to Tu Ly SFE, and then to the households. No other authorities receive funds from the company. With fewer actors, and experienced staff, Tu Ly SFE's administrative and operating costs are perceived to be less as revealed in the interviews with key stakeholders. According to Vatn (2010), experience (i.e., running systems over years) is expected to reduce transaction costs. A comparison of Tu Ly SFE's transactions with those of the Management Board of Da River Protection Forest in the district under Program 661 funds revealed that Tu Ly SFE used less staff in starting up and implementing the programme (Table 2). Because Tu Ly SFE has a global presence in the district, they disseminate information about forest programmes, monitor forest activities, and conduct other activities more easily. Provincial and district authorities in Hoa Binh acknowledged that SFEs can manage the forest better, and therefore should continue their role in the forest sector. Also, Tu Ly SFE propagates seedlings in their nursery, thereby avoiding the cost of out-sourcing.

At present, the transaction and operation costs of many implementing authorities at all levels substantially reduce the net funds available for households and communities protecting the forest in Vietnam. For instance, Thuy *et al.* (2013) reported that the PFES programme in Son La Province spends most of the 10 % of its revenue from the PFES (decree states that only 10 % of the total revenue from the PFES buyers will be retained for operating costs, including administrative and transaction costs) on checking forest protection performance and disbursing funds to 3,500 households. The programme requires more funding to reach out to all 64,000 forest owners. The costs of monitoring compliance with PFES agreements are also high (Phuc *et al.*, 2012; VNFF, 2013).

## 4 Discussion

Despite heavy criticism of SFEs, including having paid too little attention to their responsibility for protecting forests (Dang, 2001) and the inability of some to be financially sustainable (EASRD, 2005), the government of Vietnam recognised the important role played by these enterprises. Although the practices of SFEs did not fulfil all forest ecosytem services, especially on biodiversity and off-setting pressure on primary or old growth forest, the plantation forests they managed played a vital role in the provision of environmental services, such as watershed protection (Bui & Hong, 2006), when compared to agriculture and other forms of land use. Other environmental services provided by plantation forest such as reduction of soil erosion (hence reduction of sediment silted in rivers and reservoirs, regulation of water resources that provides longer inflow, rehabilitation of degraded land, and carbon sequestration), should be given importance. Thus, continued efforts should be made to reform the organisation and management of these SFEs.

### 4.1 *Criteria for Successful PES Programme*

To encourage SFEs in Vietnam to participate in the government's PFES programme as service providers, acceptable regulations, payments and incentives must be carefully considered. Innovative incentives may attract SFEs to adopt sustainable forestry practices. By recruiting SFEs as monitoring agencies of forest activities outside their administrative areas, SFEs will create additional sources of income to boost their financial viability.

#### 4.1.1 *Impact and Acceptability of PFES*

While considering SFEs as both participants in and agents of the PFES programme in Vietnam, their role in engaging poor families in their areas needs to be considered. Outreach to poor households must take precedence over any special consideration given to those individuals with special connections to SFEs. It is possible that Tu Ly SFE favours some households in the area. Those with Type 2 contracts are hired more often and received bigger loans. Similar issues were raised in the Lam Dong PES pilot study. The impact of PFES on rural poverty alleviation in the pilot study came under scrutiny when households without pre-existing contracts with the SFEs were excluded (Phuc *et al.*, 2012). In their defence, Tu Ly SFE justifies hiring households with contracts to help those households recover the high cost of the loans. The 16 % interest rate has discouraged many farmers in the area from participating in the Tu Ly SFE programme. For the period 2013 to 2014, the enterprise has entered into contracts with only three new households.

Most of the households were open-minded about the potential of forest management as an alternative livelihood, but expressed the need for more land for forest production and better contract terms. In the pilot study in Son La, the small landholdings (on average 2 ha) was one of the reasons households did not gain any significant benefit from PES payments (Phuc *et al.*, 2012).

Households also stated they would be more active in Tu Ly SFE contracts, if they were given interest-free loans. In addition, the current contracts emphasise conservation and protection, which for some, is a disincentive to participate, as there are too few livelihood opportunities with immediate gains. Although the contracts clearly state the responsibilities and benefits of the SFE and the contract farmer, there is no clear statement regarding the use of NFTPs such as honey, herbs, fruits, firewood, and bamboo. Also, tree species selection is highly centralised, with defined management rules, making the terms inflexible. There is consensus that contracts should have attractive terms and conditions, such as reduced or zero interest rates on loans, and provisions for increasing revenue through access to NTFPs.

#### 4.1.2 *Cost and financial sustainability*

In Vietnam, hydropower companies allocate a portion of their income to an environmental fund as per government directives. Whereas many private PES agreements fail over time, due to inadequate or insecure funding (Landell-Mills & Porras, 2002; Todorova *et al.*, 2013), the Vietnam directives ensure continued sources of funding for PFES in hydropower watershed areas. The PFES value has been pre-determined by the government. Hydropower operators pay 20 Vietnam dong (VND)/kwh (USD 0.001/kwh) while water supply companies pay 40 VND/m$^3$ (USD 0.002/m$^3$) and tourist organisations pay 1–2 % of their annual income. Because the programme is a legal instrument, the service users must accept the pre-determined level of payment. The government is currently revising its PFES valuation (Litzenberg, 2013). Lower transaction cost of the programme can result to acceptable payment rates.

### 4.2 *Moving forward with SFEs*

Table 3 presents a strengths, weaknesses, opportunities, and threats (SWOT) analysis of SFE as environmental service providers and using them as intermediaries for monitoring activities in a PFES programme.

As revealed in the SWOT study, implementation of PFES was faster and more effective partly because forests were managed by SFEs. Stakeholders of the PFES pilot studies in Vietnam shared similar observation. The PFES pilot study in Lam Dong province began working with SFEs in 2008. Local households with existing contracts (under Program 327 and Program 661)

**Table 3:** *SWOT analysis of SFEs as environmental service providers and as intermediaries for monitoring activities in the PFES programme.*

| *Strengths* | *Weaknesses* |
|---|---|
| • Institutional framework based on business principles to be financially viable, but remain as agencies for forest protection under government regulations<br>• Implementation of new guidelines are easily disseminated and enforced<br>• Lower transaction cost due to:<br>– Fewer parties involved in managing and monitoring their forestry programmes<br>– Many years of specialised experience in forest management<br>– Greater autonomy and outreach in the district<br>• Advantage of expertise<br>• Experience in monitoring | • Pressure to become financially independent drive SFEs to be more profitable, with less regard for forest protection.<br>• Selectiveness of SFE contracts and employment, capturing local elites with connections to political power (Sikor and Tan, 2007; Phuc *et al.*, 2012) so they are not that autonome as mentioned as strength<br>• Capital shortages and inadequate financing, due to high interest rates and stringent lending criteria imposed by banks |
| *Opportunities* | *Threats* |
| • Can create opportunities for local-level negotiations and choices regarding forest management contracts that accommodate local needs and livelihoods<br>• Possibility of contracting SFE for monitoring forest activities other than their own land<br>• The institutional arrangements of PFES in Vietnam already consider SFEs as an environmental service provider<br>• Opportunity to secure funds for the sustainability of the company<br>• Revenue from PFES is an interest-free capital for SFEs | • Recurrence of the damaging SFE-era before the reform in 1991.<br>• No clear directive from the government as to how the provinces distribute the funds. |

[1] Source: Own depiction

were given PFES contracts. To date, 3,400 households have received payments for their services from SFEs. The successful model in Lam Dong can be attributed partly to working with the 13 state organisations (SFEs and PFMBs) that own and manage most of the forest land in the area. In contrast with Lam Dong, the Son La pilot study is directly involved with local households. Forest area had already been allocated to 50,000 forest owners in the early 2000s (Phuc *et al.*, 2012). The distribution of PFES had been slow and faced high transaction costs. The SFEs and PFMBs in Lam Dong prepared the necessary documents to support the contracts with households. In return, the SFEs and PFMBs were paid to administer the contracts. Winrock International and the Center for International Forestry Research (CIFOR) have stated that monitoring of forest cover and quality is costly for the government (Thuy *et al.*, 2013; Nga, 2014) and local government agencies do not have the capacity and experience to monitor the PFES programme. Transaction costs tend to be high because of the large number of forest owners, the complexity of administrative structures, the limited capacity of public servants, conflicts of interest and week coordination and information sharing between and within government agencies.

The social and economic benefits of Vietnam's SFEs are largely ignored. Efforts are needed to reinvigorate and maximize potential of SFEs, and to realize a structure which includes SFEs' role in reducing transaction cost in PES programmes to rebuild Vietnam's forest.

To improve the financial sustainability of SFEs, innovative partnerships with communities, rather than with many individuals, can reduce transaction costs (Adhikari & Lovett, 2006; Blore *et al.*, 2013). There are success stories of communal forest management in Vietnam, especially where the social composition is heterogeneous (Sikor & Tan, 2011). It is important to increase local participation in SFE programmes to improve their financial sustainability and increase the benefits provided to local residents. This can be done through attractive terms and conditions of SFE contracts. With the potential of PFES as an interest-free capital source, SFEs can reduce interest rates on loans. The key to attracting poor farmers to participate is the identification of profitable activities. SFEs could work with households to develop alternative forms of agroforestry for adoption in forested areas. Mono-culture needs to be reduced to avoid periods without revenue. Allowing mixed forest plantations in the PFES programme, planting fruit trees with forest trees, and raising animals under forest canopies are examples of incentives for poor farmers to participate in forest management. The government could also encourage participation by studying the feasibility of in-kind payments in PES, such as the provision of materials, training and expertise.

With the devolution of the forest sector in Vietnam and the move towards a market-oriented economy, SFEs are facing financial constraints from the shortage of capital (MARD, 2012). The high interest rates imposed by banks on SFEs have reduced the activities of SFEs and limited the outreach of their forest programmes, which led to the dissolution of some enterprises. There is a need to resolve the difficulties faced by SFEs regarding access to funds and unacceptable contracts. With the proposed role of SFEs as intermediaries in PFES programme activities, payments for the services (e.g., watershed protection) could lift some of the financial burden.

Using SFEs as intermediaries for the PFES programme in Vietnam is not a novel idea. SFEs have carried out these responsibilities in past government programmes. The recruitment of SFEs to monitor forest activities outside their forest lands is logical, given their expertise and experience in collaboration with local farmers. The system has been piloted with positive outcomes, but regulations must be revisited to provide concrete guidelines. The government must issue implementation guidelines to ensure wider outreach of the programme to improve livelihoods. SFEs can achieve impacts by working with many poor households in forested areas. In 2007, more than 20,000 farmers were employed by SFEs to maintain seedling nurseries, plant and prune trees, and maintain forest firewalls (FAO, 2009).

The current discourse on transaction costs should consider the roles of institutional reform and organisational change in ensuring programme success. An effective regulatory and monitoring framework is essential to avoid repeating historical problems with SFEs. To achieve wider distributed impact, regulations regarding the acceptable terms and conditions of SFE contracts are needed to encourage local participation. Considering additional livelihood options (e.g., product development of NTFPs, bee keeping, nursery raising) in the policies is important for poor farmers, due to the long recovery period of capital and gains. In doing so, forest protection and livelihood support can be addressed together. Although the study analysed only one SFE and a modest number of its contractors or employees, results are consistent with previous pilot studies in Vietnam. It showed the strengths and weaknesses of using SFEs as environmental service providers and intermediaries for monitoring activities in the PFES programme.

*Acknowledgements:*

The authors would like to thank the project "Opportunities for economic incentives to promote sustainable land and water management in the sloping lands of South and Southeast Asia" funded by the Federal Ministry for Economic Cooperation and Development, Germany for support to the study presented in this paper.

## References

Adhikari, B. & Lovett, J. C. (2006). Transaction costs and community-based natural resource management in Nepal. *Journal of Environmental Management*, 78 (1), 5–15.

Arias, M. E., Cochrane, T. A., Lawrence, K. S., Killeen, T. J. & Farrell, T. A. (2011). Paying the forest for electricity: A modelling framework to market forest conservation as payment for ecosystem services benefiting hydropower generation. *Environmental Conservation*, 38 (4), 473–484.

Artemiev, I. (2003). *State forestry enterprise reform in Vietnam: Unlocking the potential for commercial wood growing. Technical Note May 2003.* World Bank, Washington D.C., USA.

Bac, D. V., Catacutan, D. C. & Ha, H. M. (2014). Importance of national policy and local interpretation in designing payment for forest environmental services scheme for the Ta Leng river basin in Northeast Vietnam. *Environment and Natural Resources Research*, 4 (1), 39–53.

Blore, M. L., Cundill, G. & Mkhulisi, M. (2013). Towards measuring the transaction costs of co-management in Mkambati Nature Reserve, Eastern Cape, South Africa. *Journal of Environmental Management*, 129, 444–455.

Bui, D. T. & Hong, B. N. (2006). *Payments for environmental services in Vietnam: Assessing an economic approach to sustainable forest management*. EEPSEA, Singapore.

Bulte, E. H., Lipper, L., Stringer, R. & Zilberman, D. (2008). Payments for ecosystem services and poverty reduction: Concepts, issues, and empirical perspectives. *Environment and Development Economics*, 13 (3), 245–254.

CIFOR (2013). *Rewarding the upland poor for environmental services: An innovative strategy to reward Asia's upland poor for preserving and improving our environment*. World Agroforestry Center, Bogor, Indonesia.

Clements, T., John, A., Nielsen, K., An, D., Tan, S. & Milner-Gulland, E. (2010). Payments for biodiversity conservation in the context of weak institutions: Comparison of three programs from Cambodia. *Ecological Economics*, 69 (6), 1283–1291.

Coase, R. H. (1960). The problem of social cost. *Journal of Law and Economics*, 3, 1–44.

Dang, N. V. (2001). *Lâm nghiệp Việt Nam 1945–1950 (Vietnam forestry 1945–1950)*. Ha Noi Agriculture Publishing House, Ha Noi, Vietnam.

Dunn, H. (2011). *Payments for ecosystem services. Evidence and analysis series*. Department for Environment, Food and Rural Affairs, London.

EASRD (2005). *State forest enterprise reform in Vietnam. Technical Note*. World Bank, Washington D.C., USA.

Ecosystem Marketplace (2008). *Payments for ecosystem services: Market profiles*. Forest Trends and Ecosystem Marketplace, Washington, DC, USA.

Falconer, K. (2000). Farm-level constraints on agri-environmental scheme participation: A transactional perspective. *Journal of Rural Studies*, 16 (3), 379–394.

FAO (2001). *Global forest resources assessment 2000: Main report*. FAO Forestry Paper No. 140, Rome, Italy.

FAO (2007). *The state of food and agriculture: Paying farmers for environmental services*. FAO Agricultural Series No. 38, Rome, Italy.

FAO (2009). *Vietnam forestry outlook study. Working paper No. APFSOS II/WP/2009/09*. FAO, Bangkok, Thailand.

Fuhrer, E. (2000). Forest functions, ecosystem stability and management. *Forest Ecology and Management*, 132, 29–38.

Gauvin, C., Uchida, E., Rozelle, S., Xu, J. & Zhan, J. (2010). Cost-effectiveness of payments for ecosystem services with dual goals of environment and poverty alleviation. *Environmental Management*, 45 (3), 488–501.

Government of Vietnam (1998). Decision No. 661/QD-TTg dated July 29, 1998 of the Prime Minister on objectives, tasks, policies and implementation arrangements for the Five Million Hectare Reforestation Program. Ha Noi, Vietnam.

Government of Vietnam (2010). Decree No. 99/2010/ND-CP dated September 24, 2010 of the Government on policy for payments for forest environmental services. Ha Noi, Vietnam.

de Groot, R. S. & van der Meer, P. J. (2010). Quantifying and valuing goods and services provided by plantation forests. *In:* Bauhus, J., van der Meer, P. & Kanninen, M. (eds.), *Ecosystem goods and services from plantation forests*. pp. 16–42, Earthscan, London.

de Jong, W., Sam, D. D. & Hung, T. V. (2006). *Forest rehabilitation in Vietnam: Histories, realities and future*. Center for International Forestry Research, Bogor, Indonesia.

Kile, G., Booth, T., Cromer, R., Marcar, N., Myers, B. & Polglase, P. (1998). The role of plantations and farm forests in sustainable land management. *In:* Boosting International Competitiveness in the Australian Timber and Forestry Industry, Australian Timber and Forestry Conference, Proceedings, 29–30 April, 1998, Sydney. National Association of Forest Industries, Canberra.

Kronenberg, J. & Hubacek, K. (2013). Could payments for ecosystem services create an "ecosystem service curse". *Ecology and Society*, 18 (1), 10.

Lamb, D., Erskine, P. D. & Parrotta, J. A. (2005). Restoration of degraded tropical forest landscapes. *Science*, 310 (5754), 1628–1632.

Landell-Mills, N. & Porras, I. T. (2002). *Silver bullet or fools' gold?: A global review of markets for forest environmental services and their impact on the poor*. International Institute for Environment and Development, London, UK.

Liss, B.-M. (2008). *Development and application of a pilot policy on payment for environmental services (PES) in Da river basin, Son La province*. Consultant report to the Vietnamese-German forestry programme, International consultancy services, Landsberg/Lech, Germany.

Litzenberg, I. (2013). Vietnam forest protection and development fund. Paper presented at the regional workshop on 'Challenges in benefit sharing and livelihood improvement with water storage development', March 20–21, 2014, Vientiane, Lao PDR.

MARD (2012). Forest sector development report 2011: Serving FSSP annual plenary meeting, 1 March 2012. Ministry of Agriculture and Rural Development (MARD), Ha Noi, Vietnam.

Mayrand, K. & Paquin, M. (2004). *Payments for environmental services: A survey and assessment of current schemes*. Unisfera International Centre, Montreal, Canada. 52 p.

McElwee, P. (2009). Reforesting "bare hills" in Vietnam: Social and environmental consequences of the 5 million hectare reforestation program. *Ambio: A Journal of the Human Environment*, 38 (6), 325–333.

Mettepenningen, E., Verspecht, A. & van Huylenbroeck, G. (2009). Measuring private transaction costs of European agri-environmental schemes. *Journal of Environmental Planning and Management*, 52 (5), 649–667.

Milder, J. C., Scherr, S. J. & Bracer, C. (2010). Trends and future potential of payment for ecosystem services to alleviate rural poverty in developing countries. *Ecology and Society*, 15 (2), 4.

Nambiar, E. S. (1999). Productivity and sustainability of plantation forests. *Bosque*, 20 (1), 9–21.

Nga, D. T. (2014). Payments for forest environmental services: Practice and challenges in Quang Nam, Nghe An, and Thanh Hoa. Paper presented during the Stakeholder workshop on benefit sharing and payments for environmental services, 10–11, March 2014. Ha Noi, Vietnam.

Nguyen, T. T., Bauer, S. & Uibrig, H. (2010). Land privatization and afforestation incentive of rural farms in the Northern Uplands of Vietnam. *Forest Policy and Economics*, 12 (7), 518–526.

Nguyen, T. T., Koellner, T., Le, Q. B., Lambini, C. K., Choi, I. & Shin, H.-J. (2014). An economic analysis of reforestation with a native tree species: The case of Vietnamese farmers. *Biodiversity and Conservation*, 23 (4), 811–830.

Nguyen, T. T., Pham, V. D. & Tenhunen, J. (2013). Linking regional land use and payments for forest hydrological services: A case study of Hoa Binh Reservoir in Vietnam. *Land Use Policy*, 33, 130–140.

Pagiola, S., Arcenas, A. & Platais, G. (2005). Can payments for environmental services help reduce poverty? An exploration of the issues and the evidence to date from Latin America. *World Development*, 33 (2), 237–253.

Phuc, T. X., Dressler, W. H., Mahanty, S., Thuy, P. T. & Zingerli, C. (2012). The prospects for payment for ecosystem services (PES) in Vietnam: A look at three payment schemes. *Human Ecology*, 40 (2), 237–249.

Qi, S. (2014). Research on the legal system of payment for ecosystem services under the global perspective. *Canadian Social Science*, 10 (2), 108–112.

Šálek, L. & Sloup, R. (2012). Economic evaluation of proposed pure and mixed stands in Central Vietnam highlands. *Journal of Agriculture and Rural Development in the Tropics and Subtropics*, 113 (1), 21–29.

Šálek, L. & Výlupek, O. (2012). Contribution to the restoration of mixed forests in Central Vietnam. *Journal of Sustainable Forestry*, 31 (6), 549–562.

Scherr, S. J. & Bennett, M. T. (2011). *Buyer, regulator, and enabler – The government's role in ecosystem services markets: International lessons learned for payments for ecological services in the People's Republic of China*. Asian Development Bank, Mandaluyong City, Philippines.

Scherr, S. J., Bennett, M. T., Loughney, M. & Canby, K. (2006). *Developing future ecosystem service payments in China: Lessons learned from international experience*. Forest Trends, Washington, D.C., USA.

Shelton, D., Cork, S., Banning, C., Parry, R., Hairsine, P., Verteyy, R. & Stauffacher, M. (2001). Application of an ecosystem services inventory approach to the Goulburn broken catchment. *In:* Rutherford, I., SHeldon, F., Brierly, G. & Kenyon, C. (eds.), *Proocedings of the Third Australian Stream Management Conference August 27–29, 2001*. pp. 157–162, Cooperative Research Centre for Catchment Hydrology, Brisbane.

Sikor, T. (1998). Forest policy reform: From state to household forestry. *In:* Stewards of Vietnam's upland forests. pp. 18–38, Asia Forest Network, Berkeley.

Sikor, T. & Tan, N. Q. (2007). Why may forest devolution not benefit the rural poor? Forest entitlements in Vietnam's central highlands. *World Development*, 35 (11), 2010–2025.

Sikor, T. & Tan, N. Q. (2011). *Realizing forest rights in Vietnam: Addressing issues in community forest management*. The Center for People and Forests (RECOFTC), Bangkok, Thailand.

Sommerville, M. M., Milner-Gulland, E. & Jones, J. P. (2011). The challenge of monitoring biodiversity in payment for environmental service interventions. *Biological Conservation*, 144 (12), 2832–2841.

Tallis, H., Kareiva, P., Marvier, M. & Chang, A. (2008). An ecosystem services framework to support both practical conservation and economic development. *Proceedings of the National Academy of Sciences*, 105, 9457–9464.

Tan, N. Q. (2011). *Payment for environmental services in Vietnam: An analysis of the pilot project in Lam Dong province*. Institute for Global Environmental Strategies (IGES), Hayama, Japan.

Tan, N. Q., Ngai, N. B., Thanh, T. N., Sunderlin, W. & Yasmi, Y. (2008). Forest tenure reform in Vietnam: Case studies from the Northern Upland and Central Highlands regions. The Center for People and Forests (RECOFTC), Bangkok, Thailand and The Rights and Resources Initiative (RRI), Washington D.C., USA. , unpublished Draft Report.

Thuy, P. T., Bennett, K., Phuong, V. T., Brunner, J., Dung, L. N. & Tien, N. D. (2013). *Payments for forest environmental services in Vietnam: From policy to practice. Occasional Paper 93.* Center for International Forestry Research, Bogor, Indonesia.

Todorova, M., Martini, M., Lucius, I., Grigorova, Y. & Tresierra, J. (2013). *Promoting payments for ecosystem services and related sustainable financing schemes in the Danube Basin*. WWF, Switzerland.

Vatn, A. (2010). An institutional analysis of payments for environmental services. *Ecological Economics*, 69, 1245–1252.

VNFF (2013). Proceedings workshop on payment for forest environmental services in Vietnam – Status and solutions. Vietnam National Forest Protection and Development Fund (VNFF), Ha Noi, Vietnam.

Wunder, S. (2008). Payments for environmental services and the poor: Concepts and preliminary evidence. *Environment and Development Economics*, 13 (3), 279–297.

Zhu, X., Møller, L. R., De Lopez, T. T. & Zaballa Romero, M. E. (2010). *Pathways for implementing REDD+: Experiences from carbon markets and communities*. Danmarks Tekniske Universitet, Risø Nationallaboratoriet for Bæredygtig Energi.

Zilberman, D., Lipper, L. & McCarthy, N. (2008). When could payments for environmental services benefit the poor? *Environment and Development Economics*, 13 (3), 255–278.

# 11

# Identifying the factors governing attitude towards the e-Agriservice among dairy farmers in Maharashtra, India

Sagar Kisan Wadkar [a,*], Khajan Singh [b], Asif Mohammad [b], Ravinder Malhotra [c], Rajiv Baliram Kale [d]

[a] *PMRDF Scheme, Tata Institute of Social Science, Mumbai, Maharashtra*
[b] *Division of Dairy Extension, ICAR – National Dairy Research Institute, Karnal, Haryana*
[c] *Division of Economics, Statistics and Management, ICAR – National Dairy Research Institute, Karnal, Haryana*
[d] *Zonal Project Directorate, Zone VI, CAZRI Campus, Jodhpur, Rajasthan*

**Abstract**

Information and communication technology (ICT) projects have a great potential to revolutionise the information delivery system by bridging the gap between farmers and extension personnel. aAQUA (Almost All Questions Answered) portal was launched by the Developmental Informatics Laboratory (DIL) at Indian Institute of Technology (IIT) Mumbai, Maharashtra, India in 2003 as an information providing system to deliver technology options and tailored information for the problems and queries raised by Indian dairy farmers. To measure the effectiveness of this service the attitudinal dimensions of the users of aAQUA e-Agriservice were investigated using a 22 item scale. A simple random sampling technique was used to select 120 dairy farmers from which data were collected and subjected to factor analysis to identify the underlying constructs in this research. From the attitude items, four components were extracted and named as the *pessimistic, utility, technical* and *efficacy perspective*, which influenced the development of varied level of attitudinal inclination towards the e-Agriservice. These components explained 64.40 per cent of variation in the attitude of the users towards the aAQUA e-Agriservice. This study provides a framework for technically efficient service provision that might help to reduce the pessimistic attitude of target population to adopt e-Agriservice in their farming system. The results should also be helpful for researchers, academics, ICT based service providers and policy makers to consider these perspectives while planning and implementing ICT projects.

*Keywords:* attitude scale, aAQUA e-Agriservice, Likert's scaling technique, Principal Component Analysis (PCA)

## 1 Introduction

In agriculture the integration of Information and Communication Technologies (ICTs) has been assumed to have the potential to revolutionise the information delivery system towards the farmer in many developing countries. In recent years, the government of India initiated many ICT projects in agriculture to enrich the knowledge and tackled information asymmetry problems of farming community. *The services (like weather and marketing services, input and output price information, good agricultural and dairy farming practices, inputs availability, and government subsidies and schemes) provided by ICT tools via internet in online as well as offline mode to the target group, is called as 'e-Agriservice'*. The e-Agriservice uses the power of global computer networks and telecommunication networks, which help to gather, transfer, transform, store, retrieve and disseminate information and knowl-

* Corresponding author
Email: sagarkwadkar@gmail.com

edge among different stakeholders. The service provides information on different aspects of farming like crop planning as per market demand, best management practices for crops, animal husbandry, and dairy, input and output prices, sector wise availability of subsidies, weather and market services, which save information searching time and acts as an effective medium of empowerment of farmers through efficient knowledge sharing. Thus, it helps to strengthen the linkages between research-extension-farmers by providing timely and location based information. ICT radically change the ways in which knowledge and information are constructed (Bolter, 2001; Brunner, 1992; Logan, 1995).

aAQUA e-Agriservice is one such ICT based project, launched by the Developmental Informatics Laboratory (DIL) at Indian Institute of Technology, Mumbai in collaboration with the Farm Science Centre (*Krishi Vigyan Kendra*, KVK), Baramati and Vigyan Ashram (NGO), Pabal, Maharashtra in 2003 as an information providing system to deliver technology options and tailored information for the problems and queries raised by Indian dairy farmers. It is capable of multi-lingual retrieval in 3 Languages spoken among Indian farmers – Marathi, Hindi and English, allowing the registered users (anybody can register freely) to search, ask, see and/or select agricultural keywords on the database. In addition to this, they get information on crop, livestock, government schemes and subsidies, weather and market information for proper planning and management of their farms. The field engineer prints the new queries, allocate these to the Farm Science Centre's (KVK-Baramati) extension personnel on the basis of their area of expertise, get the answers and upload these on aAQUA. It normally provides answers to farmers queries (agri-dairy-livestock and other related) within 24 to 48 hours depending on its difficulty. After the queries are answered and uploaded on aAQUA, the kiosk operators or the users can check these. With this the farmers' query resolving process is completed (Ramamritham *et al.*, 2011).

Many of such ICT initiatives are implemented in rural India, to deploy ICT based services for farming communities and thereby strengthening the research-extension system (Saravanan, 2012). However, the information must be relevant and meaningful to farmers, in addition to being packaged and delivered in a way preferred by them (Diekmann *et al.*, 2009). Context-specific information has higher impacts on the adoption of technologies and increases farm productivity for marginal and small agricultural landholders (Samaddar, 2006). So, it is of importance to investigate whether the psychological traits like attitude and perception of farmers can influence the utilisation of the e-Agriservice rendered by public and private functionaries.

An attitude refers to one's positive or negative opinion about a concrete subject. Attitudes are learnt; they are mouldable and may change with experience of the stimulus objects and with social rules or institutions (Binder & Niederle, 2006). Ajzen (1988) described an attitude as a predisposition to respond favourably or unfavourably to an object, person, or event. Attitude toward ICT usage has been defined as a person's general evaluation or feeling towards ICT and specific computer and internet related activities (Smith *et al.*, 2000). Li (2002) has pointed to a wide range of factors affecting attitude toward ICT. The variations in the factors identified by different researchers might be attributed to differences in the context, participants, and type of research. A Positive attitude towards ICTs is widely recognized as a necessary condition for their effective implementation (Woodrow, 1992). Therefore there is need to explore the attitude of farming communities towards a particular technological intervention (e.g. aAQUA e-Agriservice). Therefore a study was conceptualised to identify factors governing attitude towards the aAQUA e-Agriservice.

## 2 Materials and methods

### *2.1 Sampling*

The Maharashtra state situated in the western part of India was purposively selected for the present study as the aAQUA e-Agriservice was launched as a pilot project in this state in 2003 and still continues to deliver its services to the farmers of this state (www.aaqua.org). Four out of eight pilot districts (*Pune, Nasik, Jalna* and *Amravati*) were selected randomly. Using the list of users from the service provider of aAQUA (Presently Agrocom Software Technologies Pvt. Ltd., Mumbai), thirty users (who asked at least one query per season to the aAQUA e-Agriservice and rearing at least five dairy animals) from each district were randomly identified and surveyed using pre-tested interview schedule. Thus, a total of 120 respondents were considered for the data collection and subsequent analysis.

### *2.2 Instrument*

The items used to measure the attitude of dairy farmers towards aAQUA e-Agriservice (i.e. attitudes scale) were developed based upon discussions with the extension professionals, ICT experts, from reviewed literature as well as on author's own experience and knowledge on the basis of criteria given by Edwards (1957). These items were further assessed and modified based on the summated rating scaling technique as suggested by Likert (1932). One benefit of the summated rating scale is its ability to represent the multiple aspects of a concept in a single measure (Hair Jr. *et al.*, 2006). The items were pre-tested by administering these to 60 agricultural extension scientist, to indicate their degree of agreement or disagreement on each item using a three point continuum *'Agree'*, *'Uncertain'* and *'Disagree'* with scoring 3, 2, and 1, respectively for positive items and vice-versa for negative items. Finally, twenty two items were selected based on an item analysis of the pre-test (i.e. based on the 't' value equal to or greater than 1.75).

#### *2.2.1 Data Analysis*

The scale containing 22-items was administered to the 120 users of the aAQUA e-Agriservice. The subjects were asked to respond using a five-point scale (strongly agree, agree, neutral, disagree, and strongly disagree). The score 1 represented the option "strongly disagree" while score 5 on the scale represented the category "strongly agree". Out of 22 items, 10 of the items were negatively worded and the rest were positively worded (see Appendix 1). For the analysis of the data, all negatively worded items were reversed so that a higher numbered response on the Likert scale would represent positive attitude. Thus, the overall possible minimum and maximum scores of users ranged from 22 to 110.

### *2.3 Factor Analysis*

The item wise score of each respondent with their cumulative score was subjected to factor analysis by Principal Component Analysis (PCA) method using SPSS statistics 20.0. Factor analysis is a technique for identifying groups or clusters of variables towards any object (Field, 2005). The items were used for assessing factors governing attitude of users' towards the e-Agriservice related to one another for the $i^{\text{th}}$ respondent in a general form as follows:

$$Y_e = \sum_{f=1}^{n} a_{e,f} X_f; \quad f = 1, 2, \ldots, n$$

Where,

$Y_e$: Attitude of $e^{\text{th}}$ respondent

$a_{e,f}$: Assessment of the $f^{\text{th}}$ item by $e^{\text{th}}$ respondent

$X_f$: $f^{\text{th}}$ item

$n$: Number of items

Prior to performing factor analysis, test of sample adequacy was done through Kaiser-Meyer-Olkin (KMO) test and Bartlett's test of sphericity. The KMO test is the measure of sampling adequacy, which varies between 0 and 1. The values closer to 1 are better and the value of 0.6 is the suggested minimum. Further Bartlett's test of sphericity confirms the adequacy of the sample population by testing the null hypothesis that the variables in the population correlation matrix are uncorrelated and inadequate.

Null Hypothesis (H0): There is no statistically significant interrelationship between factors affecting the attitude of users towards e-Agriservice.

Alternate Hypothesis (H1): There may be a statistically significant interrelationship between factors affecting the attitude of users towards e-Agriservice.

The Bartlett's test of sphericity had the p-value (Sig.) of $0.000 < 0.01$, which indicated that the sample population was adequate so the null hypothesis was rejected and accepted the alternate hypothesis (H1) that there may be statistically significant interrelationship between factors. The KMO and Bartlett's test measure of sampling adequacy was used to examine the appropriateness of Factor Analysis. The approximate of Chi-square was 1767.737 with 231 degrees of freedom, which is significant at 0.01 level of significance. The KMO statistic is greater than 0.60 (0.825). Thus, the data was suitable and supports the factorability of the correlation matrix.

## 3 Results and discussion

Table 1 show the demographic characteristics of the users of the aAQUA e-Agriservice in Maharashtra, western part of India. These include age, education, occupation, farming experience, land holding, herd size, annual income, and mass media exposure. One third of the users (26.67 %) were between 36 to 40 years, 34 per cent were educated up to secondary level and 39 per cent had a dairy farming and agriculture as their main occupation. Most of the users (45 %) had 14 to 18 numbers of herd size, includes cattle and buffalo. Furthermore, 29 per cent of the users had 10 to 14 years of farming experience, and 34 per cent possessed less than 1.34 ha of land while 29 per cent had Rs. 47,814 to 65,921 of annual income. Most users had access to print and electronic media for getting farming and related information, as two third of the respondents had medium to very high level of mass media exposure.

**Table 1:** *Demographic characteristics of the users of the e-Agriservice (n=120)*

| *Sl. No.* | *Variable* | *Category* | *Percentages* |
|---|---|---|---|
| 1 | Age (Years) | Less than 35 | 24.17 |
| | | 36–40 | 26.67 |
| | | 41–45 | 25.83 |
| | | 46–53 | 13.33 |
| | | More than 54 | 10.00 |
| 2 | Education | Primary (1$^{st}$ – 5$^{th}$ Standard) | 5.00 |
| | | Middle (6$^{th}$–8$^{th}$) | 16.67 |
| | | Secondary (9$^{th}$–10$^{th}$) | 34.17 |
| | | Higher secondary(11$^{th}$–12$^{th}$) | 33.33 |
| | | Graduate and above | 10.83 |
| 3 | Occupation | Dairy farming | 32.50 |
| | | Dairy farming + Agriculture | 39.17 |
| | | Dairy farming + Business | 20.00 |
| | | Dairy farming + Agril. labourer | 8.33 |
| 4 | Farming Experience (Years) | Less than 9 | 28.33 |
| | | 10–14 | 29.17 |
| | | 15–18 | 18.33 |
| | | 19–24 | 17.50 |
| | | Above 25 | 6.67 |
| 5 | Land Holding (ha) | Less than 1.34 | 34.17 |
| | | 1.35–3.42 | 23.33 |
| | | 3.43–5.60 | 19.17 |
| | | 5.61–9.08 | 19.16 |
| | | Above 9.09 | 4.17 |
| 6 | Herd Size (Numbers) | Less than 8 | 7.50 |
| | | 9–13 | 22.50 |
| | | 14–18 | 45.00 |
| | | 19–20 | 5.83 |
| | | Above 21 | 19.17 |
| 7 | Annual Income (Rs.) | Less than 33,557 | 13.33 |
| | | 33,558 to 47,813 | 17.50 |
| | | 47,814 to 65,921 | 29.17 |
| | | 65,922 to 83,776 | 22.50 |
| | | Above 83,777 | 17.50 |
| 8 | Mass Media Exposure | Very Low (< 9) | 9.17 |
| | | Low (10–12) | 24.17 |
| | | Medium (13–15) | 30.00 |
| | | High (16–18) | 20.83 |
| | | Very High (> 19) | 15.83 |

### *3.1 Factor Analysis*

The descriptive statistics shown in Table 2 present the means, standard deviations and communalities of the rating of the attitude towards e-Agriservice for each of the items or variables (see also appendix 1). For example, "the e-Agriservice improves the efficiency of extension workers (IMPRVEFF)" and "It is the best means to collect information on market prices of agricultural and non-agricultural products (MRKTPRINFO)" were the most important items that influence the users to use this e-Agriservice. These items had the highest mean of 4.41 and 4.22 respectively. The standard deviation (SD) for "the e-Agriservice facility is a time consuming activity (TMCNSA)" found to be 1.19, which shows that users differed in their opinion about total time requirement in availing the e-Agriservice facility while for the items "MRKTPRINFO" the SD was 0.62, meaning users had similar opinion on the appropriateness of marketing information provided by the e-Agriservice.

The communalities of the attitude items show how much of the variance of the items has been accounted for by the extracted components. Items with high values are well represented in the common factor space, while items with low values (< 0.4) are not well represented (Bryant & Yarnold, 1995; Loehlin, 2004). For instance 80 per cent of the variance in "the e-Agriservice does not improve the knowledge regarding different aspects of dairy farming (NTIMPRVKNW)" is accounted for, while only 47 per cent of the variance in "the service provider helps to retain and attract new users with the efficient mobilisation of its activities (SERPROV)" is accounted for.

**Table 2:** *Descriptive statistics of raw scores on attitude scale items (n=120; 5-point scale)*

| *Sl. No.* | *Variable* † | *Mean* | *SD* | *Communalities* |
|---|---|---|---|---|
| 1 * | TMCNSA | 3.58 | 1.19 | .756 |
| 2 * | PROPAGANDA | 3.25 | 1.18 | .650 |
| 3 * | COTHERSER | 3.72 | 1.04 | .690 |
| 4 * | EAGSSTOP | 3.89 | 1.10 | .749 |
| 5 * | NTLOCSP | 3.92 | 1.09 | .731 |
| 6 | DEVESFREL | 3.91 | 0.94 | .702 |
| 7 | CONTWRK | 3.86 | 0.99 | .594 |
| 8 | MRKTPRINFO | 4.22 | 0.62 | .634 |
| 9 | GNRTEMPLY | 3.14 | 1.06 | .620 |
| 10 | IMPRVEFF | 4.41 | 0.77 | .526 |
| 11 | WEATHERSERS | 3.79 | 1.06 | .526 |
| 12 * | NTRELWRTHY | 3.52 | 1.20 | .786 |
| 13 * | NTIMPRVKNW | 3.90 | 1.08 | .802 |
| 14 * | TCHSVYPB | 3.40 | 1.26 | .548 |
| 15 | DEVECAPA | 3.73 | 1.00 | .692 |
| 16 | IMPRSTNDLV | 3.87 | 0.90 | .621 |
| 17 | SERPROV | 3.44 | 1.05 | .478 |
| 18 | ALTERNATEP | 3.68 | 1.07 | .725 |
| 19 * | ALNTSLVPRB | 3.40 | 1.23 | .620 |
| 20 | EFFECTW | 3.81 | 0.77 | .635 |
| 21 * | INTUNAOBCT | 3.83 | 0.92 | .485 |
| 22 | ANSWTM | 3.83 | 1.02 | .599 |

Note: * Negative items
† the explanation of variables is mentioned in App. 1

The eigenvalues ≥ 1 were considered for the number of components to be generated. The eigenvalues associated with each component represent the variance explained by that particular linear component (Field, 2005). Table 3 shows all the factors extractable from the analysis along with their eigenvalues, the percent of variance attributable to each factor, and the cumulative variance of the factor and the previous factors. The eigenvalues of the first four components from the principal component analysis were larger than 1 i.e. 8.88, 2.28, 1.82 and 1.19, respectively. The variance explained by these four components were 40.36, 10.37, 8.25 and 5.41, per cent respectively. All the remaining components were not significant. The first component accounted for the highest variance (and hence had the highest eigenvalue) and the next component accounted for as much of the left over variance as it could, and so on. Thus, these four components accounted for 64.40 per cent of variance which is regarded as satisfactory in social sciences (Hair Jr. *et al.*, 2006).

### *3.2 The Extracted Factor Loading by Variamax Rotation*

In order to obtain a meaningful representation of items and factor mapping along the principal axis, the resulted principal component was rotated using orthogonal transformation by varimax. The items with loadings equal to or greater than 0.4 were considered meaningful and extracted for factor analysis (Field, 2005). Table 4 reveals how the item analysis reduced the original 22 items to four independent constructs. The results show that the factor loadings ranging from 0.718 to 0.777 were substantially loaded on the component 1 subscale, from 0.477 to 0.778 on the component 2 subscale, from 0.519 to 0.800 on the component 3 subscale and from 0.478 to 0.711 on the component 4 subscale.

The behaviour of individual items in relation to others within the same subscale provides good evidence for content validity because the highest factor loading is central to the domains assessed by these subscales (Francis *et al.*, 2000). Table 4 depicts the factor loadings after VARIMAX rotation which is shown under four components.

**Table 4:** *Varimax rotated component matrix score for variables*

| *Variable* * | *Component* | | | |
|---|---|---|---|---|
| | 1 | 2 | 3 | 4 |
| TMCNSA | .777 | | | |
| PROPAGANDA | .768 | | | |
| COTHERSER | .755 | | | |
| EAGSSTOP | .743 | | | |
| NTLOCSP | .718 | | | |
| DEVESFREL | | .778 | | |
| CONTWRK | | .729 | | |
| MRKTPRINFO | | .654 | | |
| GNRTEMPLY | | .637 | | |
| IMPRVEFF | | .567 | | |
| WEATHERSERS | | .477 | | |
| NTRELWRTHY | | | .800 | |
| NTIMPRVKNW | | | .720 | |
| TCHSVYPB | | | .681 | |
| DEVECAPA | | | .620 | |
| IMPRSTNDLV | | | .587 | |
| SERPROV | | | .519 | |
| ALTERNATEP | | | | .711 |
| ALNTSLVPRB | | | | .696 |
| EFFECTW | | | | .650 |
| INTUNAOBCT | | | | .511 |
| ANSWTM | | | | .478 |

Note: * the explanation of variables is mentioned in App. 1

**Table 3:** *Total variance explained by the generated principal components*

| *Component* | *Initial Eigenvalues* | | | *Extraction Sums of Squared Loadings* | | | *Rotation Sums of Squared Loadings* | | |
|---|---|---|---|---|---|---|---|---|---|
| | *Total* | *% of Variance* | *Cumulative %* | *Total* | *% of Variance* | *Cumulative %* | *Total* | *% of Variance* | *Cumulative %* |
| 1 | 8.88 | 40.36 | 40.36 | 8.88 | 40.36 | 40.36 | 4.30 | 19.53 | 19.53 |
| 2 | 2.28 | 10.37 | 50.73 | 2.28 | 10.37 | 50.73 | 3.58 | 16.29 | 35.82 |
| 3 | 1.82 | 8.25 | 58.98 | 1.82 | 8.25 | 58.98 | 3.55 | 16.14 | 51.96 |
| 4 | 1.19 | 5.41 | 64.40 | 1.19 | 5.41 | 64.40 | 2.74 | 12.43 | 64.40 |

### *3.2.1 Component 1 – Pessimistic Perspective*

In this case the mean score for negative items found to be 3.76, indicates that dairy farmers disagreed with the negative worded items, which means this process consumes less time of users to solve the farming queries rather than searching/ finding solutions from any other sources. They did not face any problems to avail and use this service. The adoption of mobile telephony by farmers helped to reduce their information searching, transportation and transaction costs (Aker & Mbiti, 2010; Bhatnagar, 2008; de Silva & Ratnadiwakara, 2008; Jensen, 2007). Thus most of the negative items fell in component 1 having strong correlation with each other. Hence it was named *'pessimistic perspective'* (items 1 to 5 as mentioned in App. 1). It explains 40.36 per cent of the variability on the attitude of users towards the e-Agriservice. This component suggest the service provider should pay attention to loopholes of technology and make it more user friendly by providing timely & location specific solutions to the problems and establishing kiosks (computer based system to access the information through e-Agriservice) in cluster of villages to provide easy accessibility and affordability. The need based and hassles free information flow contributed immensely in developing positive attitude towards the e-Agriservice and to attract new users & retain the existing users.

### *3.2.2 Component 2 – Utility Perspective*

Most of the positive items explaining the usefulness of the e-Agriservice fell in component 2, therefore it was named *'utility perspective'* (items 6 to 11). It explains 10.37 per cent of the variability on the attitude of users towards the e-Agriservice. It provides the market and weather information to plan the farming activities, by making them self-reliant for getting timely information and thereby makes them knowledge empowered, which leads to improving the efficiency of extension workers in reaching a large number of farmers with less effort. CTA (2007) and Jenson *et al.* (2004) have shown that farmers with access to market information obtain higher farm prices. Similarly, Raj *et al.* (2011) conducted action research in Nagapattinam district in the state of Tamil Nadu, India using intervention of mobile technology and found that the net income of the intervention farmers was 15.2 % higher than that of the control group. Uphoff (2012) has also mentioned that the ICT empowers farmers by enabling them to gain access to and control over more and better resources, access to new or better technologies and link them into organizations.

### *3.2.3 Component 3 – Technical Perspective*

The items 12 to 17 (App. 1) comprises the technical perspective of the e-Agriservice. It explains 8.25 per cent of the variability on the attitude of users towards the e-Agriservice. This component suggests the service provider need to provide accurate and specific information to retain the interest of farming community in availing the e-Agriservice. David & Foray (2003) also suggested that the access to correct and relevant (and timely) information makes emancipatory and participatory action possible, which has capacity to empower people. Further this would lead to better farming output, helps farmers to improve their socio-economic level. The technicality of the service provider may be reduced and facilitator may be deployed to cater the needs of the farmers in using the service as well as enhancing e-literacy.

### *3.2.4 Component 4 – Efficacy Perspective*

The fourth component (items 18 to 22) characterises the efficacy perspective of the e-Agriservice. This component explains 5.41 per cent of the variability on the attitude of users towards the e-Agriservice. The e-Agriservice can be more effective in timely dissemination of information if uninterrupted internet facility is provided at village level. This would reduce the overdependence on face-to-face interaction with extension functionaries for getting relevant information on farming.

Thus, the service provider must focus on the utility, technical and efficacy perspective to reduce the pessimistic attitude in order to sustainable use of e-Agriservice by farming communities. The findings are in line with the Feng *et al.* (2007) who have developed the conceptual model for identifying factors affecting the success of ICT based knowledge transfer such as knowledge quality, system quality, service quality (with sub-dimensions of e-service quality and extension quality, use, user satisfaction and net benefits. Rogers (2003) also indicated that "trialability" and "observability" are the two attributes of an innovation that might increase the rate of adoption of innovations. Further to confirm the findings, Cronbach alpha coefficients as a measure of reliability focuses on the internal consistency of the set of items forming the sub-scale was calculated. The Cronbach alpha value for four components viz. pessimistic, utility, technical and efficacy subscales were 0.90, 0.81, 0.84 and 0.77, respectively and it was calculated to be 0.91 for the entire items. Thus, the results show that the alpha coefficients for all components were significantly high, suggesting the internal reliability index was adequate. The same set of items could be useful for assessing the attitude of farmers towards other ICT based services as well.

In this psychological analysis, the overall attitude towards e-Agriservice may not give clear representation of the qualitative data. Therefore, a factor analysis was carried out to reduce the number of variables into a meaningful component, which could help in identifying factors responsible for developing the favourable attitude towards e-Agriservice. The outcome of this study could be helpful in planning and implementing rural ICT projects in developing countries. The findings could also be helpful in successful penetration of ICT services in rural areas and sustaining interest of target population.

## 4 Conclusion

Many ICT projects in developing countries are involved in providing timely and relevant information to the farming community. Most of them are funded by the external agencies and donor agencies in the selected area on pilot basis. However, the need, preferences and attitude of the target group is perceived as key indicator in predicting acceptability of any technological intervention among farming community. In this paper an attempt was made to identify the factors, which govern the attitude of dairy farmers towards the aAQUA e-Agriservice. The factor analysis by principal components has been used for the evaluation of the attitude items towards the e-Agriservice. The results obtained revealed four components viz. pessimistic perspective, utility perspective, technical perspective and efficacy perspective, which governed the attitude of farmers towards the aAQUA e-Agriservice. The aforementioned perspectives explained the majority of the variability in the dependent variable, i.e. attitude towards the aAQUA e-Agriservice. The service provider must focus on the utility, technical and efficacy perspective of e-Agriservice to reduce the pessimistic attitude of target population for developing and sustaining favourable attitude towards the aAQUA e-Agriservice. Therefore, policy makers and service providers must consider these four components, in order to assure end-users acceptance of technology and its sustainability.

### Acknowledgement

The author acknowledges the University Grant Commission (UGC), Government of India for financial support to carry out research work.

## References

Ajzen, I. (1988). *Attitudes, personality and behavior*. Open University Press, Bristol, UK.

Aker, J. C. & Mbiti, I. M. (2010). Mobile phones and economic development in Africa. *Journal of Economic Perspectives*, 24 (3), 207–232.

Bhatnagar, S. (2008). Benefits from Rural ICT applications in India: Reducing transaction costs and enhancing transparency? LIRNE-Asia presentation at public lecture on ICT in Agriculture, Colombo, Sri Lanka. URL http://www.lirneasia.net/wp-content/uploads/2008/02/bhatnagar_public_lecture.pdf.

Binder, M. & Niederle, U. (2006). Institutions as determinants of preference change- A one way relation? Papers on Economics and Evolution # 0607, Max Planck Institute of Economics, Jena, Germany. URL http://www.econ.mpg.de/deutsch/research/EVO/discuss.php#p2006.

Bolter, J. D. (2001). *Writing space: Computers, hypertext and the remediation of the print*. M. J. Lawrence Erlbaum Associates, Mahwah.

Brunner, C. (1992). *Integrating technology into the curriculum: Teaching the teachers*. Centre for Technology in Education, New York (ERIC document reproduction service No. EJ 553816).

Bryant, G. & Yarnold, H. (1995). Principal components analysis and exploratory and confirmatory factor analysis. *In:* Grimm, L. G. & Yarnold, P. R. (eds.), *Reading and understanding multivariate analysis*. American Psychological Association Books.

CTA (2007). Annual Report 2006. Technical Centre for Agricultural and Rural Cooperation ACP-EU (CTA), Wageningen, The Netherlands.

David, P. A. & Foray, D. (2003). Economic Fundamentals of the Knowledge Society. *Policy Futures in Education*, 1 (1), 20–49.

Diekmann, F., Loibl, C. & Batte, M. T. (2009). The economics of agricultural information: factors affecting commercial farmers' information strategies in Ohio. *Review of Agricultural Economics*, 31 (4), 853–872.

Edwards, A. L. (1957). *Techniques of attitude scale construction*. Vakils, Feffer and Simons Private Ltd. Ballard Estate, Mumbai.

Feng, W. Z., Duan, Y. Q., Mathews, B. & Fu, Z. T. (2007). Measuring ICT based agricultural knowledge transfer under a sender advantage asymmetric information structure. *New Zealand Journal of Agricultural Research*, 50 (5), 1307–1314.

Field, P. (2005). *Discovering statistics using SPSS for Windows: Advanced techniques for the beginner*. Sage, London, UK.

Francis, L., Katz, Y. & Jones, S. (2000). The reliability and validity of the Hebrew version of the Computer Attitude Scale. *Computers & Education*, 35 (2), 149–159.

Hair Jr., J. F., Black, W. C., Babin, B. J., Anderson, R. E. & Tatham, R. L. (2006). *Multivariate Data Analysis, 6th Edition*. Pearson Education, Inc., Upper Saddle River, New Jersey.

Jensen, R. (2007). The digital provide: Information (Technology), market performance and welfare in the south Indian fisheries sector. *Quarterly Journal of Economics*, 122 (3), 879–924.

Jenson, M., Myers, M. & Southwood, R. (2004). The impact of ICT in Africa. Paper Prepared for the Commission for Africa.

Li, N. (2002). *Culture and gender aspects of students' information searching behaviour using the Internet: a two culture study of China and the United Kingdom*. Ph.D. thesis, Open University United Kingdom, ProQuest Digital Dissertations.

Likert, R. A. (1932). *A technique for the measurement of attitude*. Archives of Psychology, New York.

Loehlin, J. C. (2004). *Latent variable models: An Introduction to factor, path, and structural equation analysis, 4th Edition*. Erlbaum Associates, Hillsdale NJ.

Logan, R. K. (1995). *The fifth language: Learning a living in the computer age*. Stoddart, Toronto, ON.

Raj, D. A., Poo Murugesan, A. V., Aditya, V. P. S., laganathan, S. O. & Sasikumar, K. (2011). A crop nutrient management decision support system: India. *In:* Grimshaw, D. J. & Kala, S. (eds.), *Strengthening Rural Livelihoods. The impact of information and communication technologies in Asia*. Practical Action Publishing Ltd., pp. 33–52.

Ramamritham, K., Bahuman, C. & Bahuman, A. (2011). aAQUA: ICT-enabled Knowledge Services to Farmers in India. *In:* Hazelman, M. & Attaluri, S. (eds.), *Success Stories on ICT/ICM in AR4D in Asia and the Pacific Region*. FAO.

Rogers, E. M. (2003). *Diffusion of Innovations (5th Ed.)*. Free Press, New York, London.

Samaddar, A. (2006). Traditional and Post-Traditional: A Study of Agricultural Rituals in Relation to Technological Complexity among Rice Producers in Two Zones of West Bengal, India. *Journal of Culture and Agriculture*, 28 (2), 108–121.

Saravanan, R. (2012). ICTs for Agricultural Extension in India: Policy Implications for Developing Countries. Proceedings of the AFITA/WCCA 2012 - 8th Asian Conference for Information Technology in Agriculture (AFITA) and World Conference on Computer in Agriculture (WCCA) during September, 3-6, 2012, at Taipei, Taiwan.

de Silva, H. & Ratnadiwakara, D. (2008). Using ICT to reduce transaction costs in agriculture through better communication: A Case-study from Sri Lanka. LIRNEasia, Colombo 08, Sri Lanka. URL `http://lirneasia.net/wp-content/uploads/2008/11/transactioncosts.pdf`.

Smith, B., Caputi, P. & Rawstone, L. (2000). Differentiating computer experience and attitude towards computers: An empirical investigation. *Computers in Human Behavior*, 16, 59–81.

Uphoff, N. (2012). Empowerment of farmers through ICT. Draft paper for ECOSOC expert group meeting on promoting empowerment of people in advancing poverty eradication, social integration, and decent work for all, UN HQ, NYC.

Woodrow, J. E. (1992). Locus of control and student teacher computer attitudes. *Computers in Education*, 14 (5), 421–432.

**Appendix 1:** *Variable names to the items*

| *Sl. No.* | *Variable* | *Items* |
|---|---|---|
| 1* | TMCNSA | The e-Agriservice facility is a time consuming activity |
| 2* | PROPAGANDA | The e-Agriservice is more of propaganda & less usage for dairy farming |
| 3* | COTHERSER | I could have contacted other sources for dairy related queries |
| 4* | EAGSSTOP | The aAQUA e-Agriservice should be stopped |
| 5* | NTLOCSP | The e-Agriservice cannot meet location specific needs of the farmers |
| 6 | DEVESFREL | The e-Agriservice helps to develop self-reliance among farming community |
| 7 | CONTWRK | The e-Agriservice empowers me to have control over farming activities |
| 8 | MRKTPRINFO | It is the best means to collect information on market prices of agricultural and non-agricultural products |
| 9 | GNRTEMPLY | It helps to generate employment opportunities among farming community |
| 10 | IMPRVEFF | The e-Agriservice improves efficiency of experts and extension workers in reaching a large number of farmers with less effort |
| 11 | WEATHERSERS | The weather services provided by the e-Agriservice are satisfactory |
| 12* | NTRELWRTHY | The services provided by the e-Agriservice are not realistic and worthwhile |
| 13* | NTIMPRVKNW | The e-Agriservice does not improve the knowledge regarding different aspects of dairy farming |
| 14* | TCHSVVYPB | The tech-savvy people can benefit more from the aAQUA e-Agriservice |
| 15 | DEVECAPA | It is not just the agro-advisory service but also develop my capability in dairy farming |
| 16 | IMPRSTNDLV | It aids to increase income which leads to enhance standard of living |
| 17 | SERPROV | The service provider helps to retain and attract new users with the efficient mobilisation of its activities |
| 18 | ALTERNATEP | The aAQUA e-Agriservice is an alternative to the present dairy extension system |
| 19* | ALNTSLVPRB | e-Agriservice alone would solve the problems of farmers |
| 20 | EFFECTW | It enhances users effectiveness about dairy farming |
| 21* | INTUNAOBCT | The internet unavailability obstruct the accessibility of the e-Agriservice by the farmers |
| 22 | ANSWTM | It provides appropriate answers to farmers queries within a short period of time |

Note: * Represents negative items.

# 12

# Index-based agricultural insurance products: challenges, opportunities and prospects for uptake in sub-Sahara Africa

Nepomuscene Ntukamazina [a,*], Richard N. Onwonga [a], Rolf Sommer [b], Jean Claude Rubyogo [c], Clare M. Mukankusi [d], John Mburu [e], Rahab Kariuki [f]

[a]*Department of Land Resource Management and Agricultural Technology, University of Nairobi, P.O. Box 29053-00625, Nairobi, Kenya*

[b]*International Center for Tropical Agriculture (CIAT), Nairobi, Kenya*

[c]*International Center for Tropical Agriculture (CIAT), Arusha, Tanzania*

[d]*International Center for Tropical Agriculture (CIAT), Kampala, Uganda*

[e]*Department of Agricultural and Applied Economics, University of Nairobi, Nairobi, Kenya*

[f]*Agriculture and Climate Risk Enterprise, Nairobi, Kenya*

**Abstract**

Agricultural insurance products have been piloted in Sub-Saharan Africa (SSA) to address climate related risks faced by farmers. However, these products in general face low rates of adoption in SSA. Factors and challenges that may explain the low uptake of index-based insurance products in SSA are reviewed in this paper with the objective of assessing and documenting (i) the insurance products available to farmers, (ii) factors influencing farmers to purchase insurance products, (iii) challenges limiting farmers accessing to insurance products and (iv) opportunities that can positively enhance uptake in SSA. This review reveals that area yield index insurance, index-based crop insurance and index-based livestock insurance have been piloted or implemented in the region. The uptake of these products was found to be positively correlated with on-farm income/savings, literacy, and family size with estimated coefficients of 0.211, 0.292 and 0.018, respectively; and negatively correlated with premium rate (−0.183), age of farmer (−0.058), land tenure (−0.800) and farm size (−0.167). Challenges that impede uptake of index-based products include weakness of regulatory environment and financial facilities, basis risk, quality and availability of weather data, capacity building of stakeholders (farmer, insurer, and regulator), and lack of innovation for local adaptation and scalability. The current gap between high promise and low uptake calls for farmer-driven product design, strong public-private partnerships and improved quality and availability of weather data.

*Keywords:* adoption driving factor, crop and livestock insurance, developing countries, weather risk management

## 1 Introduction

Even with the introduction of improved management practices, most of the agricultural activities in Sub-Sahara Africa (SSA) remain susceptible to adverse weather events that can severely impact the quality and yield of a crop (Dick *et al.*, 2011).Traditionally, smallholders in SSA have managed these risks by diversifying production activities on-farm and income generating activities off-farm. While these mechanisms work well for low-magnitude losses, they often prove to be inadequate for risk that is infrequent but severe (Hazell, 1992; Aidoo *et al.*, 2014). Therefore, efficient management of available resources with variable weather conditions is essential to increase productivity of agriculture

* Corresponding author
Email: ndabanepo@gmail.com

in order to feed an increasing population (Kumar *et al.*, 2006). In other words, innovative conservation methods and adaptation measures need to be implemented to reduce some of the negative impacts of climate change and variability. Some of these measures include diversification, hedging, planting of robust crops on marginal arable lands, contract farming and agricultural insurance.

Although not so common in the developing world compared to the other strategies mentioned above, agricultural insurance is one way by which farmers can stabilize farm income and investment and guard against disastrous effects of losses due to natural hazards or market prices variability (Amador-Ramirez *et al.*, 2007; Aidoo *et al.*, 2014). Agricultural insurance not only stabilizes the farm income but also helps the farmers to initiate production activity after experiencing crop failure (Raju & Chand, 2008). In addition, agricultural insurance models have demonstrated their important role to address crop production risks and climate related disasters (Carter *et al.*, 2014).

While positive impacts have been recorded where index-based insurance products have been adopted, their uptake has generally been low. Reasons of the low uptake of insurance products are often not clear cut; whether it is due to product design aspects, basis risk, lack of demand, or barriers to demand linked to liquidity, financial literacy or lack of trust (Giné & Yang, 2009; Dick *et al.* 2011; Clarke *et al.* 2012; Cole *et al.*, 2012; Norton *et al.*, 2014; Takahashi *et al.*, 2016; Mensah *et al.*, 2017). It is therefore of central importance to evaluate available literature on index-based insurance models, to understand why SSA region lags behind in uptake of such insurance products.

The overall objective of this review is to identify key elements that allow for or hinder to increase the penetration of agricultural insurance in SSA. Specifically, this review was initiated to identify challenges, opportunities and factors that explain the low uptake of index-based agricultural insurance products in SSA. The review was guided by the following questions: (i) what are the insurance products available to farmers, (ii) what are the factors that influence farmers to adopt insurance products, (iii) what are the major challenges limiting farmers accessing to insurance products, and (iv) which opportunities exist that can increase uptake. The review was limited to index-based insurance products as they are the most piloted and implemented in developing countries (Mahul *et al.*, 2012).

## 2 Approach used in the review

The implementation of index-based insurance in developing countries is a very recent practice. In order to maximize the number of relevant studies, the review targeted studies conducted in the period from 1990 to date. The study identification focused on studies conducted in Sub-Saharan Africa countries. In terms of the study scope and design, the review included studies that examined designing, implantation or promotion of index-based insurance products. Particularly, the identification singled out studies on index-based insurance products, factors and challenges influencing adoption of these products, and opportunities that might change the uptake of these products. Exclusion criteria, used to filter studies are presented in Table 1.

*Literature search and screening strategy*

Iterative search strategy was used to identify potential studies from online databases such as Google search, Google scholar, Scopus, Science Direct and websites of international organisations. Peer review journals and publishers deemed to be important sources were searched for published studies. This process generated over one hundred (121) studies, which were considered for screening. These studies were then subjected to the title, abstract and full-document screening, respectively. The screening process reduced the number of relevant papers to ninety six (96) papers for in-depth review and inclusion in the synthesis (Fig. 1).

## 3 Synthesis results

### *3.1 Index-based agricultural insurance products*

A total of 96 studies dealt with a survey population from Benin, Burkina Faso, Cameroon, Ethiopia, Ghana, Kenya, Malawi, Mali, Morocco, Mozambique, Niger, Nigeria, Rwanda, Senegal, Syria, Tanzania and Zimbabwe. The review has identified various agricultural insurance products that are either piloted or implemented in SSA region (Table 2). In 2015, Agriculture and Climate Risk Enterprise Ltd. (ACRE) provided insurance to around 400,000 farmers in Kenya, Rwanda and Tanzania, resulting in a significant business (GIIF, 2016). According to Greatrex *et al.* (2015), this positive trend in insurance uptake is attributed to the wide range of products offered by ACRE, its role as an intermediary between insurance companies, reinsurers and dis-

**Table 1:** *Exclusion criteria.*

| *Category* | *Exclusion criteria* |
|---|---|
| Time | – Study conducted before 1990 |
| Geographical location | – Study conducted outside Sub-Saharan Africa countries |
| Study scope and study design | – Study discussing risk mitigation (e.g., irrigation or crop diversification) rather than risk management<br>– Study presenting information not related to (i) index based insurance products, (ii) challenges/factors influencing their uptake, (iii) possible opportunities to increase adoption of index-based insurance products. |

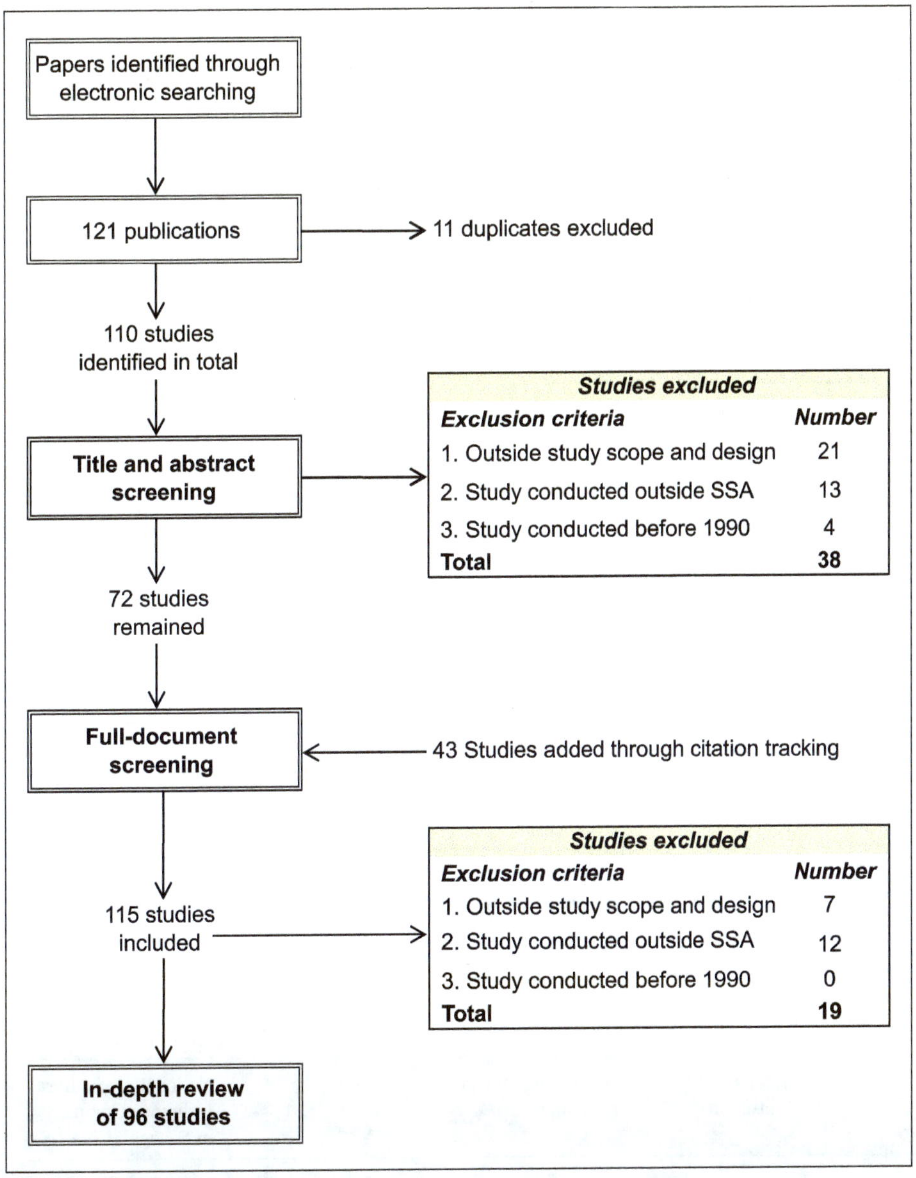


**Fig. 1:** *Filtering of papers from searching to map to synthesis (Authors).*

**Table 2:** *Summary of agricultural products piloted/implemented in SSA.*

| Country | Period/Year | Insurance products | Insured crops/livestock | Insured perils | Insurance programme / project | Estimated clients |
|---|---|---|---|---|---|---|
| Benin | 2012–2014 | – Satellite-based index insurance | Maize and Cotton | Drought | PlaNet Guarantee | 1,099 |
| Burkina Faso | 2011–2014 | – Satellite-based index insurance<br>– Area-yield index insurance | Maize, cotton | Drought Decreased yield | PlaNet Guarantee | 8,281 |
| Ethiopia | 2009–2011 | – Satellite-based index insurance | Maize, Sorghum | Drought | Horn of Africa Risk Transfer for Adaptation (HARITA) Project | 13,000 |
| Ethiopia | 2009 | – Satellite-based index insurance | Haricot beans, teff, and other cereals | Drought | Nyala Insurance Share Company (NISCO) | 22,200 |
| Kenya Ethiopia | 2010–2013 | – Index-based livestock insurance (IBLI) | Camels, Cattle, Goats, Sheep | Drought | The International Livestock Research Institute (ILRI) | 10,067 |
| Kenya, Rwanda, Tanzania | 2011–2015 | – Weather station based index insurance<br>– Satellite-based index insurance<br>– Dairy livestock insurance | Maize, beans, wheat, sorghum, coffee, potatoes, and dairy cows | Drought, excess rain and storms, pregnancy losses for calving cows | Agriculture and Climate Risk Enterprise (ACRE-Africa) | 394,426 |
| Malawi | 2004–2011 | – Weather index insurance | Groundnut, maize, tobacco | Drought | Opportunity International Bank of Malawi (OIBM) | 13,092 |
| Malawi | 2012–2013 | – Relative evapo-transpiration (RE) index | Maize | Drought | COINRe (Dutch-based re-insurance company), Local insurance companies | 1282 |
| Mali | 2011–2014 | – Satellite-based index insurance | Maize | Drought | Allianz | 17,481 |
| Mozambique | 2011–2013 | – Weather index insurance | Maize, cotton | Drought, temperature | Hollard Mozambique, EMOSE; Cotton Institute of Mozambique | 43,000 |
| Rwanda Zambia | 2010–2013 | – Satellite-based index insurance | Irish potatoes, Maize, rice, cotton | Drought and excess rainfall | SORAS (Rwanda); Focus Insurance (Zambia) | 35,134 |
| Senegal | 2011–2012 | – Weather station-based index insurance | Maize, groundnut | Drought | PlaNet Guarantee | 6,600 |
| Senegal | | – Area yield insurance | Groundnut and millet | Drought | National Agricultural Insurance Company of Senegal (CNAAS) | |

Source: Authors

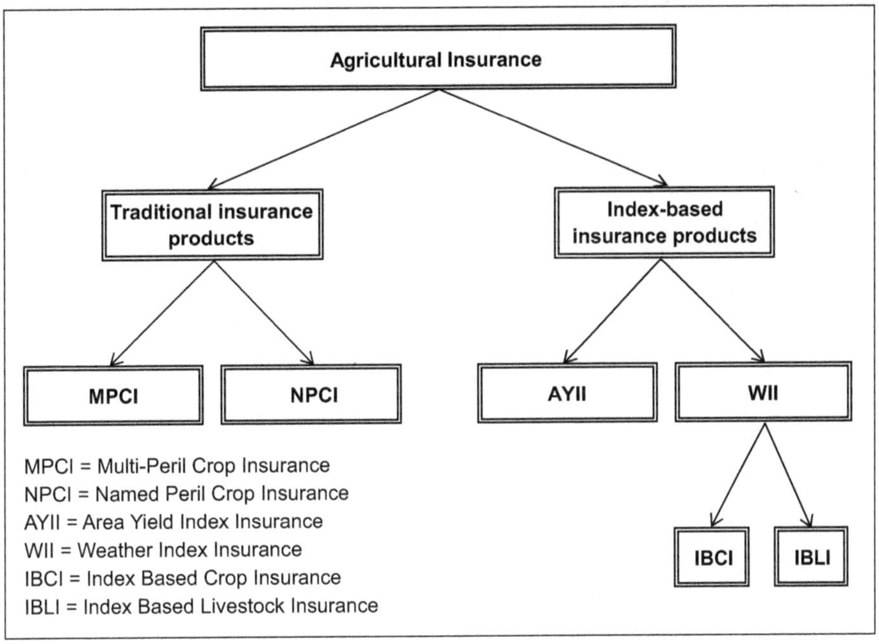


**Fig. 2:** *Agricultural insurance models (Source: Authors).*

tribution channels/aggregators and its link to the mobile money providers.

Three categories of index-based insurance products are piloted or implemented in SSA (Fig. 2). These products include area yield index insurance (AYII) and weather index insurance (WII) presented as index based crop insurance (IBCI) and index based livestock insurance (IBLI). In the surveyed population, these three types of insurance models (AYII, IBCI, and IBLI) were found being piloted and implemented at estimated rates of 12, 69 and 19 percent, respectively. This is consistent with the study of Binswanger-Mkhize (2012), namely that since the 1990s the focus has moved from individually assessed insurance (traditional insurance) to index-based insurance.

### *3.2 Factors influencing uptake of index-based insurance products*

Studies conducted in SSA indicate that socio-demographic and socio-economic factors are considered as driving factors for farmers to adopt index-based insurance products, in addition to premium rates and delivery channels (Table 3). As expected, the higher the premium rate, the lower the farmers' willingness to purchase index-based insurance. Literacy, family size and on-farm income/savings have a positive impact on farmers' willingness to adopt insurance with estimated coefficients of 0.292, 0.018, and 0.211, respectively. While age of farmer, land ownership and increase in farm size decreases the willingness to adopt insurance products by 0.058, 0.8, and 0.167, respectively.

### *3.3 Challenges for index-based insurance products*

Despite the apparent advantages of the index based insurance products, pilots and feasibility studies have shown challenges inherent with index products (World Bank, 2010; FSD Kenya, 2013). As presented by the International Finance Corporation (IFC), weakness of insurance regulatory environment and poor financial facilities are considered as country/programme specific challenges that impede development of insurance markets in SSA. In addition, the review has identified cross cutting challenges such as basis risk, quality and availability of historical weather and yield data, capacity building of stakeholders (farmer, insurer and regulator), limited product options for different weather risks, and lack of innovation for local adaptation and scalability.

### *3.4 Opportunities for index-based insurance products*

African Risk Capacity program (ARC), Government and public sector support, use of mobile network operators, public-private partnership, and interlinking insurance with safety net programs are presented in this review as opportunities to speed up the uptake of index-based agricultural insurance in SSA.

**Table 3:** *Summary results from probit models to explain adoption of index insurance products.*

| *Independent variables* | *Relationship with dependant variable*† (estimated coefficient) | | | | | |
|---|---|---|---|---|---|---|
| | A | B | C | D | E | Average |
| Premium rates/bid | negative | negative | negative (−0.125) | negative (−0.24) | negative | negative (−0.122) |
| Socio-demographic factors | | | | | | |
| Years of education | positive (0.063) | positive (0.807) | positive (0.012) | positive (0.09) | positive (0.490) | positive (0.292) |
| Age of farmer | negative (−0.172) | * | negative (−0.009) | negative (−0.003) | negative (−0.048) | negative (−0.058) |
| Family size | negative (−0.126) | positive (0.222) | negative (−0.023) | positive (0.0001) | * (0.018) | positive |
| Socio-economic factors | | | | | | |
| On-farm income and savings | positive (0.803) | positive (0.242) | positive (0.008) | positive (0.001) | positive (0.0001) | positive (0.211) |
| Land ownership | negative (−0.194) | negative | * | positive (0.002) | negative (−2.207) | negative (−0.560) |
| Farm/herd size | negative (−0.433) | negative (−0.091) | negative (−0.012) | * | negative (−0.131) | negative (−0.167) |

*A*: Wairimu *et al.* (2016), *B*: Koloma (2015), *C*: Takahashi *et al.* (2016), *D*: Gallenstein *et al.* (2015), *E*: Aidoo *et al.* (2014).
* Independent variable not included in the model;
† Willingness of farmers to adopt index-based insurance product;
Source: Authors

## 4 Discussion

### 4.1 *Index-based agricultural insurance products*

#### 4.1.1 *Area yield index insurance*

Area yield index insurance (AYII) represents an alternative approach to address and overcome many of the drawbacks of traditional multi-peril crop insurance (MPCI). The key feature of this product is that it does not indemnify crop yield losses at the individual field or grower level. Rather, an area-yield index product makes indemnity payments to growers according to yield loss or shortfall against an average area-yield (the index) in a defined geographical area (e.g. county, district, province or department). The insured yield is expressed as a percentage (coverage level) of the historical average yield for a defined crop in the defined geographical region, considered as insured unit. The holder of an area yield insurance policy receives an indemnity whenever the realized county yield falls below some specified critical yield (i.e., strike), regardless of the realized yield on his or her farm (World Bank, 2009).

While traditional MPCI is often constrained by a lack of reliable historical yield data at the individual farm level, the required 10 years' historical data at country-level, district-level or provincial-level are usually available to determine the coverage level for area yield index insurance contracts (Mahul *et al.*, 2009; Rao, 2010).

As the index is based on yield, the insurance covers risks encountered from planting to harvesting, pre-planting and post-harvest losses are not reflected in the area yield index. In addition, the basis risk is an important factor affecting the efficacy of area yield insurance. The higher (lower) the positive correlation between the farm and county yield, the lower (higher) the basis risk (Barnett *et al.*, 2005). Lowering the size of insured unit and double or triple trigger mechanism were presented by World Bank (2009) and Stoeffler *et al.* (2016) as ways to minimize the basis risk. The Burkinabé index insurance pay-out occurred only if both the cooperative yield is below the cooperative strike-point (e.g. 750 7kg $ha^{-1}$) and the district yield is below the district strike-point (e.g. 1000 7kg $ha^{-1}$). Whereas in Mali, the cotton area-yield insurance provided three level payments: small pay-out, medium pay-out and big pay-out when yields were below 20, 8, and 4 % of the yield distribution (Stoeffler *et al.*, 2016).

#### 4.1.2 *Index based crop insurance (IBCI)*

The index-based crop insurance product is an innovative form of index insurance that covers farmers against weather-related extreme events. The product uses a proxy (or index) – such as the amount of rainfall, temperature, wind speed, relative evapotranspiration, or biomass index delivered from Normalized Difference Vegetation Index (NDVI) – to trigger indemnity payouts to farmers. For example, the rainfall index derivative for wheat in Morocco, the *Kilimo Salama* insurance in Kenya, Tanzania and Rwanda, and the Nyala Insurance Share Company (NISCO) in Ethiopia, indemnity payments are made, for the selected crop, when actual rainfall in the cropping season, recorded in the nearest weather station, falls below pre-defined threshold levels (Dercon *et al.*, 2014; Wairimu *et al.*, 2016). West African grain farmers found the most promising contract to be one based on the NDVI, a remotely sensed, satellite-based measure of the greenness of the vegetation, and as such a proxy for its biomass and/or density (Hill, 2010). Whereas in Malawi, the COINRe reinsurance organisation in collaboration with local insurance companies piloted the use of relative evapotranspiration (RE) as an index instead of using the rainfall index (Leblois *et al.*, 2014).

The defined index helps to determine whether farmers have suffered losses from the insured peril and hence need to be compensated. Therefore, the index is set so as to correlate, as accurately as possible, with the crop losses suffered by the policyholder (World Bank, 2011; Tadesse *et al.*, 2015; Wairimu *et al.*, 2016).

For example, a maize drought contract offered by Agriculture and Climate Risks Enterprise (ACRE) in Kenya consists of three phase contract, where for each phase different minimum rainfall requirements apply. When the rainfall measures below the defined minimum threshold in a block of 5 to 10 days, a pay-out is triggered. The length of each phase, its relative importance, and the minimum thresholds are determined using the FAO's water requirement satisfaction index (WRSI) with the local historical climate data, crop variety characteristics and farmer feedback. An example is shown for a medium to long maturing maize variety in a location in central Kenya in Fig. 3.

#### 4.1.3 *Index based livestock insurance (IBLI)*

The lack of a comprehensive 100-year mortality database has led the IBLI team to explore the use of satellite images in designing index based livestock insurance. Mude *et al.* (2011) found the NDVI is highly correlated with forage availability and therefore can be linked to animal mortality. In addition, NDVI data are publicly available in near-real time and objectively verifiable, also widely used, as indicator of vegetative cover in drought monitoring programs in Africa (Chantarat *et al.*, 2009).

A predicted livestock mortality index is established from a statistical relationship between satellite-generated vegetation imagery and historical records of community level livestock losses. This process generated a parameter objectively, cost effectively measured and non-human manipulable as an index that triggers insurance pay-out index (McPeak *et al.*, 2010; Greatrex *et al.*, 2015).

In Kenya and Ethiopia, remotely sensed NDVI measures were used to set up an IBLI based on the relationship between predicted livestock mortality and forage availability (Chantarat *et al.*, 2011; Greatrex *et al.*, 2015). The insurance product covers the short rains short dry season (SRSD) or the long rains long dry season (LRLD). The contract is specific at the location level, based on the predicted mortality rate as a function of the vegetation index specific to the grazing range of that location (Chantarat, 2009). The IBLI contracts are sold just before the start of rainy season and are assessed at the end dry period to determine whether indemnity payments are to be made (Fig. 4).

The initial launch of IBLI and associated commercialisation and outreach was met with robust demand for the product. In the sales periods following the launch, there is a continued upward trend in cumulative adoption but there is also a substantial rate of dis-adoption. The trust of pastoralist clients in the underwriter and logistical complications dampened the product demand (Chantarat, 2009; Jensen *et al.*, 2015).

### 4.2 *Factors influencing uptake of index-based insurance products*

#### 4.2.1 *Premium rates*

Higher premium rates (or lower expected returns) result in substantially lower levels of participation in agricultural insurance programs (Smith & Watts, 2009). For example, Arshad *et al.* (2015) reported that the increase in premium rate decreases the levels of participation in agricultural insurance programs by 0.03.

#### 4.2.2 *Socio-demographic factors*

*Literacy* has a positive relationship with the willingness of farmers to adopt agricultural insurance scheme (Aidoo *et al.*, 2014; Arshad *et al.*, 2015; Koloma, 2015; Lin *et al.*, 2015). Index based insurance products can be difficult to understand especially for populations with

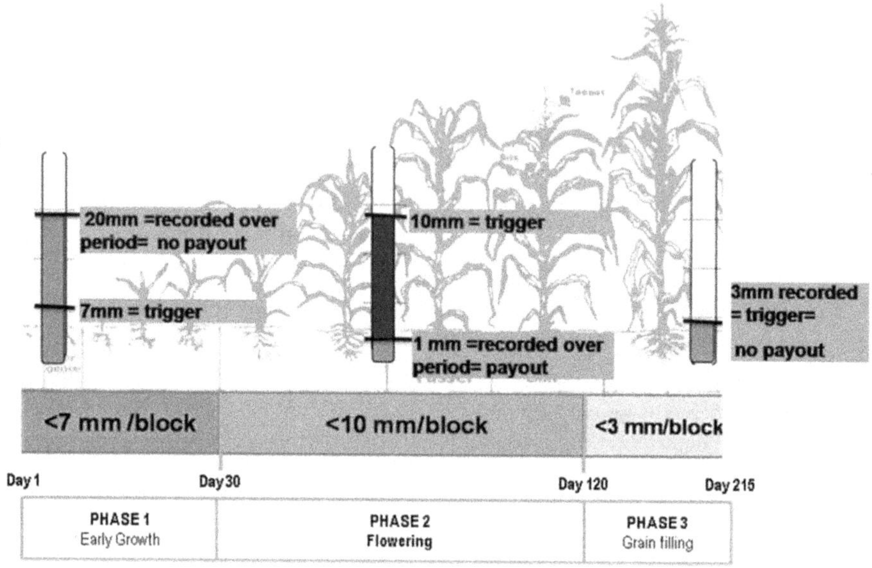


**Fig. 3:** *Example of a three phase maize contract (adapted from Nganga, 2013).*

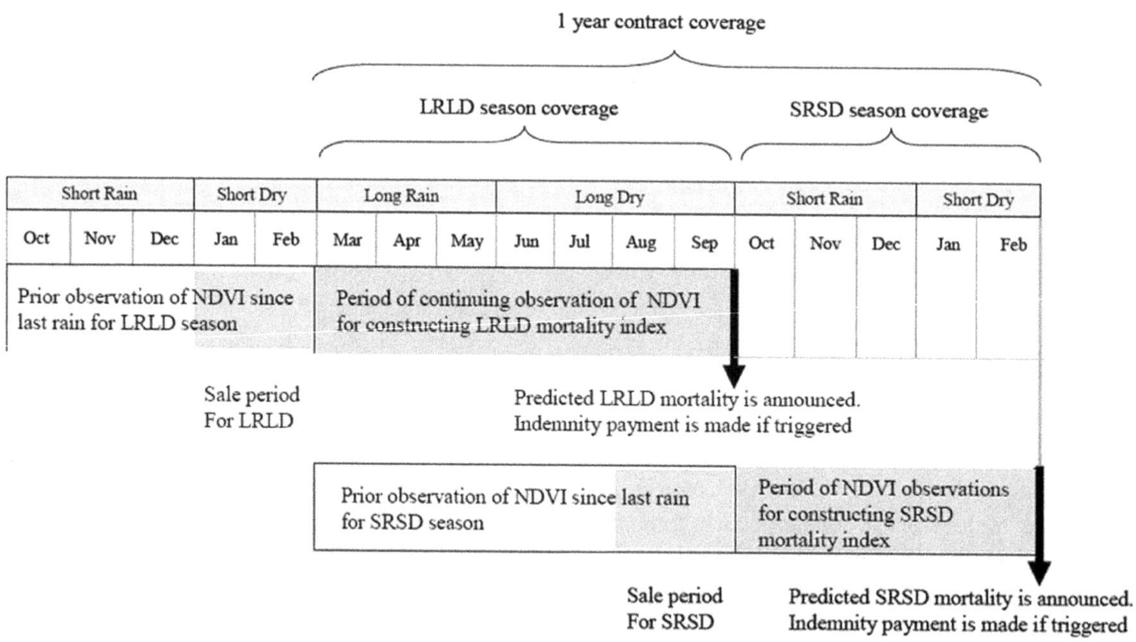


**Fig. 4:** *Temporal structure of IBLI contract (Chantarat, 2009).*

low literacy rates and little or no previous insurance experience. Therefore, education may facilitate the diffusion of new technology and as such has a positive relation with innovation adoption and the payment of accompanying charges. While studying the willingness to pay for crop insurance, Smith & Watts (2009) and De Angelis (2013) also reported that farmers with more literacy rates were more interested in rainfall insurance and willing to pay higher amounts. More educated farmers are likely to appreciate crop insurance issues better than their less educated counterparts.

Concerning family human resources, an increase in *family size* increases the probability of having access to micro-insurance. The higher the family workforce is, the higher the probability of becoming a micro-insurance beneficiary (Koloma, 2015). Family size also positively affected the willingness to pay, exposing a potential market for insurance among households having a large number of family members. The findings indicated also that the joint family system in rural areas can positively influence the decision on making investments like purchasing insurance contracts (Arshad *et al.*, 2015).

Aidoo *et al.* (2014) pointed out a negative influence of *age of farmer* on willingness to adopt crop insurance. As farmers grow older they gain more experience in farming through learning by doing, and are more likely to accept risks than younger farmers. In addition, older farmers lack receptivity towards newly introduced technologies. Dercon *et al.* (2014) found in Ethiopia that households with younger household heads who hold official positions are more likely to purchase crop insurance.

### 4.2.3 *Socio-economic factors*

The on-farm income and *savings* were found to have a positive effect (coefficient = 0.041) on payment of insurance premium. Both Giné & Young (2009) and Cole *et al.* (2012) found, as we do in this review, that insurance uptake is correlated with farmers' wealth. Indeed, insurance premium is usually paid from current income or accumulated income (represented by savings). According to Gine & Yang (2009), the lack of access to credit has traditionally been considered a major obstacle to technology adoption and development. In addition, since the agricultural insurance policies are purchased at the onset of the season, coinciding with the purchase of other agricultural inputs (labour for land preparation, seeds, fertiliser, etc.) only the better-off can afford the policy (Gene & Yang, 2009).

*On-farm income* is positively correlated with the amount farmers are willing to pay as insurance premium. Indeed, premiums are paid with income and hence farmers with high farm income tend to have higher payment capacity than those with low farm income, ceteris paribus. Skees & Barnet (2006) reported that many of the poorest farmers in Tanzania indicated that they simply could not afford to pay any insurance premiums (at least prior to harvest) because their cash flow situation was so dire and their incomes and wealth were so low. Similarly, Smith & Watts (2009) reported that Moroccan farmers with relatively high incomes were more likely to consider purchasing rainfall insurance than farmers with low incomes (quite possibly also because of cash flow problems).

Aidoo *et al.* (2014) found farmers who *own lands* are less willing to adopt crop insurance compared to tenants and sharecroppers. Such farmers have the capacity to diversify into other crops and enterprises since they have easy access to land. In addition, farmers who own lands do not have to pay anything to anybody in times of crop failure but rather manage the little at their disposal. It is therefore not surprising that tenants and sharecroppers tend to be more willing to adopt new innovations such as crop insurance to cope with production risk. However, for land conservation technologies that enhance land fertility and the overall value of the land, land tenure has a positive relationship with willingness to adopt such innovations (Kong *et al.*, 2011).

*Farm size* was found to have a negative impact on the amount a farmer was willing to pay as premium for crop insurance. Since in the insurance business payments are made on per acre basis (the larger the farm the higher the amount paid as premium), farmers with larger farm sizes will tend to pay less as premium per acre (Aidoo *et al.*, 2014). This is quite understandable since the total premium they will pay for their total farm size will be far higher than their counterparts with smaller farms. However, in other adoption studies a positive correlation was found between amount farmers are willing to pay for an agricultural technology and farm size. This was because larger farm sizes tend to have more advantage from adoption of innovations due to economies of scale (Osipenko *et al.*, 2015).

### 4.2.4 *Delivery channels*

As insurers normally have limited or no business (or offices) in rural areas, distribution is best organized through existing links to farmers or farmer groups (Dick *et al.*, 2011). The insurance product distribution through existing services or networks operating in rural areas is important to increase coverage, reduce transaction costs, and reach more clients. Complementary support for agricultural insurance operations could include the promotion of "aggregators"; that is, farmers associations, cooperatives, producer associations, rural banks, and microfinance institutions as delivery channels for agricultural insurance (World Bank, 2010).

For instance in Kenya, *Kilimo Salama* insurance is distributed using local stockists at the time of purchasing inputs, making it easier for the customer to adopt the new product. This distribution channel capitalizes on existing relationships since farmers are more likely to take advice from someone they know and trust (Kilimo Salama, 2011). Dercon *et al.* (2014) and Tadesse *et al.* (2015) found the uptake of weather index insurance higher in Ethiopia when insurance is channelled through group-based informal insurance schemes *iddir* (a funeral society in Ethiopia) with appropriate training for group leaders.

## 4.3 *Challenges for index-based insurance products*

### 4.3.1 *Regulatory environment and financial facilities*

Poor regulatory environment and collaboration with financial institutions are reported as country/programme specific challenges to implementing agricultural insurance in SSA (IFC, 2017). These challenges include weakness of insurance regulatory environment, reluctance of banks and micro-finance institutes to finance agriculture sector, disbursement of loans too late for the planting season leading to a late sowing phase for farmers and higher risk exposure, absence of financial institutions in rural areas (IFC, 2017). Mensah *et al.* 2017 found lack of agricultural insurance legislation and low collaboration with financial institutions among the most pressing constraints faced by the development of agricultural insurance for cashew crop farmers in Ghana. While promoting private sector approaches to help farmers to access index insurance in Kenya, Global Index Insurance Facility (GIIF), Syngenta Foundation for Sustainable Agriculture (SFSA) and International Livestock Research Institute (ILRI) found the need to address restrictive regulations (Kilimo Salama, 2011). Fortunately, Insurance Regulatory Authority (IRA) has been established in Kenya and Uganda in addition to the regional body of the insurance industry for 14 countries in Francophone Africa. In addition, GIIF has defined a strategy of providing legal and regulatory assistance to these bodies for public policy dialogue and regulatory environment facilitation to address insurance market failures (GIIF, 2016; IFC, 2017).

### *4.3.2 Basis risk*

Basis risk is the most problematic feature of index based insurance products, which means that pay-outs may not be fully correlated with crop losses. It represents the difference between the pay-out, as measured by the index, and the actual loss incurred by the policyholder. Because no field loss assessment is made under index insurance, the pay-out is based entirely on the index measurement and may be either higher or lower than the actual loss (World Bank, 2010). Microclimates and uneven topography may affect the yields greatly and these aspects are sometimes not accurately factored into the product design (Bageant & Barrett, 2017).

There has been significant research aiming at addressing basis risk by increasing the density of automatic weather stations (every 10–15 km) or designing hybrid index insurance products using a combination of satellite-rainfall estimates and vegetation indices (Greatrex *et al.*, 2015). Although NDVI can be more effectively used for monitoring pastoral forage and livestock losses, its use for crops like coffee and bananas would be limited, because losses often do not correlate with extent of vegetation (FSD Kenya, 2013). In addition, accuracy of NDVI is limited below 100 $km^2$ area due to the quality of imaging; areas of that size still contain a wide range of diverse weather.

### *4.3.3 Quality and availability of weather and yield data*

The development of index based insurance products requires accurate and complete historical data on weather and crop yield. The amount of required data depends on the frequency of the risk to insure. Twenty years of data may be sufficient to set initial premium rates for relatively frequent weather events, while thirty or forty years of data may not be sufficient for infrequent but potentially catastrophic events (Barnett & Mahul, 2007; World Bank, 2010). The scarcity of these data may entail model risk and additional premium loadings that make crop insurance unattractive to potential buyers, despite the huge potential demand for yield risk reduction (Odening & Shen, 2014).

In many countries, weather data have public goods characteristics, they are unlikely to be collected, cleaned and archived. In addition, these data are not freely available, either as a result of restrictive use policies and fees being charged, or poor data coverage and quality. Consequently, data quality and access remain an important unresolved challenge in the implementation of weather index insurance at larger scale (Barrett *et al.*, 2007).

Some of the suggested options to mitigate the problem of data scarcity include the use of daily observations of temperature and/or rainfall to construct a weather index or simulate synthetic yield-data series through plant-growth models for area-yield index (Dick *et al.*, 2011; Odening & Shen, 2014). In Ethiopia, agronomist and weather experts developed the Livelihoods, Early Assessment and Protection (LEAP) software application which uses ground and satellite rainfall data to map the whole of Ethiopia with ability of covering areas without weather stations (Hazell *et al.* 2010). As reported by GIIF (2016), where both historical yield and weather data are not available, ACRE-Africa relied on satellite data and testing analysis techniques to generate the most accurate proxy for the farmer experience.

### *4.3.4 Capacity building of stakeholders*

Index insurance is a complex concept which requires substantial investment in training of stakeholders along the implementation scheme (Miranda & Milangu, 2016). Particularly, potential policyholders need to understand the basis risk inherent with index insurance to make an informed purchase decision (World Bank, 2010). In Ethiopia, weather index insurance for famine prevention tested by World Bank and World Food Program (WFP) in 2005/2006 was later discontinued by farmers after one year with good rainfall. Farmers and policymakers were not sufficiently educated on how weather index insurance principles operate and become hesitant after a good harvest to pay for the insurance coverage in the following season (Tadesse *et al.*, 2015). Therefore any rollout of the product requires intense education programs to strengthen them to understand the principles of the entire delivery system.

To date, experience with capacity building and education of stakeholders has provided positive and convincing results (Barnett & Mahul, 2007). While explaining the index insurance, McPeak *et al.* 2010 designed an illustrative and playing game through which pastoralists in Northern Kenya were able to understand how it works, what it does and does not cover. For a successfully publicize an insurance product and prepare extension effort, Mude *et al.* 2011 suggested to train first master trainers (MT) followed by another training, run with MTs of Village Insurance Promoters (VIP) recruited from the targeted villages. After the product launch, MTs and VIPs continued to offer their extension services and sales agents began, for the first time, to sell IBLI to clients across Marsabit district in Kenya. Dercon *et al.* (2014) reported that the demand and uptake for insurance products among trained policyholders increased when groups were exposed to training and other capacity building opportunities. While investigating the demand for insurance in Ethiopia, Dercon *et al.*, 2014 found a higher uptake among farmers who had heard about the insurance policy (22 %) or trained (36 %) against only 2 % among those that were not trained.

#### *4.3.5 Lack of innovation for local adaptation and scalability*

While the insurers have shown considerable interest in selling indexed products, their ability to innovate is limited. Until there is commercial success, there is little incentive for private companies to invest adequate time and resources in building internal capacity and funding experiments for setting up new models (Carballo & Reis, 2013; FSD Kenya, 2013). However, on-going annual reviews of the trigger levels are advisable, especially in the first years of a program. The lack of this technical work limits the speed at which the scaling up of a pilot program to a regional or national levels (World Bank, 2010).

### *4.4 Opportunities for index-based insurance products*

#### *4.4.1 African Risk Capacity program*

The African Risk Capacity (ARC) program is a specialized agency of the African Union designed to improve the capacity of African Union Member States to manage natural disaster risk, adapt to climate change and protect food insecure populations. To date (February 2017), sixteen countries have signed the Memorandum of understanding with ARC. These countries include Malawi, Kenya, Niger, Lesotho, Senegal, Burkina Faso, Mozambique, Mauritania, Zimbabwe, Ghana, The Gambia, Mali, Comoros, Chad, Madagascar, and Ethiopia (ARC, 2016).

Voluntarily, member states subscribe to the ARC risk pool and based on the WRSI calculations, Africa RiskView software estimates the number of people potentially affected by drought for each country participating in the insurance pool. Due to drought stress observed during 2014 and 2015 agricultural seasons, Governments of Senegal and Malawi benefited from an ARC pay-out of USD 16 million, and 8.1 million, respectively (ARC, 2016). With support from the German and UK governments, ARC Ltd issued nearly $130 million in drought coverage to Kenya, Mauritania, Niger, Senegal, The Gambia, Malawi and Mali for the risk pool in 2014–2016 (ARC, 2016).

#### *4.4.2 Government and public sector support*

Government can be a catalyst by ensuring that insurers target small-holders, particularly if a publicly owned insurer involved in index insurance contracts. Governments and their regulatory agencies also play a central role in properly positioning index insurance programs within the existing insurance and financial regulatory framework (Miranda & Milangu, 2016).

Arshad *et al.* (2015) reported that governmentally subsidized crop insurance schemes are needed to attract the small farmers to purchase insurance contracts. However, the insurer should be financially responsible for its own affairs, free of government manipulation and not accessing to government funds. If needed, subsidies should be set as some fixed percentage of the total premium. The insurance provider is more likely to succeed if it is an autonomous public institution with its own board of directors, and not a department within the Ministry of Agriculture (Hazell, 1992; World Bank, 2007).

#### *4.4.3 Use of mobile network operators*

The largest challenge in developing any financial product is its distribution, especially if the product is targeting to reach small-scale farmers. One of the solutions to this barrier is partnering with mobile network operators. Under "community based health insurance" in Rwanda and "mi-life" micro-insurance in Ghana, MTN subscribers were able to buy life insurance products, pay premiums and make claims through their mobile phones (International Growth Centre, 2016). Collaborating with Safaricom, the largest mobile network operator in Kenya, ACRE Africa sold its insurance products to over 390,000 farmers in Kenya and Rwanda, by the end of 2015 (Kilimo Salama, 2011; Tadesse *et al.*, 2015). In Ethiopia, M-Birr, a mobile money channel targeting rural residents, enabled almost 50,000 account holders to transfer, deposit or withdraw money without leaving the comfort of their homes (Mugambi, 2015). Initiated in 2015, the mobile money interoperability between different mobile network operators (MNOs) is also presented by as a winning formula to increase insurance penetration within Africa.

#### *4.4.4 Public and private partnerships*

The development of agricultural insurance markets requires public and private sectors to overcome institutional, technical and financial challenges (World Bank, 2010). For example in East Africa (Kenya, Tanzania and Rwanda), Agriculture and Climate Risk Enterprise (ACRE) is demonstrating positive development impact with index based crop insurance. ACRE recognizes the wide range of partners as a major reason behind their rapid scaling and demand. Partners include banks and micro-finance institutes (MFIs), mobile network operators, seed companies, government agencies, research institutions, insurance and reinsurance companies, and global donors like Global Index Insurance Fund "GIIF" (Greatrex *et al.*, 2015).

#### *4.4.5 Interlinking weather index insurance with safety net programs*

The Horn of Africa Risk Transfer and Adaptation project (HARITA) developed a more participatory weather index insurance product in Ethiopia. Through the creation of employment opportunities, HARITA project integrated the Productive Safety Nets Program (PSNP) activities of the Ethiopian government (tree planting or other public goods) with the so-called insurance-for-work (IFW) model (Bageant & Barrett, 2017). Resource-poor farmers were able to pay insurance premiums in kind and receive insurance certificate to guarantee pay-outs in the event of drought affecting crop production. This approach has been tested in northern Ethiopia by Oxfam America, and about 60 % of the households chose to participate in the insurance-for-work program to get coverage for their most important staple cereal crop, *teff*. In 2012, about 19,000 farmers were insured over 76 villages in northern parts of Ethiopia (Greatrex *et al.*, 2015). This approach resolves the cash constraints of the poor to invest in risk transfer instruments and could contribute to enhancing wider uptake if the index is appropriate (Tadesse *et al.*, 2015).

## 5 Conclusion

The purpose of this paper was to review the index based insurance products that have been exposed to farmers in Sub-Sahara Africa (SSA), factors influencing their uptake as well as challenges and opportunities associated to the provision of these insurance products.

Even if the implementation of index-based insurance products in SSA is too recent, the review found that index based insurance products have potential for uptake by farmers. In line with this, farmers have been exposed to three types of index based insurance products such as area yield index insurance (AYII), index based crop insurance (IBCI) and index based livestock insurance (IBLI). These products have also demonstrated their potentiality to replace traditional agricultural insurances, because they might be provided at lower costs to be even affordable to farmers with lower and mid on-farm income. For this reason, together with other advantages of index insurance, a further growth of opportunities for index-based insurance products is expected.

Several factors appeared to affect demand for index based insurance products. On-farm income, savings, educational level, and family size were found to influence positively the trend of uptake, while premium rates, age of farmers, land tenure and farm size have negative impact on the uptake of these products. A part from these factors, there are challenges that need to be addressed while designing, piloting, implementing and promoting index based insurance products. These challenges include regulatory environment and financial facilities, basis risk, data quality and availability, capacity building of stakeholders (farmer, insurer, and regulator), and lack of innovation for local adaptation and scalability.

The provision of agricultural insurance in SSA has potential opportunities that can be considered as positive catalysts in increasing the trend of adoption of index-based insurance products. These opportunities include Africa Risk Capacity (ARC) program, government subsidy and public sector support, use of mobile network operators, public-private partnerships, and interlinking weather index insurance with safety net programs.

The findings from this review do contribute to fill the gaps related to promoting uptake of index based insurance products in SSA. In this respect, insurers should collaborate more closely with economists, agro-meteorologists, agro-dealers, microfinance, banks, researchers, and farmers' organisations in order to more effectively develop successful applications of index based insurance products in SSA.

**Acknowledgement**

This review is one component of PhD training, jointly supported by CIAT, Syngenta Foundation for Sustainable Agriculture and University of Nairobi. Their technical and financial supports are greatly acknowledged. Authors are also very grateful to the editor and anonymous reviewers for their comments.

## References

Aidoo, R., Mensah, O. J., Wie, P. & Awunyo-Vitor, D. (2014). Prospects of crop insurance as a risk management tool among arable crop farmers in Ghana. *Asian Economic and Financial Review*, 4 (3), 341–354.

Amador-Ramirez, M. D., Acosta-Diaz, E., Medina-Garcia, G. & Guetierrez-Luna, R. (2007). An empirical model to predict yield of rain-fed dry bean with Multi-year data. *Revista Fitotecnia Mexicana*, 30 (3), 311–319.

ARC (2016). Africa Risk View End of Season Report, Senegal. Africa Risk Capacity (ARC). Available at: http://www.africanriskcapacity.org (accessed on: 15.01.2017).

Arshad, M., Amjath-Babu, T. S., Kächele, H. & Müller, K. (2015). What drives the willingness to pay for crop insurance against extreme weather events (flood and drought) in Pakistan? A hypothetical market approach. *Climate and Development*, 8 (3), 234–244.

Bageant, E. R. & Barret, C. B. (2017). Are There Gender Differences in Demand for Index-Based Livestock Insurance? *The Journal of Development Studies*, 53 (6), 932–952.

Barnett, B. J., Black, J. R., Hu, Y. & Skees, J. R. (2005). Is Area Yield Insurance Competitive with Farm Yield Insurance? *Journal of Agricultural and Resource Economics*, 30 (2), 285–301.

Barnett, B. J. & Mahul, O. (2007). Weather index insurance for agriculture and rural areas in lower-income countries. *American Journal of Agricultural Economics*, 89 (5), 1241–1247.

Barrett, C. B., Barnett, B. J., Carter, M. R., Chantarat, S., Hansen, J. W., Mude, A. G., Osgood, D. E., Skees, J. R., Turvey, C. G. & Ward, M. N. (2007). Index Insurance for Climate Risk Management and Poverty Reduction. IRI Technical Report 07-03.

Binswanger-Mkhize, H. P. (2012). Is There Too Much Hype about Index-based Agricultural Insurance? *The Journal of Development Studies*, 489 (2), 187–200.

Carballo, A. D. & Reis, L. (2013). The Agricultural Insurance. Market in the Caribbean. En breve no. 183. World Bank, Washington, DC.

Carter, M., De Janvry, A., Sadoulet, E. & Sarris, A. (2014). Index-based weather insurance for developing countries: A review of evidence and a set of propositions for up-scaling. Background document for the workshop "Microfinance products for weather risk management in developing countries: State of the arts and perspectives", Paris, June 25 2014.

Chantarat, S. (2009). *Pro-Poor Risk Management: Essays on the economics of index-based risk transfer Products*. Ph.D. thesis, Cornell University, Ithaca, NY (USA).

Chantarat, S., Mude, G. A. & Barrett, B. C. (2009). *Willingness to pay for index based livestock insurance: Results from a field experiment in Northern Keny.* Cornell University, NY (USA).

Chantarat, S., Mude, G. A., Barrett, B. C. & Carter, R. M. (2011). Designing index based livestock insurance for managing asset risk in northern Kenya. *The Journal of Risk and Insurance*, 80 (1), 205–237.

Clarke, D. J., Mukerji, S. & Dercon, S. (2012). *A theory of rational demand for index insurance*. Ph.D. thesis, University of Oxford, England.

Cole, S., Bastian, G., Vyas, S., Wendel, C. & Stein, D. (2012). *The effectiveness of index-based micro-insurance in helping smallholders manage weather-related risks*. EPPI-Centre, Social Science Research Unit, Institute of Education, University of London.

De Angelis, K. (2013). Index-based livestock insurance: The case of Mongolia. World Resources Institute. 5 pp.

Dercon, S., Hill, R. V., Clarke, D., Outes-Leon, I. & Taffesse, A. S. (2014). Offering rainfall insurance to informal insurance groups: Evidence from a field experiment in Ethiopia. *Journal of Development Economics*, 106, 132–143.

Dick, W., Stoppa, A., Anderson, J., Coleman, E. & Rispoli, F. (2011). Weather Index-based Insurance in Agricultural Development. A Technical Guide. International Fund for Agricultural Development (IFAD).

FSD Kenya (2013). Review of FSD's index based weather insurance initiatives. Financial Sector Deepening Kenya (FSD), Nairobi, Kenya.

Gallenstein, R., Mishra, K., Sam, A. & Miranda, M. (2015). Willingness to Pay for Insured Loans in Northern Ghana. Agricultural & Applied Economics Association and Western Agricultural Economics Association, Annual Meeting, San Francisco, CA, July 26–28, 2015.

GIIF (2016). Achievements in ACP Countries; Phase 1 (2010–2015). Global Index Insurance Facility (GIIF). Available at: http://www.indexinsuranceforum.org (accessed on: 12 May 2017).

Giné, X. & Yang, D. (2009). Insurance, credit and technology adoption: Field experimental evidence from Malawi. *Journal of Development Economics*, 89, 1–11.

Greatrex, H., Hansen, J. W., Garvin, S., Diro, R., Blakeley, S., Le Guen, M., Rao, K. N. & Osgood, D. E. (2015). Scaling up index insurance for smallholder farmers: Recent evidence and insights. CCAFS Report No. 14, Copenhagen.

Hazell, P., Anderson, J., Balzer, N., Clemmensed, H. A., Hess, U. & Rispoli, F. (2010). *The Potential for Scale and Sustainability in Weather Index Insurance for Agriculture and Rural Livelihoods*. IFAD and World Food Programme, Rome. 151 pp.

Hazell, P. R. (1992). The appropriate role of agricultural insurance in developing countries. *Journal of International Development*, 4 (6), 567–581.

Hill, R. V. (2010). Agricultural insurance in Sub-Saharan Africa: can it work? Paper prepared for the Fourth African Agricultural Markets Program (AAMP) policy symposium, Lilongwe, Malawi, September 6–10, 2010.

IFC (2017). Index Insurance: Frequently Asked Questions. International Finance Corporation – IFC. Available at: http://www.ifc.org (accessed on: 9 January 2017).

International Growth Centre (2016). Mobile money: Lessons for West Africa. International Growth Centre (IGC), London School of Economic and Political Science, Houghton Street, London WC2A 2AE.

Jensen, N. D., Barret, C. B. & Mude, A. G. (2015). The favorable impacts of index-based livestock insurance: Evaluation results from Ethiopia & Kenya. ILRI Research Brief–May 2015, Nairobi, Kenya. 5 pp.

Kilimo Salama (2011). Index-based Agriculture Insurance: A Product Design Case Study. International Finance Corporation (IFC). World Bank Group. Available at: http://hdl.handle.net/10986/21682 (accessed on: 6 December 2015).

Koloma, Y. (2015). Crop Microinsurance for Maize Farmers in Burkina Faso: Access and Agriculture Performance in the Dandé Village. *Strategic Change*, 24, 115–129.

Kong, R., Turvey, C. G., Rong, K., Guangwen, H., Jiujie, M. & Patrick, M. (2011). Factors influencing Shaanxi and Gansu farmers' willingness to purchase weather insurance. *China Agricultural Economic Review*, 3 (4), 423–440.

Kumar, R. N., Sailaja, B. & Voleti, S. R. (2006). Crop Modelling with Special Reference to Rice Crop. Rice Knowledge Management Portal (RKMP), Hyderabad, India. 24 pp.

Leblois, A., Quirion, P. & Sultan, B. (2014). Price vs. weather shock hedging for cash crops: Ex ante evaluation for cotton producers in Cameroon. *Ecological Economics*, 101, 67–80.

Lin, J., Boyd, M., Pai, J., Porth, L., Zhang, Q. & Wang, K. (2015). Factors affecting farmers' willingness to purchase weather index insurance in the Hainan Province of China. *Agricultural Finance Review*, 75 (1), 103–113.

Mahul, O., Belete, N. & Goodland, A. (2009). Index-based Livestock Insurance in Mongolia. Innovations in Insuring the Poor, No 17, 2020 vision focus, International Food Policy Research Institute (IFPRI).

Mahul, O., Verma, N. & Clarke, J. D. (2012). *Improving Farmers' Access to Agricultural Insurance in India*. The World Bank.

McPeak, J., Chantarat, S. & A., M. (2010). Explaining index-based livestock insurance to pastoralists. *Agricultural Finance Review*, 70 (3), 333–352.

Mensah, N. O., Fialor, S. C. & Yeboah, E. (2017). Evaluating the Constraints to Development of Agricultural Insurance for Cashew Crop Farmers in Brong-Ahafo Region of Ghana. *Open Journal of Business and Management*, 5, 215–229.

Miranda, M. J. & Mulangu, F. M. (2016). Index Insurance for Agricultural Transformation in Africa. African Transformation Report 2016: Transforming Africa's Agriculture.

Mude, G. A., Chantarat, S., Barrett, C. B., Carter, R. M., Ikegami, M. & McPeak, G. J. (2011). Insuring Against Drought Related Livestock Mortality: Piloting Index Based Livestock Insurance in Northern Kenya. *In:* Towards priority actions for market development for African farmers. Proceedings of an international conference, Nairobi, Kenya, 13–15 May 2009. pp. 175–188, AGRA and International Livestock Research Institute (ILRI). Available at: http: //hdl.handle.net/10568/16493 (accessed on: 2 Octo-ber 2015).

Mugambi, C. (2015). Mobile money and insurance in Africa: Whats the connection? KPMG's Africa blog. Available at: http://www.blog.kpmgafrica.com (accessed on: 05.05.2017).

Nganga, R. (2013). Public Private Partnership model of insurance in Africa. Presentation to the CCAFS Workshop on Institutions and Policies to Scale out Climate Smart Agriculture, Colombo, Sri Lanka.

Norton, M., Osgood, D., Madajewicz, M., Holthaus, E., Peterson, E., Diro, R., Mullally, C., Teh, T. & Gebremichael, M. (2014). Evidence of Demand for Index Insurance: Experimental Games and Commercial Transactions in Ethiopia. *The Journal of Development Studies*, 50 (5), 630–648. doi:10.1080/00220388.2014.887685.

Odening, M. & Shen, Z. (2014). Challenges of insuring weather risk in agriculture. *Agricultural Finance Review*, 74 (2), 188–199.

Osipenko, M., Shen, Z. & Odening, M. (2015). Is there a demand for multi-year crop insurance? *Agricultural Finance Review*, 75 (1), 92–102.

Raju, S. S. & Chand, R. (2008). Agricultural Insurance in India Problems and Prospects. National Centre for Agricultural Economics and Policy Research (NCAP), Working Paper No. 8, New Delhi, India.

Rao, K. N. (2010). Index Based Crop Insurance. *Agriculture and Agricultural Science Procedia*, 1, 193–203.

Skees, J. R. & Barnett, B. J. (2006). Enhancing microfinance using index-based risk-transfer products. *Agricultural Finance Review*, 66 (2), 235–250.

Smith, H. V. & Watts, M. (2009). Index Based Agricultural Insurance in Developing Countries: Feasibility, Scalability and Sustainability. Paper prepared for the Gates Foundation.

Stoeffler, Q., Gelade, W., Guirkinger, C. & Carter, M. (2016). Indirect protection: the impact of cotton insurance on farmers' income portfolio in Burkina Faso. Agricultural and Applied Economics Association, Annual Meeting, Boston, Massachusetts, July 31–August 2, 2016.

Tadesse, M. A., Shiferaw, B. A. & Erenstein, O. (2015). Weather index insurance for managing drought risk in smallholder agriculture: lessons and policy implications for sub-Saharan Africa. *Agricultural and Food Economics*, 3, 1–26. doi:10.1186/s40100-015-0044-3.

Takahashi, K., Ikegami, M., Sheahan, M. & Barret, C. B. (2016). Experimental Evidence on the Drivers of Index-Based Livestock Insurance Demand in Southern Ethiopia. *World Development*, 78, 324–340.

Wairimu, E., Obare, G. & Odendo, M. (2016). Factors affecting weather index-based crop insurance in Laikipia County, Kenya. *Academic Journals*, 8 (7), 111–121.

World Bank (2007). *China: Innovations in Agricultural Insurance Promoting Access to Agricultural Insurance for Small Farmers*. The World Bank, Washington, DC. 92 pp.

World Bank (2009). *Index-based Crop Insurance in Senegal Promoting Access to Agricultural Insurance for Small Farmers*. The World Bank, Washington, DC. 104 pp.

World Bank (2010). *Agricultural Insurance in Latin America: Developing the Market. Report no. 61963*. The Worldbank, Washington, DC. 152 pp.

World Bank (2011). *Nigeria: Crop Weather Index Insurance Pre-Feasibility Study Report*. The Worldbank, Washington, DC. 136 pp.

# 13

# Multi-methodological characterisation of Costa Rican biochars from small-scale retort and top-lit updraft stoves and inter-methodological comparison

Joeri Kaal [a,b,*], Roberto Calvelo [c], Art Donnelly [d], Anna McBeath [e], Antonio Martínez Cortizas [b], Hugh McLaughlin [f]

[a]*Institute for Heritage Science (Incipit), Spanish National Research Council (CSIC), Santiago de Compostela, Spain*

[b]*Departamento de Edafoloxía e Química Agrícola, Fac. Bioloxía, Universidade de Santiago de Compostela, Campus Sur, Santiago de Compostela, Spain*

[c]*New Zealand Biochar Research Centre, Soil and Earth Sciences Group, Institute of Agriculture and Environment, Massey University, Palmerston North, New Zealand*

[d]*SeaChar.Org, University Place, Seattle, USA*

[e]*College of Science, Technology and Engineering, James Cook University, Cairns, Australia*

[f]*NextChar, LLC, Amherst, MA, USA*

---

**Abstract**

We applied common (pH, elemental analysis, thermogravimetry) and less-common (infrared spectroscopy, GACS adsorption test, pyrolysis-GC-MS, hydrogen pyrolysis) analytical procedures to a set of biochars from Costa Rica (bamboo stalk, cacao chaff, sawmill scrap, coconut husk and orchard prunings feedstocks). The biochars were produced by high temperature combustion in a top-lit updraft stove (TLUD) and low temperature anaerobic charring in a retort (RET), the latter of which was heated by the gas that evolved from the TLUD. The RET biochars exhibit a smaller adsorption capacity, higher molecular diversity and larger proportion of thermolabile materials, because of the lower degree of thermochemical alteration (DTA) and therefore limited formation of the microporous polycondensed aromatic matrix typical of the TLUD biochars. Multivariate statistics showed that DTA, not feedstock composition, controls biochar organic chemistry. The TLUD biochars might be better candidates for soil amendment because of their adsorption capacities and will probably exert a more prolonged effect because of their chemical stability. The cross-comparison of the methods showed the complementarity of especially elemental analysis, GACS, thermogravimetry, hypy and pyrolysis-GC-MS.

*Keywords:* biochar, characterisation, Costa Rica, rural communities

---

## 1 Introduction

Biochar is thermally modified biomass, such as charcoal obtained from urban, industrial, agricultural or forestry residues, destined for soil amendment. Applications using biochar are increasingly adopted to enhance soil productivity and promote carbon (C) sequestration (e.g. Lehmann *et al.*, 2006; Fowles, 2007; Laird, 2008; Sohi *et al.*, 2008; Kookana *et al.*, 2011). In soil, biochar acts as a sponge-like material enhancing water and nutrient retention, reducing erodibility, providing a habitat for soil microorganisms while it also has a liming effect as a result of its ash content. These properties often culminate into improved soil productivity and reduced loss of topsoil in agricultural land (Carrión & Badal, 2004; Jeffery *et al.*, 2011; Lehmann *et al.*, 2011). The effects

* Corresponding author
Email: joeri@samage.net

of biochar are reportedly stronger in soils with low adsorption capacity and pH, and in poor tropical soils with fast organic matter turnover rates (Glaser *et al.*, 2002; Jien & Wang, 2013). However, these effects are diverse, depending not only on feedstock type and production conditions (e.g. temperature) (Bruun *et al.*, 2008; Peng *et al.*, 2011; Preston & Schmidt, 2006; Zimmerman, 2010; Zhao *et al.*, 2013), but also on the targeted crop and soil properties (Masiello, 2004; Haefele *et al.*, 2011), giving rise to the need for optimising biochar production conditions to the specific requirements of each soil type and land use (Abiven *et al.*, 2014). Obviously, a thorough analysis of the properties of such biochars is a prerequisite to evaluate their suitability for different soil amendment applications.

Among the non-industrial biochar manufacturing systems, often used in rural areas and small farms, is the "Jolly Roger Oven", which consists of two adapted oil drums, the bottom drum of which is the top-lit updraft (TLUD) gasifier where intense burning converts the feedstock into charcoal (Kearns, 2012; McLaughlin & Clayton, 2012). The top drum, which is sealed, is heated by the exhaust rising from the TLUD and therefore acts like a retort (RET) oven in which the biomass is charred under oxygen-depleted conditions. In Costa Rica, several field studies, mostly in banana, coffee and cacao agroforestry contexts have demonstrated the potential of biochar to enhance soil properties and plant growth. For example, cacao biochar improved cacao seed production, especially when biochar was combined with a fertiliser such as poultry manure (Acosta Buitrago, 2013; Hojah da Silva, 2013). These studies have been performed with the focus on socio-economic benefits for indigenous communities, who sell the biochar produced in essentially smoke-free cook-stoves (Schultz, 2013). The directly appreciable utility of cleaner cookstoves, for biochar production, is an essential ingredient of small-scale biochar applications in rural areas where migrating Ngöbe-Buglé indigenous populations (with very basic household equipment) account for a significant part of the harvesting labour of the aforementioned crops. Current applications in Costa Rica include the use of coffee prunings, again in TLUD cookstoves, for biochar production (Scholz *et al.*, 2014). In addition, there is considerable interest in the potential alleviation of the damaging effects on cacao pods of the pathogens "mazorca negra" (*Phytophthora* sp.) and "molinia" (*Moniliophthora roreri*), by biochar application. The characterisation of the biochars applied had hitherto been limited to elemental composition and nutrient retention assays. Knowledge on the properties of these biochars, which are currently under field trial experimentation, is needed to optimise their implementation strategy.

This study addresses the properties of the organic matter of TLUD and RET biochars obtained from several important feedstocks in Costa Rica: bamboo stalk, cacao chaff and sawmill scrap. In addition, we studied TLUD biochars of orchard prunings and coconut husk. This is not a typical characterisation study of small amounts of biochar produced in the laboratory, but of biochars that are being created in rural areas of Costa Rica and are currently used as a soil amendment. We analysed these biochars for ash content, elemental composition, pH-$H_2O$, Fourier-transform infrared spectroscopy (FTIR), pyrolysis-gas chromatography-mass spectrometry (pyrolysis-GC-MS), hydrogen pyrolysis (hypy), thermogravimetry (TG) and gravimetric adsorption capacity scan (GACS). For the first time, hypy residues were analysed by pyrolysis-GC-MS to identify the nature of the residual volatiles. The assessment of general, molecular and adsorption properties is not only expected to shed light on the properties of these biochars and make cautious predictions on their likely effects on soil conditions, but also generate knowledge on biochar chemistry and on the complementarities of the methodologies applied.

## 2 Materials and methods

### 2.1 Feedstocks

Cacao chaff was obtained from fresh cacao pods at APPTA Fair Trade (Talamanca, Costa Rica). Bamboo stalk and coconut husk were obtained from their direct surroundings. The sawmill scrap was obtained from the nearby mill and the mixed orchard prunings were taken from the branches of isolated trees in a nearby banana field in the surroundings of the plant. The orchard prunings are composed mainly of, but not limited to, branches of *Inga edulis* (guabo) and *Erythrina* sp. (poró).

### 2.2 Biochar production

The Jolly Roger Oven setup is as described in detail by McLaughlin & Clayton (2012). For each biochar production, a 12" K-type probe was used to verify the temperature at various levels in the TLUD gasifier drum as the combustion front migrates through the barrel, which ranged between 500 and 800 °C over the course of one run with typical maximum heating temperature of 650 °C. The retort (RET) drum, set on top of the

TLUD, was sealed using local wet clay. The maximum temperature within a run was close to 450 °C (coinciding with the moment of maximum temperature in the TLUD). The coconut husk and orchard prunings biochars were produced in an Estufa Finca cookstove, which is a TLUD-system with a smaller volume than the TLUD from the Jolly Roger Oven. No RET biochars are available for these feedstocks. The Estufa Finca is used for cooking by rural communities and more than 4500 kg of the resultant biochar has already been sold and implemented into numerous agricultural fields. The temperature profile in the Estufa Finca is 550–650 °C. Biochars were homogenised and ball-milled to powder before analysis.

### 2.3 *Biochar characterisation*

#### 2.3.1 *Chemical characterisation and thermal analysis*

The elemental composition was determined in three different laboratories. Proportions of C, H, N, O and S were determined on a Vario MACRO cube CHNS elemental analyser (Elementar Analysensysteme GmbH, Hanau, Germany). In addition, C and N content were determined by dry combustion using a LECO analyser (model CHN-1000), and C content was measured on a Costech 4010 elemental analyser. Values for C content (three laboratories) were almost identical, with differences below 1 % and linear determination coefficients ($r^2$) between 0.996 and 0.999 ($P < 0.0001$). Nitrogen content showed larger differences (15 ± 11 %) because of the low N content of the biochars that were produced from materials other than cacao chaff, but were also strongly correlated ($r^2 = 0.998$, $P < 0.0001$). We only present data for the more complete elemental analysis using the Vario MACRO cube CHNS and those data are used for inter-methodological comparisons.

The pH in water was measured in 1:10 w:v in deionised water, after 60 minutes of equilibration in an orbital shaker.

Thermogravimetric TG and derivative (DTG) curves were obtained using an SDT Q600 instrument. The weight loss between 110 and 900 °C in a $N_2$ atmosphere was considered as the biochar volatile matter content, whereas the loss of weight at 900 °C after the introduction of air current was considered as the stable, thermo-resistant fraction or fixed C ($C_{fixed}$). The fraction of volatile matter with respect to the sum of volatile matter content and $C_{fixed}$ is considered as the thermolabile fraction of biochar ($C_{thermo}$, ash-free, dry matter basis; expressed as %). Ash content is the fraction of the sample material remaining after combustion of $C_{fixed}$ (Calvelo Pereira *et al.*, 2011). This method is also known as proximate analysis.

The inorganic C ($C_{inorg}$) content of the biochars was estimated by measuring the weight loss associated with the endothermic peak at ca. 600–850 °C (Calvelo Pereira *et al.*, 2011; Wang *et al.*, 2014). Weight loss and/or peak area changes were assessed as $CO_2$ mass loss but corrected to calculate $C_{inorg}$.

#### 2.3.2 *Gravimetric Adsorption Capacity Scan (GACS)*

The GACS adsorption test was applied using a custom-built modified TG analyser as described by McLaughlin (2010). Briefly, biochar is placed in 1.5 mL wire basket, suspended below an analytical balance in a heated $N_2$-purged chamber, with a thermocouple located above the sample inside the wire basket. The balance is tared at 100 °C, after which the chamber is heated to 300 °C at 13.3 °C min$^{-1}$. At that point, the purge gas is replaced by 1,1,1,2-tetra-fluoro-ethane (a refrigerant known as R134a) and the chamber allowed to cool to ambient temperature at 4 °C min$^{-1}$. The weight gain of the sample during cooling, caused by adsorption of R134a, is a measure of the adsorption capacity.

#### 2.3.3 *Fourier-Transform Infrared spectroscopy (FTIR)*

Attenuated total reflectance FTIR was performed using a Gladi-ATR (Pike Technologies) spectrometer, scanning in the 4000–400 cm$^{-1}$ region. The weak signal implied that adequate baseline subtraction was not always feasible and that the FTIR spectra could not be reliably quantified.

#### 2.3.4 *Pyrolysis-GC-MS*

For pyrolysis-GC-MS, biochar samples were pyrolysed at 750 °C for 10 seconds, with a heating rate of 10 °C ms$^{-1}$. The relatively high analytical pyrolysis temperature proved the most suitable temperature for biochar analysis using pyrolysis-GC-MS (Kaal *et al.*, 2009). Briefly, 1–1.5 mg of sample was placed in fire-polished quartz tubes with quartz wool on both ends and pyrolysed using a resistive heating Pt-filament CDS Pyroprobe 5250. The pyrolysis products were transferred into a 6890N gas chromatograph (Agilent Technologies) by He flow (1 mL min$^{-1}$), separated on a HP-5MS polysiloxane-based column (temperature program 60–325 °C at 20 °C min$^{-1}$, 5 min dwell time) and identified using an Agilent 5975B mass spectrometer (70 eV electron ionization). The peaks in the pyrolysis chromatograms were identified on the basis of pyrolysis-GC-MS literature of biochar and the NIST ’05 library. This resulted in a list of 108 pyrolysis products, which was reduced to 101 compounds by combining unresolved isomers and eliminating potential contamination products (phthalates and styrenes). The 101 compounds were quantified using their primary ion fragments (*m/z*). Relative proportions of each pyrolysis product are expressed as percentage of total quantified peak area (TQPA). TQPA is a rough measure of signal intensity and therefore sample pyrolysability (Kaal *et al.*, 2009).

### 2.3.5 *Hydrogen pyrolysis (hypy)*

The proportion of stable polycyclic aromatic carbon (SPAC) of biochar, expressed as the percentage of total C ($C_t$), was determined by hypy following (Meredith *et al.*, 2012; Wurster *et al.*, 2012; McBeath *et al.*, 2015). Briefly, a Mo catalyst was added to ca. 100 mg of sample using an aqueous/methanol solution of ammonium dioxydithiomolybdate. Catalyst loaded samples were placed in the hypy reactor, which was pressurised with $H_2$ to 150 bar with a purge gas flow of 5 L min$^{-1}$. The reactor was heated initially at 300 °C min$^{-1}$ to 250 °C and then at 8 °C min$^{-1}$ to 550 °C, and then held at maximum temperature for 2 min. The $C_t$ content of the sample residue after hypy was determined by elemental analysis using a Costech 4010. The error associated with the estimation of SPAC content is better than ±4 % of the measured value (McBeath *et al.*, 2015). The hypy residues were analysed by pyrolysis-GC-MS (see above) to assess the presence and properties of volatile components.

### 2.4 *Data evaluation*

Differences between TLUD and RET biochars were tested using One-way ANOVA (SPSS 20), and considered significant when the probability of the null hypothesis (no difference between the groups) was below 0.05. Relationships between continuous variables were tested by Pearson linear determination coefficients ($r^2$). A Principal Component Analysis (PCA) was performed in order to illustrate the relative similarities and differences between selected parameters, after standardisation (Z-scores) of the data.

## 3 Results

### 3.1 *Elemental composition and pH*

The $C_t$ content of the biochars ranged between 550 and 870 g kg$^{-1}$ (Table 1). The cacao chaff biochars (RET and TLUD) have lowest $C_t$ content, also if the $C_t$ content is corrected for ash (not shown). The (atomic) ratios from elemental analysis are corrected for the minor contents of $C_{inorg}$ as determined by thermogravimetry (see section 3.2). Cacao chaff biochars have the highest N content and lowest $C_{org}$/N ratio, reflecting the nature of this protein-rich feedstock. The N content and $C_{org}$/N ratios of the other biochars are < 10 g kg$^{-1}$ and > 115, respectively (Table 1). The S content was below the detection limit except for the biochars from cacao chaff and bamboo stalk. The H/$C_{org}$ ratio ranged from 0.21 to 0.88 and the O/$C_{org}$ ratio from 0.03 to 0.18. They are higher for the RET biochars than for the TLUD biochars ($P<0.01$ and $P<0.05$, respectively). The Van Krevelen plot (Fig. 1) shows that the cacao chaff TLUD biochar is closer to the range of the RET biochars.

The pH in water of the biochars ranged from 8.3 to 10.5 (Table 2). TLUD biochars seem to have slightly higher pH (9.9–10.5) than the RET biochars (8.3–10.0), but this difference is insignificant ($P=0.054$).

**Table 1:** *Elemental composition of the biochars analysed (all values on dry matter basis)*

| *Type* | *Feedstock* | $C_t$ [g kg$^{-1}$] | $C_{org}$ [g kg$^{-1}$] | *N* [g kg$^{-1}$] | *H* [g kg$^{-1}$] | *S* [g kg$^{-1}$] | *O* [g kg$^{-1}$] | $C_{org}$/N [–]* | O/$C_{org}$ [–]* | H/$C_{org}$ [–]* |
|---|---|---|---|---|---|---|---|---|---|---|
| Retort | Cacao chaff | 552.5 | 549.9 | 48.9 | 40.8 | 2.3 | 132.6 | 13.1 | 0.181 | 0.884 |
| | Sawmill scrap | 741.3 | 734.7 | 6.2 | 32.7 | < DL† | 103.9 | 138.9 | 0.106 | 0.531 |
| | Bamboo stalk | 668.1 | 668.1 | 6.8 | 40.7 | 1.8 | 141.6 | 115.3 | 0.159 | 0.726 |
| TLUD | Cacao chaff | 571.6 | 570.0 | 24.2 | 24.0 | 3.6 | 93.4 | 27.4 | 0.123 | 0.501 |
| | Sawmill scrap | 836.6 | 833.8 | 8.2 | 14.9 | < DL† | 49.8 | 118.0 | 0.045 | 0.213 |
| | Bamboo stalk | 705.4 | 705.4 | 5.5 | 17.9 | 0.5 | 32.7 | 150.4 | 0.035 | 0.303 |
| | Coconut husk | 867.4 | 865.4 | 7.0 | 19.7 | < DL† | 52.9 | 143.8 | 0.046 | 0.272 |
| | Orchard prunings | 855.6 | 850.5 | 7.9 | 17.8 | < DL† | 35.6 | 125.3 | 0.031 | 0.250 |

* Atomic ratio; † Below detection limit

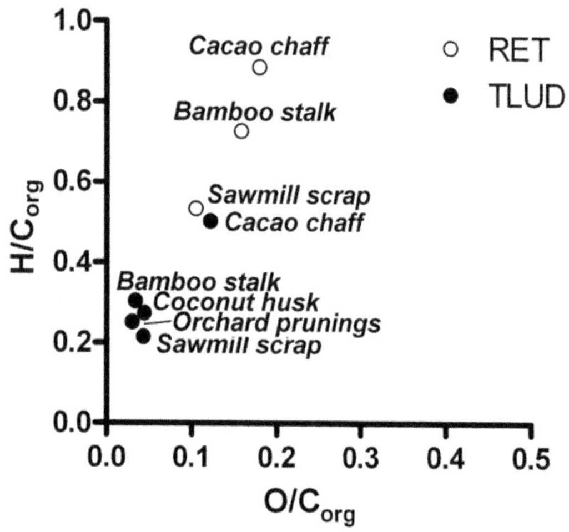


**Fig. 1:** *Van Krevelen plot of the biochars analysed (atomic ratios of O/$C_{org}$ and H/$C_{org}$).*

## 3.2 *Thermogravimetry*

The ash content (on a dry weight basis) ranged from 50 to 285 g $kg^{-1}$ (Table 2). Highest ash contents were measured for the cacao chaff biochars (TLUD and RET). There are no significant differences between the ash contents of the TLUD and RET biochars. The RET biochars (35 ± 10 %) have a larger proportion of thermolabile C ($C_{thermo}$) than the TLUD biochars (18 ± 8 %) ($P < 0.05$). The proportion of $C_{inorg}$, which can be recognised from a peak in the differential mass loss curve around 650 °C (Fig. 2), was up to 6.6 g $kg^{-1}$ with no apparent differences between feedstocks or production types. The differential mass loss curve obtained of the RET biochars (Fig. 2a) shows that the cacao biochar produced a broad (composed) peak between 200 and 500 °C. The mass loss of the bamboo RET biochar is concentrated in a very broad band in the range 350–600 °C, while the sawmill scrap RET biochar shows minor and gradually increasing weight loss in the same region. The differential mass loss curve of the TLUD biochars (Fig. 2b) shows few peaks, which is indicative of their thermal stability. In fact, the most prominent peak at ca. 650 °C corresponds to $C_{inorg}$.

## 3.3 *GACS adsorption capacity*

The GACS test showed that the RET samples have much lower adsorption capacity for R134a (20–45 g $kg^{-1}$) than the TLUD equivalents (50–100 g $kg^{-1}$) ($n$=6; $P < 0.05$) (Table 2). The sawmill scrap TLUD biochar has the highest GACS value, showing an adsorption isotherm very similar to that of the CC#14 reference sample (McLaughlin, 2010), whereas the sawmill scrap RET biochar has the highest adsorption capacity of the RET biochars, which can be explained by the generally larger surface areas of biochars from wood feedstock (Zhao *et al.*, 2013). From comparison with the GACS analyses of a reference set of pine- and oak-derived biochars produced at 350, 450 and 550 °C, in general the RET biochars show adsorption behaviour similar to that of the 450 °C chars, except for the Bamboo biochar which is more similar to pine-350 °C, while the TLUD isotherms more closely resemble that of the 550 °C biochars (McLaughlin, 2010).

Previous comparative tests of TLUD biochars using other measures of adsorption capacity showed that the GACS test is linearly related to e.g. the $N_2$-BET surface area (McLaughlin, 2010). The advantage of the GACS test using R134a over BET-surface analysis is that the latter requires more costly equipment and that R134a is larger and heavier than $N_2$ or $CO_2$, more relevant to the adsorption of soil constituents such as nutrients and contaminants.

## 3.4 *Fourier-Transform Infrared spectroscopy (FTIR)*

FTIR absorbance spectra are presented in Fig. 3. Several TLUD samples (sawmill scrap, bamboo stalk, coconut husk, orchard prunings) showed very weak absorbance, indicative of the virtual absence of any functional group other than hybridised C bonds in polynuclear aromatic clusters. The cacao chaff TLUD biochar showed slightly better signal, mainly of the O-H stretch (3400 $cm^{-1}$), C-O stretch (1030 $cm^{-1}$), phenolic O-H bend (1430 $cm^{-1}$) and aromatic ring stretch at ca. 1610 $cm^{-1}$, with minor aromatic bands at 1720 $cm^{-1}$. These groups can all be traced back to aromatic groups with partial substitution by hydroxyl groups, typical of high-temperature charcoal (Sharma *et al.*, 2004; Keiluweit *et al.*, 2010; Rutherford *et al.*, 2012).

Qualitatively, the RET samples showed some more functional group diversity. Especially the cacao chaff sample exhibited significant peaks of symmetric and asymmetric methyl ($CH_3$) and methylene ($CH_2$) stretches at 2920 and 2850 $cm^{-1}$, traces of which can also be recognised in the bamboo RET sample. Furthermore, the band at 1030 $cm^{-1}$ (aromatic C-O) is relatively narrow in the cacao RET and TLUD biochar and shifted towards 1000 $cm^{-1}$, which most likely corresponds to thermally altered protein (C-N stretch; Coates, 2000), and a slight broadening of the band at 1650 $cm^{-1}$ might reflect overlap between aromatic C=C and C=N bands (cf. Apaydın-Varol & Eren Pütün, 2012). All RET biochars showed significant absorbance of the aforementioned O-H stretch, C-O stretch, phenolic O-H bend and the aromatic ring stretch. These results show that the biochars consist of a combination of aromatic and phenolic moieties, with traces of aliphatic groups in methyl and methylene groups.

**Table 2:** *Proximate analysis, pH, adsorption capacity (GACS) and stable polycondensed aromatic carbon (SPAC, from hypy analysis) (all values on dry matter basis).*

| *Type* | *Feedstock* | *ash* [g kg$^{-1}$] | $C_{inorg}$ [g kg$^{-1}$] | $C_{volatile}$ [g kg$^{-1}$] | $C_{fixed}$ [g kg$^{-1}$] | $C_{thermo}$* [–] | *pH-$H_2O$* [–] | *GACS* [g kg$^{-1}$] |
|---|---|---|---|---|---|---|---|---|
| Retort | Cacao chaff | 223 | 2.7 | 350 | 428 | 44.98 | 10.03 | 20.71 |
| | Sawmill scrap | 116 | 6.6 | 223 | 661 | 25.19 | 9.43 | 41.4 |
| | Bamboo stalk | 141 | < DL† | 304 | 555 | 35.35 | 8.33 | 26.85 |
| TLUD | Cacao chaff | 283 | 1.6 | 237 | 480 | 33.09 | 10.15 | N/A |
| | Sawmill scrap | 91 | 2.8 | 140 | 769 | 15.44 | 10.49 | 91.18 |
| | Bamboo stalk | 238 | < DL† | 120 | 642 | 15.77 | 10.13 | 56.22 |
| | Coconut husk | 53 | 1.9 | 121 | 826 | 12.8 | 10.33 | 70.62 |
| | Orchard prunings | 83 | 5.1 | 132 | 785 | 14.38 | 9.87 | 50.85 |

* Thermolabile C (volatile matter / (volatile matter + Fixed C)*100 %); † Below detection limit

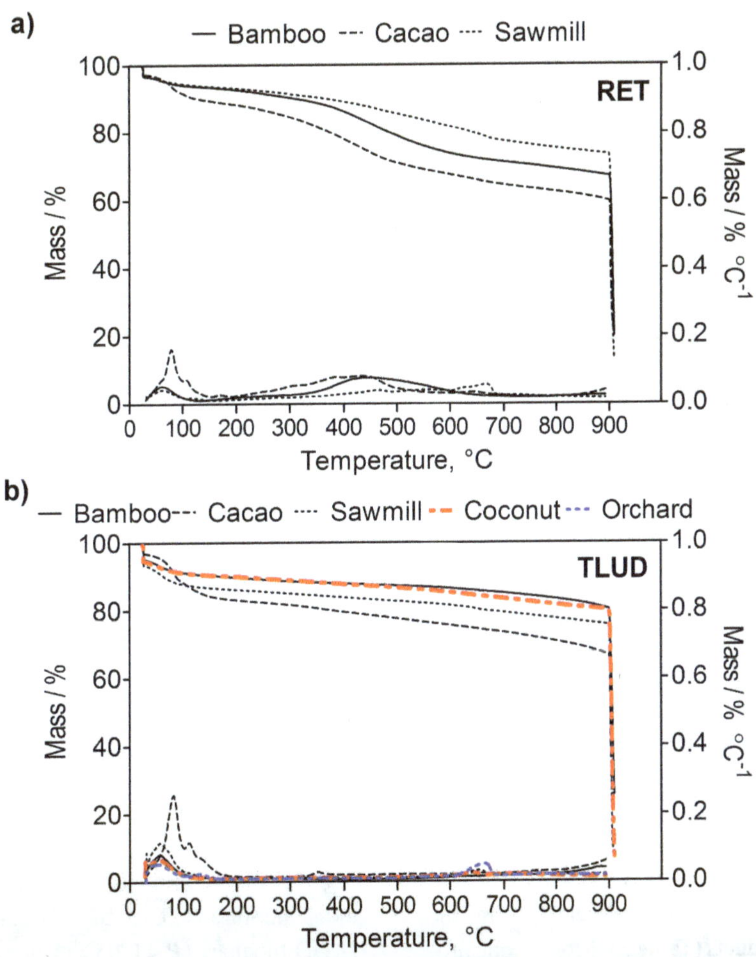


**Fig. 2:** *Proximate of the biochars analysed: thermogravimetric (TG; mass change, %) and derivative (DTG; mass change with temperature, % °C$^{-1}$) curves of retort (RET) (a) and TLUD (b) biochars.*

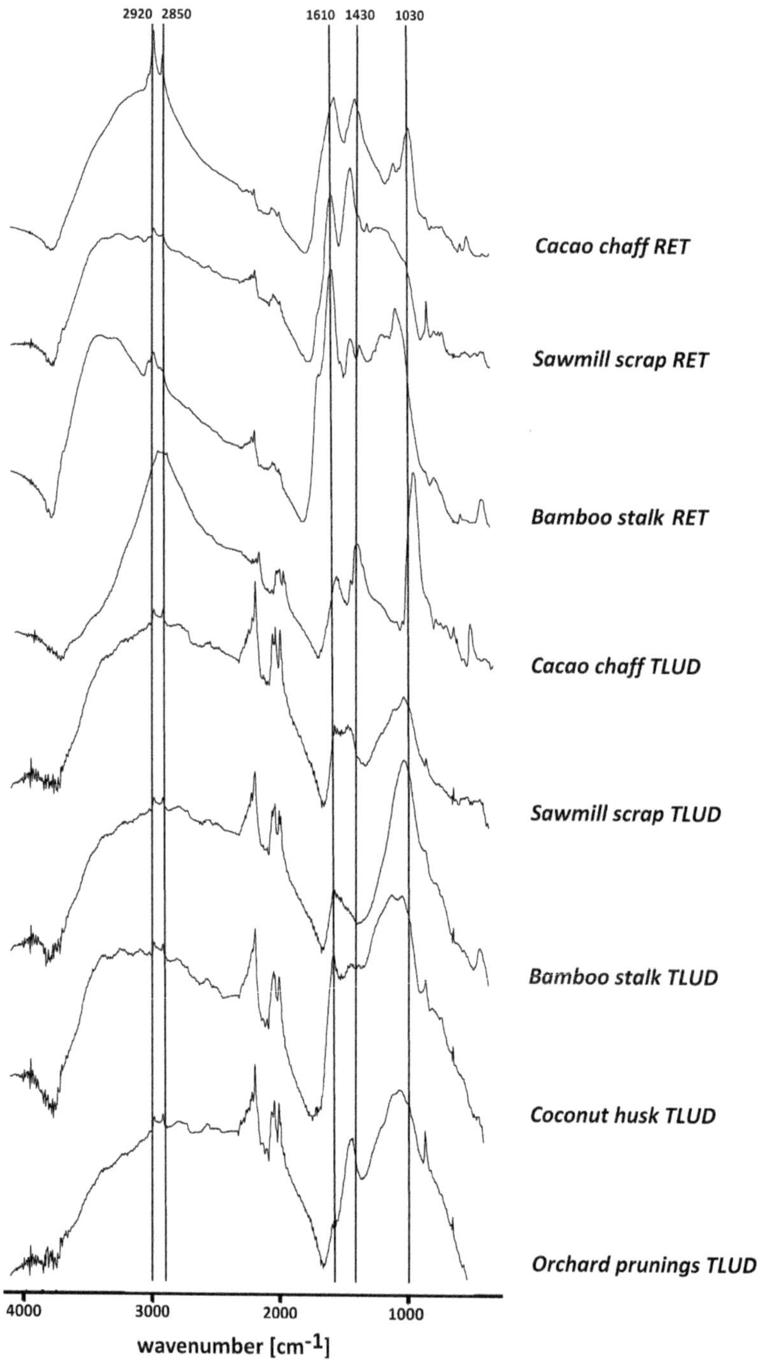


**Fig. 3:** *FTIR spectroscopy spectra of the biochars analysed. Wavenumbers of selected bands are indicated (see text).*

### *3.5 Pyrolysis-GC-MS*

Example pyrolysis chromatograms are shown in Figure 4a. In all samples, aromatic hydrocarbons are the dominant pyrolysis products (Table 3). Monocyclic aromatics hydrocarbons (MAHs) with one to three alkyl C substitutions account for $59 \pm 25\,\%$ of TQPA, and are more abundant in TLUD ($67 \pm 29\,\%$) than in RET pyrolyzates ($47 \pm 14\,\%$) ($P < 0.05$). Polycyclic aromatic hydrocarbons (PAHs) account for $9 \pm 5\,\%$. The sum of MAHs, PAHs and benzonitriles, which represent the more intensely charred fraction of Black C (Kaal *et al.*, 2009), account for $70 \pm 21\,\%$. The RET biochars show larger proportions of indenes (indene and $C_1$-indenes) and benzofurans ($C_1$- and $C_2$-benzofurans), than the TLUD biochars ($P < 0.05$). A homologous series of n-alkane/n-alkene pairs was detected in all samples except for the sawmill scrap TLUD biochar ($11 \pm 9\,\%$). Other compounds that are based on methylene chain units, i.e. *n*-alkanoic acids and *n*-alkanoic acid methyl esters ($3.5 \pm 7.0\,\%$), are particularly abundant in the TLUD biochar from bamboo stalk feedstock (20 %).

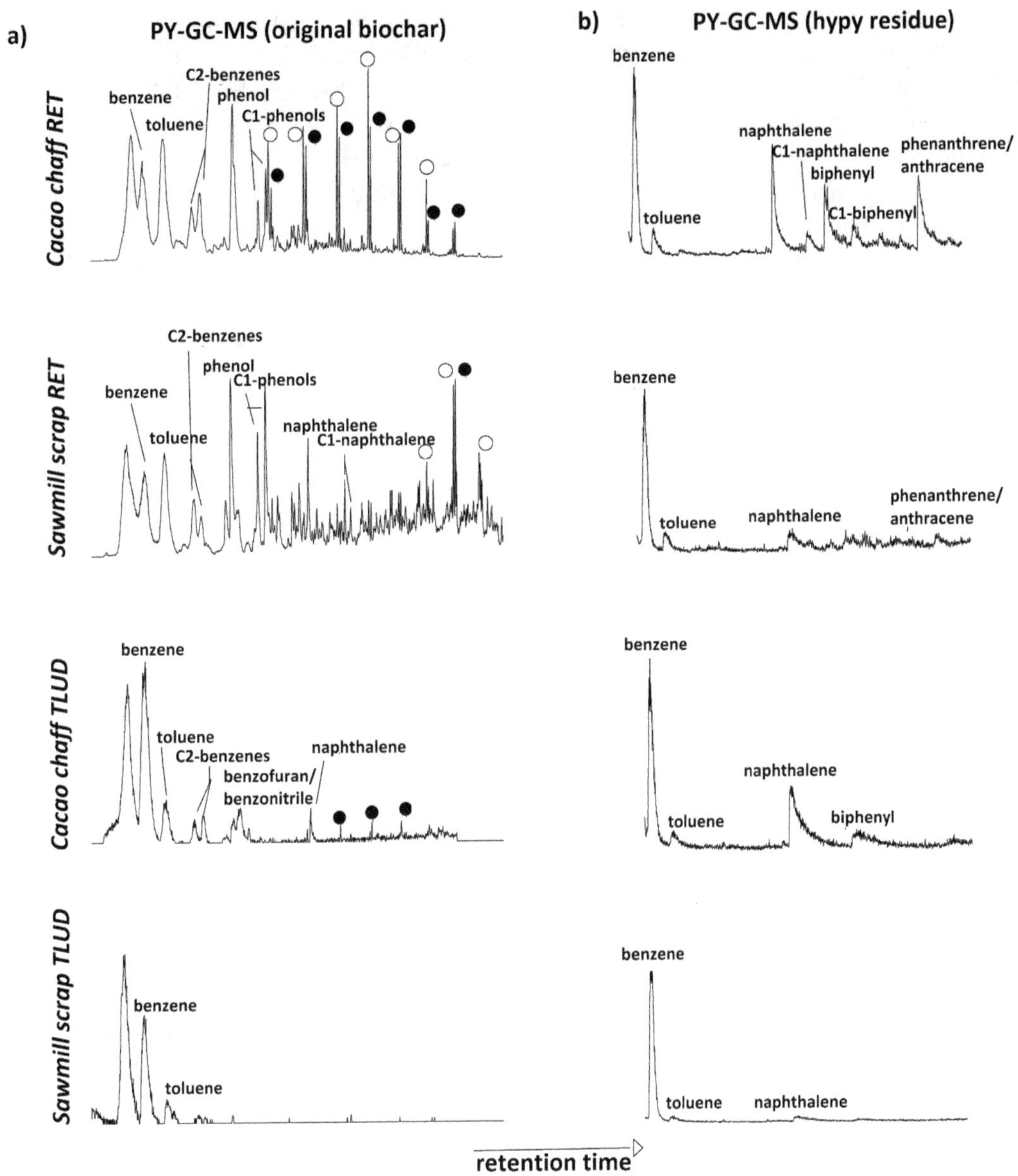


**Fig. 4:** *Total ion current chromatograms from pyrolysis-GC-MS of selected biochars. a) Original biochars. b) Hypy residues. Closed circles = n-alkanes, open circles = n-alkenes.*

**Table 3:** *Pyrolysis-GC-MS product groups from the biochars analysed and their relative proportions (%) of total quantified peak area (TQPA). TQPA and the benzene/toluene ratio are presented as well.*

| *Type* | *Feedstock* | *MAH* [%] | *PAH* [%] | *ALK* [%] | *FA* [%] | *PHE** [%] | *NC* [%] | *CARB* [%] | *LIG* [%] | *OTHER* [%] | *TQPA* [–] | *B/T* [–] |
|---|---|---|---|---|---|---|---|---|---|---|---|---|
| Retort | Cacao chaff | 38.6 | 4.3 | 25.6 | 0.6 | 12.9 | 14.5 | 0.8 | 0.3 | 2.4 | 2.5E+07 | 0.33 |
| | Sawmill scrap | 38.5 | 11.3 | 9.6 | 4.7 | 23.6 | 3.3 | 2.1 | 2.2 | 4.7 | 2.8E+06 | 0.54 |
| | Bamboo stalk | 63.0 | 9.9 | 2.1 | 2.1 | 9.9 | 3.7 | 2.9 | 4.0 | 2.3 | 3.9E+06 | 0.94 |
| TLUD | Cacao chaff | 81.2 | 5.7 | 4.8 | 0.0 | 0.0 | 4.0 | 0.0 | 0.0 | 4.4 | 2.5E+05 | 3.58 |
| | Sawmill scrap | 100.0 | 0.0 | 0.0 | 0.0 | 0.0 | 0.0 | 0.0 | 0.0 | 0.0 | 4.9E+04 | 3.10 |
| | Bamboo stalk | 22.4 | 17.9 | 15.9 | 20.3 | 10.8 | 2.7 | 2.6 | 3.3 | 4.1 | 5.6E+05 | 1.25 |
| | Coconut husk | 61.6 | 10.9 | 18.2 | 0.1 | 1.0 | 1.8 | 3.2 | 0.0 | 3.3 | 6.6E+05 | 2.48 |
| | Orchard prunings | 70.1 | 8.8 | 15.4 | 0.0 | 0.0 | 3.1 | 0.0 | 0.0 | 2.5 | 2.3E+05 | 1.51 |

MAH = monocyclic aromatic hydrocarbon, PAH = polycyclic aromatic hydrocarbon, ALK = n-alkanes and n-alkenes, FA = n-alkanoic acids and n-alkanoic acid methyl esters, PHE = phenols, NC = N containing compounds, CARB = carbohydrate products, LIG = lignin products, OTHER = other compounds

Pyrolysis products of lignin, which can be recognised from the methoxylic groups in guaiacols and syringols, account for 1.2 ± 1.7 %. The RET biochars have significantly larger relative proportions of guaiacols (guaiacol, 4-ethylguaiacol and $C_3$-guaiacol). Among the RET biochars, this intact lignin is more abundant in the biochars from lignocellulosic feedstock (bamboo, 4.0 %; sawmill scrap 2.2 %) than in the cacao chaff (0.3 %). In the TLUD biochars, lignin could only be detected in the bamboo biochar (3.3 %). Considering that the methoxylic group of lignin is completely eliminated at 400–450 °C even under anaerobic conditions (Kaal *et al.*, 2012a), the RET biochars were at least partially subjected to highest treatment temperatures (HTT) below that range. The presence of lignin in the bamboo stalk TLUD biochar suggests that some of the biomass was incompletely charred, as the HTT of that sample was well-above 600 °C.

Phenols account for 7.3 ± 8.6 %, and are more abundant in the RET biochars ($P < 0.05$). Nonetheless, the bamboo TLUD biochar also produced a significant proportion of phenols. Phenols can have multiple origins, including uncharred lignin and protein. Therefore, in the RET biochars, their precursor is unknown. Considering the presence of methoxyphenols in the bamboo TLUD biochar, the phenols probably originate from partially rearranged lignin as well (Kaal *et al.*, 2012b).

Nitrogen-containing compounds (4.1 ± 4.4 %) are more abundant in the cacao chaff biochars than in the other biochars, which reflects the abundance of protein in this feedstock. Benzonitrile is the unambiguous principal N-containing compound of strongly charred biomass (Kaal *et al.*, 2009). The $C_1$-benzonitriles and indole, on the other hand, represent more weakly altered N-containing precursors, and they are enriched in the RET biochars. The same N-containing compounds were detected in the pyrolysates of chicken manure biochar (Schnitzer *et al.*, 2007).

Recognisable products of carbohydrates (1.5 ± 1.4 %) originate from charred polysaccharides such as cellulose and hemicellulose. Their low abundance is indicative of the thorough rearrangement of polysaccharides, which is generally known to be the most thermolabile of the main constituents of plant biomass. This also suggests that there are no uncarbonised cores in the biochars analysed (Skjemstad *et al.*, 1996). The RET biochars have significantly larger relative proportions of acetylfuran ($P<0.05$).

The benzene/toluene (B/T) ratio (Table 3) represents the degree of polycondensation of Black C samples, with high values corresponding to a higher abundance of polycondensed aromatic clusters. The TLUD samples consist of more ($P<0.05$) polycondensed moieties represented by high B/T ratios (2.4 ± 1.0), even though the bamboo TLUD sample has a B/T ratio of 1.3, which is closer to the range of the RET biochar B/T ratios (0.6 ± 0.3). The naphthalene/methylnaphthalenes (N/$C_1$ N) ratio (Kaal *et al.*, 2009) cannot be applied as two TLUD biochars did not produce naphthalene (sawmill scrap) or methylnaphthalenes (cacao chaff).

The measure of signal intensity TQPA, varied considerably between 5 × 104 and 3 × 107 (Table 3). Retort biochars have significantly ($P < 0.001$) higher TQPA (3 × 106–3 × 107) than TLUD biochars (5 × 103–7 × 104), which indicates that the fraction of non-pyrolysable polycondensed aromatic clusters is higher in the TLUD biochars. The theoretical relation between TQPA and the percentage of mass loss between 110 °C and 750 °C by thermogravimetry was confirmed by linear ($r^2$=0.58, $P < 0.01$) and exponential ($r^2$=0.70, $P < 0.05$) correlations.

### *3.6 Hypy and pyrolysis-GC-MS of hypy residues*

The proportion of the organic C that remained after hypy treatment, i.e. stable polycyclic aromatic carbon (% SPAC), ranged between 16 and 50 % for the RET biochars and 90–100 % for the TLUD biochars (Table 4), being significantly higher in the latter ($P < 0.001$). The proportion of SPAC of the sawmill scrap RET biochar is much larger (50 %) than that of cacao chaff and bamboo stalk RET biochars (17–19 %).

For the first time, hypy residues were analysed by pyrolysis-GC-MS to identify the nature of the residual volatiles (Fig. 4b), even though this might reflect only a minor portion of the C present after hypy. Benzene was more abundant ($P < 0.05$) among the pyrolysis products of the hypy residues of TLUD biochars (11–43 % RET, 43–87 % TLUD), whereas $C_2$-benzenes were more abundant in the RET biochars ($P < 0.05$) (Table 4). The PAHs identified were naphthalene, methylnaphthalene, biphenyl, methylbiphenyl, phenanthrene/anthracene, fluorene, dihydroanthracene, phenylnaphthalene and pyrene. Traces of benzonitrile could be identified in the bamboo (1.7 %) and cacao (2.6 %) RET biochars only. Alkylbenzenes and alkylnaphthalenes were very scarce. No other compounds were detected. These results show that the hypy efficiently removed all labile components (carbohydrates, phenols and aliphatic structures) from the biochars. In comparison with the pyrolysate compositions before hypy treatment, after hypy the proportions of benzene and PAHs, in particular that of 3 and 4 ring PAHs, increased considerably. The

B/T ratio also increased 2–18-fold. Clearly, the more condensed nature of the TLUD biochars can still be recognised after hypy treatment.

## 4 Discussion

### *4.1 Inter-method comparison*

Figure 5a shows the Factor 1–Factor 2 plot of the loadings of the selection of parameters from the different methodologies: elemental analysis ($C_{org}$, $C_{org}$/N, H/$C_{org}$, O/$C_{org}$), pH-$H_2O$, GACS adsorption capacity, thermogravimetry (ash content, $C_{thermo}$), pyrolysis-GC-MS (∑(MAHs, PAHs, benzonitrile), ∑(lignin, carbohydrates, phenols), TQPA, B/T ratio) and hypy (% SPAC, B/T post-hypy). Principal component 1 (PC1) explains 57 % of total variance. High positive loadings (> 0.71, i.e. > 50 % of variance explained) are observed for H/$C_{org}$, O/$C_{org}$, $C_{thermo}$ , ∑(lignin, carbohydrates, phenols) and TQPA. This set of parameters reflects the abundance of labile components of the biochars. Strong negative loadings are observed for $C_{org}$, GACS adsorption capacity, B/T (pre- and post-hypy), ∑(MAHs, PAHs, benzonitrile) and % SPAC. These parameters reflect the degree of aromatisation, aromatic polycondensation and surface area. Clearly, PC1 is indicative of the degree of thermochemical alteration (DTA), with negative values corresponding to more strongly altered biomass. With increasing DTA, there is a progressive conversion into a microporous network of polyaromatic clusters, independent of the feedstock used. The rationale for the link between this process and the outcome of the methodologies applied is as follows: (1) with increasing DTA, there is a progressive elimination of functional groups and accumulation of the C-dominated aromatic clusters, which explains the positive loadings of H/$C_{org}$/N and O/$C_{org}$ and negative loading of $C_{org}$ and, qualitatively, the elimination of non-aromatic bands in the FTIR spectra. (2) Obviously, with increasing DTA, the thermal stability measured by thermogravimetry increases, hence the positive loading of $C_{thermo}$. (3) The polyaromatic clusters that accumulate with increasing DTA are well-known to be responsible for the surface area of biochars, which explains the negative loading of the GACS adsorption capacity. Furthermore, more complete burnout of volatiles with increasing DTA will have contributed to the enhanced adsorption capacity. (4) With increasing DTA, the susceptibility of the biochar to analytical pyrolysis decreases, and therefore TQPA has a positive loading. The B/T ratio, on the other hand, is inverse to the abundance of alkyl-cross bridges between aromatic groups, which are progressively eliminated with increasing DTA, hence B/T has a negative loading on PC1. Considering the large differences in DTA between the biochars, it is not surprising that most variables are related to DTA, which also turned out to control most biochar properties in other characterisation studies (Zhao *et al.*, 2013, and references therein). Moreover, the degree of aromatic polycondensation, reflected by DTA, was the controlling factor in biochar stability after soil amendment (Kuzyakov *et al.*, 2014) and laboratory ageing experiments (Mašek *et al.*, 2013).

The main source of variation in DTA is the difference between the RET and TLUD biochars: the RET samples have positive scores on PC1 (1.5–5.0), while the TLUD biochars have negative scores (-0.1– -3.6) (Fig. 5b). The RET-derived biochars more closely resemble the nature of the feedstock than the TLUD-derived biochars, be-

**Table 4:** *Proportion of hydrogen pyrolysis (hypy)-resistant stable polycyclic aromatic carbon (SPAC) and the composition of the pyrolysates obtained PY-GC-MS of the hypy residue.*

| *Type* | *Feedstock* | *SPAC* [% $C_t$] | *MAH* [%] | *PAH* [%] | *BN* [%] | *B/T* [-] |
|---|---|---|---|---|---|---|
| Retort | Cacao chaff | 19.2 | 43.0 | 54.4 | 2.6 | 5.6 |
| | Sawmill scrap | 49.9 | 54.4 | 45.6 | 0.0 | 5.3 |
| | Bamboo stalk | 16.8 | 20.7 | 77.6 | 1.7 | 1.7 |
| TLUD | Cacao chaff | 97.6 | 53.1 | 46.9 | 0.0 | 6.7 |
| | Sawmill scrap | 103.2 | 86.1 | 13.9 | 0.0 | 22.1 |
| | Bamboo stalk | 89.6 | 69.2 | 30.8 | 0.0 | 5.7 |
| | Coconut husk | 91.8 | 72.7 | 27.2 | 0.1 | 13.2 |
| | Orchard prunings | 102.2 | 91.0 | 9.0 | 0.0 | 27.3 |

MAH= monocyclic aromatic hydrocarbon, PAH = polycyclic aromatic hydrocarbon, BN = benzonitrile, B/T = benzene/toluene ratio.

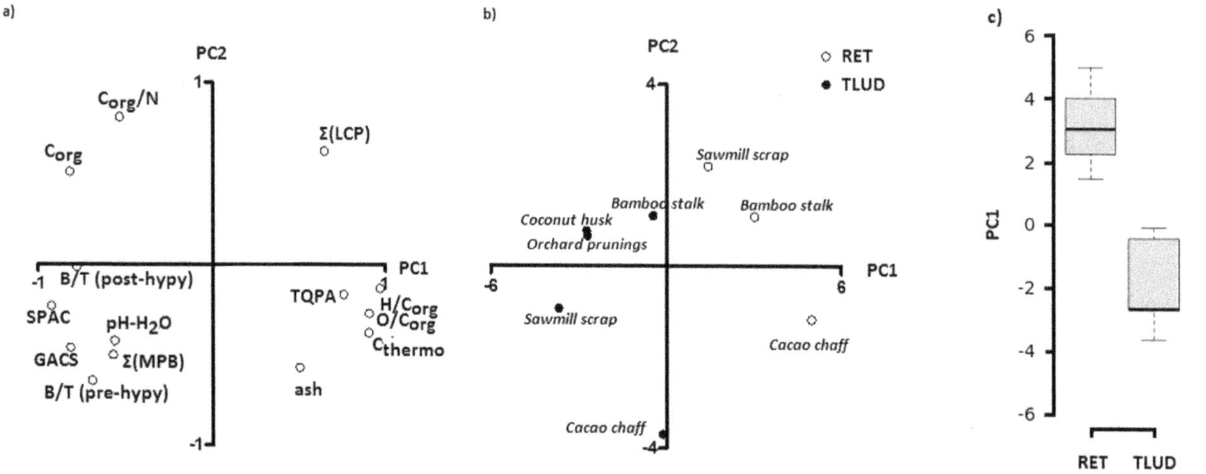


**Fig. 5:** *Principal component analysis of selected parameters. a) PC1 and PC2 loadings (B/T = benzene toluene ratio; before and after hypy); ∑(MPB) = sum monocyclic aromatic hydrocarbons, polycyclic aromatic hydrocarbons and benzonitrile; ∑(LCP) = sum lignin, carbohydrates and phenols; GACS = adsorption capacity; SPAC = stable polycyclic aromatic carbon; $C_{thermo}$= thermolabile C. b) PC1 and PC2 scores of the retort and TLUD biochars. c) Boxplot of the PC1 scores of the retort and TLUD biochars.*

cause the TLUD stove conditions induce a higher DTA to the feedstock. As such, using thermogravimetry, the RET biochars have larger proportions of $C_{thermo}$ and, using FTIR, the RET biochars show more extensive O-containing functional groups (mostly -OH and -$OCH_3$), which is corroborated by the presence of phenols from moderately charred lignin and methoxyphenols from non- and weakly-charred lignin in lignocellulose-derived biochars, and evidence of intact peptide material in the cacao chaff-derived RET biochar. On the other hand, the TLUD biochars are composed primarily of clusters of polycondensed aromatic structures, which explain their higher SPAC content, GACS adsorption capacity, B/T ratio and low $C_{thermo}$ and TQPA values. Apart from the difference in PC1 scores between RET and TLUD biochars, it is worth mentioning that the sawmill scrap RET biochar seems to have been subjected to higher DTA than the other RET biochars, as evidenced by the intermediate PC1 scores, H/$C_{org}$ and % SPAC levels.

PC2 (22 % of total variance) has a large positive loading for $C_{org}$/N and moderate positive loadings for $C_{org}$ and ∑(lignin, carbohydrates, phenols), whereas negative loadings correspond to ash content, GACS adsorption capacity and the B/T ratio. The cacao-derived biochars have the highest negative scores on PC2. In combination with the strong loadings of $C_{org}$/N, this suggests that PC2 reflects the influence of feedstock type, in particular its N content.

There are several observations that cannot be fully explained from DTA and the contrast between TLUD and RET biochars. Firstly, the pyrolysis-GC-MS pyrolysate of the bamboo TLUD biochar is a clear outlier as it contains traces of lignin and n-alkanoic acids, which are thermolabile above 400–450 °C (Kaal *et al.*, 2012a,b), yet this sample is predominantly composed of polycondensed aromatic structures, evidenced by the high GACS adsorption capacity and SPAC content. Hence, this biochar must contain a fraction of biomass that was incompletely charred, possibly isolated in a corner of the TLUD barrel or within larger fragments of bamboo feedstock. This minor proportion of weakly and non-charred biomass will cause strongly biased results by pyrolysis-GC-MS because of the different thermal stabilities (and therefore pyrolysabilities) of the two components. It is probably reflected in the corresponding DTG curve at 250 °C (Fig. 2b).

Secondly, the N content of the biochars is very low, with the exception of the cacao chaff biochars, which can be explained by their high protein content. As expected, the sum of N-containing pyrolysis products by pyrolysis-GC-MS is correlated to N content ($P < 0.001$) and inverse to the C/N ratio ($P < 0.05$), but also inverse to $C_{thermo}$ ($P < 0.05$), which might suggest that N functionalities in general and/or the cacao chaff biochars have a higher proportion of thermolabile material. The relatively thermolabile nature of even strongly carbonised N moieties (evidenced by benzonitrile) is in agreement with Sigua *et al.* (2014) who found that biochars from N-rich feedstocks (manures) were miner-alised faster than lignocellulose-derived biochars.

The methodologies applied allowed for a detailed assessment of the organic matter present in the biochars analysed. All evidence combined, the $H/C_{org}$ and SPAC seem to be the most reliable parameters to determine DTA of biochar, which is related directly to biochar stability. The pyrolysis-GC-MS results were in agreement with these parameters, but overestimate the abundance of labile biochar components for biochars with apparently heterogeneous HTT, which calls for cautious interpretation. Nonetheless, pyrolysis-GC-MS is the method that gives the most detailed information on the molecular properties of the biochars. Thermogravimetry allowed for correcting the elemental analysis data for $C_{inorg}$, which is essential to calculate the $H/C_{org}$. In addition, it provides an indication of the abundance of labile structures. FTIR data could not be quantified and in this case FTIR appeared less useful for biochar characterisation. The GACS adsorption test provides a useful and cost-efficient estimation of effective surface area.

### 4.2 *Detailed biochar characterisation*

*RET biochars*. The cacao chaff biochar has a high N content some of which corresponds to intact proteins, which suggests that it may have beneficial short-term impacts on soil productivity (Jeffery *et al*., 2011). Because of its high ash content it could be suitable for local application in cacao plantations, which require ash-rich biochars (Chepote *et al*., 2007; Hojah da Silva, 2013). However, this biochar has very low DTA and adsorption capacity, which implies that it will probably be degraded to a significant extent after implementation into soil and that it is probably of limited use for water and nutrient retention as well. The sawmill scrap biochar has the highest C and SPAC contents, the lowest proportion of $C_{thermo}$ and strongest GACS adsorption capacity, which implies that it is probably the most stable biochar after soil amendment and has a strong water and nutrient retention. The large proportion of $C_{thermo}$, small SPAC and low DTA suggest that the bamboo stalk biochar consists mainly of labile components which limits its potential as a long-term soil enhancement. This biochar has the lowest adsorption capacity, the lowest pH-$H_2O$ and high $C_{org}/N$ ratio, suggesting that this biochar is not efficient in liming or improvement of the adsorption complex neither.

*TLUD biochars*. The biochar from cacao chaff has a very large SPAC content (98 %) and a pyrolysis-GC-MS fingerprint typical of biochars with high DTA (dominated by MAHs and PAHS). By contrast, the relatively large proportion of $C_{thermo}$ (33 %) suggests that a significant fraction of this biochar might be susceptible to degradation, which implies that there remains uncertainty on its stability of this biochar. The N-rich nature of the feedstock is reflected by the low $C_{org}/N$ and abundance of benzonitrile. The TLUD biochars from sawmill scrap, coconut husk and mixed orchard prunings have high GACS adsorption capacity and consist almost completely of polyaromatic domains, with very little thermolabile material ($C_{thermo}$ 12–15 %, SPAC 92–103 %). These biochars may be expected to be resistant to degradation and increase a soil's adsorption capacity on the long-term. Finally, the high SPAC, $C_{org}$, B/T and adsorption capacity, in combination with the small proportion of $C_{thermo}$ (16 %), suggests that the bamboo biochar has high stability and adsorption capacity. However, this biochar also contains incompletely charred organic matter (aliphatic material and traces of lignin and carbohydrates). It has the highest $C_{org}/N$ ratio of the biochars analysed (150) and, following Haefele et al. (2011), such biochars should be preferably applied to poor and sandy soils. This would also apply to the other biochars, except for the cacao RET and TLUD biochars.

## 5 Conclusions

The methods applied here, and particularly the combination of GACS, thermogravimetry, elemental analysis, hypy and pyrolysis-GC-MS, are complementary as their cross-comparison allowed for a better understanding of the results from each of them individually. The retort biochars were subjected to a lower degree of thermochemical alteration (DTA) than the TLUD biochars. From the statistical assessment of all data by PCA, the properties that reflect DTA become more pronounced in the order: Cacao RET < Bamboo RET < Sawmill scrap RET < Cacao TLUD < Bamboo TLUD < Coconut TLUD < Orchard prunings TLUD < Sawmill scrap TLUD. In general, the TLUD biochars have greater stability and adsorption capacity, suggesting that their potential as a C sink and long-term sponge for water and nutrients exceed that of the RET biochars. On the other hand, and in addition to a limited retention capacity, the RET biochars contain labile material that might act as a nutrient source. In addition, the RET biochars have appeared excellent cooking fuels.

**Acknowledgements**

We thank Carmen Pérez Llaguno (Department of Soil Science, University of Santiago de Compostela) and Aida González Pazos (IR-Raman facility of the RIAIDT, University of Santiago de Compostela) for analytical services. We thank Walter Rodríguez for the facilities offered at the APPTA fair trade plant (Talamanca, Costa Rica), Gabriela Soto, Tamara Benjamin (CATIE) and Kate Selting (Seachar) for their generous support, and two anonymous reviewers for their valuable comments.

## References

Abiven, S., Schmidt, M. W. I. & Lehmann, J. (2014). Biochar by design. *Nature Geoscience*, 7, 326–327.

Acosta Buitrago, J. O. (2013). *Evaluación del uso de biochar y microorganismos benéficos en cacao (Theobroma cacao L.) y banano (Gros Michel AAA "Cocos") en sistemas agroforestales de Costa Rica*. Master's thesis, Centro Agronómico Tropical de Investigación y Enseñanza (CATIE), Turrialba, Costa Rica.

Apaydın-Varol, E. & Eren Pütün, A. (2012). Preparation and characterization of pyrolytic chars from different biomass samples. *Journal of Analytical and Applied Pyrolysis*, 98, 29–36.

Bruun, S., Jensen, E. S. & Jensen, L. S. (2008). Microbial mineralization and assimilation of black carbon: dependency on degree of thermal alteration. *Organic Geochemistry*, 39, 839–845.

Calvelo Pereira, R., Kaal, J., Camps-Arbestain, M., Pardo Lorenzo, R., Aitkenhead, W., Hedley, M., Macías, F., Hindmarsh, J. & Maciá-Agulló, J. A. (2011). Contribution to characterisation of biochar to estimate the labile fraction of carbon. *Organic Geochemistry*, 42, 1331–1342.

Carrión, I. & Badal, E. (2004). La presencia de hongos e insectos xilófagos en el carbón arqueológico: propuestas de interpretación. *In:* Avances en arqueometría : 2003. pp. 98–106, Universidad de Cádiz.

Chepote, R. E., Santana, S. D., Araujo, Q. R., Sodré, G. A., Reis, E. L., Pacheco, R. G., Marrocos, P. C. L., Serôdio, M. H. C. F., Menezes, A. A. & Valle, R. R. (2007). Aptidão agrícola e fertilidade de solos para a cultura do cacaueiro. *In:* Valle, R. R. (ed.), *Ciência, tecnologia e manejo do cacaueiro*. Ilhéus, Bahia, Brazil. Pp. 467.

Coates, J. (2000). Interpretation of Infrared Spectra, A Practical Approach. *In:* Meyers, R. A. (ed.), *Encyclopedia of Analytical Chemistry*. pp. 10815–10837, John Wiley & Sons, Ltd, Chichester.

Fowles, M. (2007). Black carbon sequestration as an alternative to bioenergy. *Biomass & Bioenergy*, 31, 426–432.

Glaser, B., Lehmann, J. & Zech, W. (2002). Ameliorating physical and chemical properties of highly weathered soils in the tropics with charcoal: A review. *Biology and Fertility of Soils*, 35, 219—230.

Haefele, S. M., Konboon, Y., Wongboon, W., Amarante, S., Maarifat, A. A., Pfeiffer, E. M. & Knoblauch, C. (2011). Effects and fate of biochar from rice residues in rice-based systems. *Field Crops Research*, 121, 430–440.

Hojah da Silva, J. (2013). *Impacto del uso de biocarbón sobre la calidad de suelos y producción de cacao (Theobroma cacao L.) en sistemas agroforestales, Reserva Indígena Bribri, Talamanca, Costa Rica*. Master's thesis, Centro Agronómico Tropical de Investigación y Enseñanza (CATIE), Turrialba, Costa Rica.

Jeffery, S., Verheijen, F. G. A., van der Velde, M. & Bastos, A. C. (2011). A quantitative review of the effects of biochar application to soils on crop productivity using meta-analysis. *Agriculture, Ecosystems & Environment*, 144, 175–187.

Jien, S.-H. & Wang, C.-S. (2013). Effects of biochar on soil properties and erosion potential in a highly weathered soil. *Catena*, 110, 225–233.

Kaal, J., Martínez Cortizas, A., Reyes, O. & Soliño, M. (2012a). Molecular characterization of Ulex europaeus biochar obtained from laboratory heat treatment experiments – A pyrolysis–GC/MS study. *Journal of Analytical and Applied Pyrolysis*, 95, 205–212.

Kaal, J., Nierop, K. G. J. & Martínez Cortizas, A. (2009). Characterisation of aged charcoal using a coil probe pyrolysis-GC/MS method optimised for black carbon. *Journal of Analytical and Applied Pyrolysis*, 85, 408–416.

Kaal, J., Schneider, M. P. W. & Schmidt, M. W. I. (2012b). Rapid molecular screening of black carbon (biochar) thermosequences obtained from chestnut wood and rice straw: A pyrolysis-GC/MS study. *Biomass & Bioenergy*, 45, 115–129.

Kearns, J. (2012). Gasifier and Retort Char System. Aqueous Solutions publishers, pp. 1–6. Available at: http://www.aqsolutions.org/images/2010/06/JRO-handbook.pdf (last accessed: 12.02.2016).

Keiluweit, M., Nico, P. S., Johnson, M. G. & Kleber, M. (2010). Dynamic molecular structure of plant biomass-derived black carbon compounds (biochar). *Environmental Science & Technology*, 44, 1247–1253.

Kookana, R. S., Sarmah, A. K., Van Zwieten, L., Krull, E. & Singh, B. (2011). Biochar application to soil: Agronomic and environmental benefits and unintended consequences. *Advances in Agronomy*, 112, 103–143.

Kuzyakov, Y., Bogomolova, I. & Glaser, B. (2014). Biochar stability in soil: Decomposition during eight years and transformation as assessed by compound-specific $^{14}C$ analysis. *Soil Biology & Biochemistry*, 70, 229–236.

Laird, D. A. (2008). The charcoal vision: A win-win-win scenario for simultaneously producing bioenergy, permanently sequestering carbon, while improving soil and water quality. *Agronomy Journal*, 100, 178–181.

Lehmann, J., Gaunt, J. & Rondon, M. (2006). Biochar sequestration in terrestrial ecosystems – a review. *Mitigation and Adaptation Strategies for Global Change*, 11, 403–427.

Lehmann, J., Rillig, M. C., Thies, J., Masiello, C. A., Hockaday, W. C. & Crowley, D. (2011). Biochar effects on soil biota: A review. *Soil Biology & Biochemistry*, 43, 1812–1836.

Mašek, O., Brownsort, P., Cross, A. & Sohi, S. (2013). Influence of production conditions on the yield and environmental stability of biochar. *Fuel*, 103, 151–155.

Masiello, C. A. (2004). New directions in black carbon organic geochemistry. *Marine Chemistry*, 92, 201–213.

McBeath, A. V., Wurster, C. M. & Bird, M. I. (2015). Influence of feedstock properties and pyrolysis conditions on biochar carbon stability as determined by hydrogen pyrolysis. *Biomass & Bioenergy*, 73, 155–173.

McLaughlin, H. (2010). Schenkel and Shenxue revisited - implications on char production and biochar properties. Appendix B: GACS assay for measuring adsorption capacity. Biochar2010 Conference, Ames, Iowa, June 2010.

McLaughlin, H. & Clayton, D. (2012). The "Jolly Roger Ovens" family of biochar-making devices. Available at: http://www.biochar-international.org/node/3064 (last accessed: 12.02.2016).

Meredith, W., Ascough, P. L., Bird, M. I., Large, D. J., Snape, C. E., Sun, Y. & Tilston, E. L. (2012). Assessment of hydropyrolysis as a method for the quantification of black carbon using standard reference materials. *Geochimica et Cosmochimica Acta*, 97, 131–147.

Peng, X., Ye, L. L., Wang, C. H., Zhou, H. & Sun, B. (2011). Temperature- and duration dependent rice straw-derived biochar: characteristics and its effects on soil properties of an Ultisol in southern China. *Soil & Tillage Research*, 112, 159–166.

Preston, C. M. & Schmidt, M. W. I. (2006). Black (pyrogenic) carbon: a synthesis of current knowledge and uncertainties with special consideration of boreal regions. *Biogeosciences*, 3, 397–420.

Rutherford, D. W., Wershaw, R. L., Rostad, C. E. & Kelly, C. N. (2012). Effect of formation conditions on biochars: Compositional and structural properties of cellulose, lignin, and pine biochars. *Biomass & Bioenergy*, 46, 693–701.

Schnitzer, M. I., Monreal, C. M., Jandl, G., Leinweber, P. & Fransham, P. B. (2007). The conversion of chicken manure to biooil by fast pyrolysis II. Analysis of chicken manure, biooils, and char by curie-point pyrolysis-gas chromatography/mass spectrometry (Cp Py-GC/MS). *Journal of Environmental Science and Health Part B*, 42, 79–95.

Scholz, S. M., Sembres, T., Roberts, K., Whitman, T., Wilson, K. & Lehmann, J. (2014). *Biochar Systems for Smallholders in Developing Countries: Leveraging Current Knowledge and Exploring Future Potential for Climate-Smart Agriculture*. World Bank Studies. Washington, DC: World Bank.

Schultz, S. (2013). Biochar cookstoves boost health for people and crops. National Geographic News, January 29. Available at: http://news.nationalgeographic.com/news/energy/2013/01/130129-biochar-clean-cookstoves/ (last accessed: 12.02.2016).

Sharma, R. K., Wooten, J. B., Baliga, V. L., Lin, X., Geoffrey Chan, W. & Hajaligol, M. R. (2004). Characterization of chars from pyrolysis of lignin. *Fuel*, 83, 1469–1482.

Sigua, G. C., Novak, J. M., Watts, D. W., Cantrell, K. B., Shumaker, P. D., Szögi, A. A. & Johnson, M. G. (2014). Carbon mineralization in two ultisols amended with different sources and particle sizes of pyrolyzed biochar. *Chemosphere*, 103, 313–321.

Skjemstad, J. O., Clarke, P., Taylor, J. A., Oades, J. M. & McClure, S. G. (1996). The chemistry and nature of protected carbon in soil. *Australian Journal of Soil Research*, 34, 251–271.

Sohi, S., Lopez-Capel, E., Krull, E. & Bol, R. (2008). Biochar, climate change and soil: A review to guide future research. CSIRO Land and Water Science Report 05/09.

Wang, T., Camps-Arbestain, M., Hedley, M., Singh, B. P., Calvelo-Pereira, R. & Wang, C. (2014). Determination of carbonate-C in biochars. *Soil Research*, 52 (5), 495–504.

Wurster, C. M., Lloyd, J., Goodrick, I., Saiz, G. & Bird, M. I. (2012). Quantifying the abundance and stable isotope composition of pyrogenic carbon using hydrogen pyrolysis. *Rapid Communications in Mass Spectrometry*, 26, 2690–2696.

Zhao, L., Cao, X., Mašek, O. & Zimmerman, A. (2013). Heterogeneity of biochar properties as a function of feedstock sources and production temperatures. *Journal of Hazardous Materials*, 256–257, 1–9.

Zimmerman, A. R. (2010). Abiotic and microbial oxidation of laboratory-produced black carbon (biochar). *Environmental Science & Technology*, 44, 1295–1301.

# 14

# Exploring the potential of cassava in promoting agricultural growth in Nigeria

Sanzidur Rahman *, Brodrick O. Awerije

*School of Geography, Earth and Environmental Sciences, University of Plymouth, Drake Circus, Plymouth, United Kingdom*

**Abstract**

Cassava is one of the major food crops in Nigeria, with multiple uses from human consumption to industrial applications. This study explores the potential of cassava in Nigerian agriculture based on a review of cassava development policies; performs a trend analysis of the cultivation area, production, productivity, and real price of cassava and other competing crops for the period 1961–2013; identifies the sources of growth in production; and examines the production constraints at the local level based on a survey of 315 farmers/processors and 105 marketers from Delta State. The results revealed that several policies and programmes were implemented to develop the cassava sector with mixed outcomes. Although cassava productivity grew at 1.5 % per annum (p.a.) during the post-structural adjustment programme period (1993–2013), its real price declined at a rate of 3.5 % p.a. The effect of yield is the main source of growth in production, contributing 76.4 % of the total growth followed by the area effect (28.2 %). The cassava sector is constrained by inadequate market infrastructure, processing facilities, and lack of information and unstable prices at the local level. The widespread diffusion of improved tropical manioc selection technologies and investments in market and marketing infrastructure, processing technologies, irrigation/water provision and information dissemination are recommended to enhance the potential of the cassava sector to support agricultural growth in Nigeria.

*Keywords:* cassava sector, constraints on cassava, Nigeria, sources of growth decomposition, trend analysis

## 1 Introduction

Although Nigeria is an oil-rich economy, agriculture remains an important economic sector that serves as a major source of raw materials, food and foreign exchange, and it employs over 70 % of the labour force (Liverpool-Tasie *et al.*, 2011). Cassava is an important crop that has great potential to support agricultural growth in Nigeria due to its wide range of uses from human consumption to industrial applications (CMP, 2006). Africa produces 40–50 % of the global cassava output (FAO, 2005; Nang'ayo *et al.*, 2007), and Nigeria is one of the leading producers (Nweke, 2004; Knipscheer *et al.*, 2007; Ayoade & Adeola, 2009). Demand for cassava derivatives, such as starch, gari (a type of processed cassava), tapioca, etc., have doubled over the last two decades (Nweke, 2004), and in recent years, the establishment of the Cassava Bread Development Fund (CBDF) and the Cassava Transformation Project by the Nigerian government have further raised hopes for improving the cassava sector (Natsa, 2014).

However, the current cassava yield is only 12.3 t ha$^{-1}$, whereas the potential yield is 28.0 t ha$^{-1}$ (Nkonya *et al.*, 2010), and the yield of improved varieties at research stations range from 13–40 t ha$^{-1}$ (Eke-okoro & Njoku, 2012). Furthermore, the results of the trends in production of cassava and other major competing crops in Nigeria are mixed. For example, both Nweke (2004) and Echebiri & Edaba (2008) noted that cassava production in Nigeria has increased substantially since the 1990s (i.e., 1986–2003), and they emphasized that cassava

* Corresponding author
School of Geography, Earth and Environmental Sciences, University of Plymouth, Drake Circus, Plymouth, PL4 8AA, United Kingdom; Email: srahman@plymouth.ac.uk

should be given priority because of its high starch content, which provides a greater proportion of energy to low income households than any of the other food crops in Africa. Liverpool-Tasie *et al.* (2011) noted that the production of millet, yam, maize and rice were stagnant or declining while cassava saw a modest increase during the period from 1994–2006 and claimed that food crop production in Nigeria is far below its potential with demand outstripping domestic supply. In contrast, Ojiako *et al.* (2007), examining production trends of cassava, sweet potato, yam, cocoyam and ginger in Nigeria for the period of 1961–2005, noted that the output of all crops increased substantially during the Structural Adjustment Program (SAP) period (1986–1998); cassava experienced its lowest growth rate, and the rate of the growth of production declined for all crops during the post-SAP period (1999–2005).

Given the contrasting evidence presented above, an accurate systematic analysis of the cassava sector in Nigeria covering a wide range of issues does not currently exist (e.g., cassava development programmes and their outcomes; trends in cultivation area, production, productivity, and prices of cassava and other major crops over a long period of time; sources of the growth in production; constraints of the cassava sector). All of the aforementioned studies analysed only one or a few of the issues outlined above, and none considered trends in the real prices of these crops, which could be a major incentive for farmers to improve productivity. Furthermore, none of the existing research has attempted to decompose and explain the sources of the observed growth in the production of major crops. On methodological grounds, all of these studies ignored the cyclical nature of time-series data (e.g., Gupta *et al.*, 2009) and instead fitted growth regressions to the actual annual production, area and/or yield data. If there is a high level of annual fluctuation in the time-series, which is actually the case, the result could be inaccurate growth rate estimates, as seen from the figures presented in Nweke (2004), Echebiri & Edaba (2008), Ojiako *et al.* (2007), and Liverpool-Tasie *et al.* (2011).

In this study, we aim to overcome the aforementioned shortcomings and present a systematic analysis of the cassava sector including the constraints faced at the local level by the cassava farmers/processors and marketers, which will be a valuable source of information for policy makers, academics, development agencies and other key stakeholders, because there is great potential to develop the cassava sector to support agricultural growth in Nigeria. For example, cassava production is expected to double and reach 76 million t by 2020 (CMP, 2006). Eke-okoro & Njoku (2012) noted that 33 improved varieties of cassava, which are high yielding/early maturing/pest and disease resistant and whose mean yields range from 13–40 t/ha, were released by 2010 through collaborations between the International Institute of Tropical Agriculture (IITA) and the National Root Crops Research Institute (NRCRI). Similarly, Nweke (2004) noted that cassava is a powerful tool to fight poverty in Africa, but enhancing this role faces numerous challenges.

Given this backdrop, the specific objectives of this study are to (a) conduct a review of the various cassava development policies and programmes undertaken in Nigeria from 1961–2013; (b) examine the trends in the area, production, productivity and real prices of cassava and four major, competing crops (i.e., groundnut, maize, millet and sorghum), the government outlay for agriculture and the share of agriculture in Nigeria's national income in Nigeria over a 52-year-period (1961–2013) separated into pre-SAP, SAP and post-SAP stages; (c) identify the sources of the growth in the production of cassava and other crops (i.e., yam, cocoyam, sweet potato, potato, and ginger); and (d) identify the constraints faced by the farmers, processors and marketers of cassava and its products at the grassroots level based on an in-depth farm level survey conducted in 2008 in three regions of Delta State, Nigeria, a major cassava growing area.

The paper is organised as follows: Section 2 presents a review of the agricultural and cassava development policies and programmes in Nigeria; Section 3 presents the methodology, including a description of the data; Section 4 presents the results, and Section 5 discusses the results, provides conclusions and outlines policy implications.

## 2 A brief on the agricultural policy developments in Nigeria

The importance of agricultural policy in boosting food production and accelerating agricultural development cannot be over emphasised (OECD, 2010; Olowu, 2011; Akpan, 2012). However, the observation of declining gross receipts for farmers from 28 % in 2005 to 23 % in 2009, which arose from agricultural policy support in the OECD countries, drove developing economies to discourage agricultural growth (Walkenhorst, 2007; Awoyinka, 2009; OECD, 2010). Adubi & Okunmadewa (1999), Okoh & Dominic (2004), and Nweke (2004) traced agricultural policy and programme implementation as a product of both pre- and post-colonial administrations and concluded that Nigerian agricultural policy evolved in overlapping phases that were not explicitly delineated but could be inferred from the various programmes that were implemented. However, Ugwu & Kanu (2012) noted that agricultural policies in Nigeria followed a series of strategies: (a) an 'exploitative strategy' during the 1950s that mainly taxed farmers producing export crops, such as cocoa, groundnut, palm and cotton; (b) an 'agricultural project strategy' until 1968 that aimed to develop backward and forward linkages in the sector, including the establishment of agricultural extension and research institutions; (c) a 'direct production strategy' during the early 1970s that primarily deepened government intervention and investment in agriculture; and (d) an 'integrated rural development strategy' from the mid-1970s onward that realised agriculture to be a part of the overall rural sector and that embarked on multipurpose development schemes and the establishment of a number of implementing institutions. Ugwu & Kanu (2012) also identified phases of 'minimum government intervention' (1960–1969), 'maximum government intervention' (1970–1985), 'SAP and post-SAP periods' (1986–1998), and the 'new millennium agricultural policies' (1999–2009) and concluded that the effects of all of these economic reforms on the agricultural sector have been minimal and unsatisfactory due to instability, inconsistency, and a lack of transparency and coordination of the policies.

### *2.1 Development of cassava in Nigeria*

Cassava development in Nigeria is an outcome of several activities undertaken through government policies, research institutions, agricultural development agencies, donors and non-governmental organisations. Eke-okoro & Njoku (2012) divided cassava development into four stages: the incipient cassava development period (1940–1953); the medieval cassava development period (1954–1967); the national and international collaboration of cassava development period (1970–2010); and an overlapping pre-emptive Cassava Mosaic Disease (CMD) cassava development period (1995 to date). Within the incipient cassava development period, the first cassava hybrid, called the Gold Coast Hybrid, was developed in 1942 and generated a mean yield of 8.9 t $ha^{-1}$. It was followed by another, called Oloronto, in 1953 with a mean yield of 13.6 t $ha^{-1}$ (Eke-okoro & Njoku, 2012). During the medieval period (1954–1967), three crosses of these aforementioned varieties were released with mean yields of 15.0–16.4 t $ha^{-1}$. However, during the 1970s, research was focused on developing disease-resistant varieties due to the severe cassava bacterial blight (CBB) crisis, and the IITA released a number of Tropical Manioc Selection (TMS) varieties during this period, which are disease resistant and high-yielding (Akoroda *et al.*, 1985, cited in Eke-okoro & Njoku, 2012; Nweke, 2004). During the pre-emptive-CMD cassava development period, the thrust was on improving productivity as well as disease resistance when five new TMS varieties were released, each of which contained high levels of starch and were characterised by a very high average yield of 35–40 t $ha^{-1}$; they were developed jointly by the IITA and the NRCRI (Eke-okoro & Njoku, 2012). Similarly, Nweke (2004) classified the development of cassava into four distinct periods: mechanisation of the cassava grater (1961–71); development and release of the TMS varieties (1972–1983); diffusion of the TMS varieties (1984–1992); and hand harvesting of the high-yielding TMS varieties (1993–2001).

Cassava received continued policy support from the government from as early as 1975 onward, which has led to diversified uses. The Presidential Initiative on Cassava, launched in 2003 and continued for five years, aimed at increasing cassava production to meet domestic demand and export, and it focused its activities on the development of production, processing, and the marketing of the processed products (IITA, 2009 cited in Asante-Pok, 2013; CMP, 2006). The Cassava Transformation Project/Agenda (2011–2015), which had a tentative budget of Naira 4 billion for four years, was launched in 2011 with the goal of doubling average cassava productivity from 12.5 t $ha^{-1}$ to 25.0 t $ha^{-1}$, raising the incomes of 1.8 million farmers by USD 450 per year, generating 1 million jobs per year in rural areas, and strengthening value chains and market institutions by involving the private sector (CTA, 2011). The government also initiated the Cassava Bread Development Fund with an investment of Naira 10 billion during 2013, and it aimed to substitute wheat flour with cassava flour, which in turn is expected to save billions of Naira by reducing the importation of wheat flour (Natsa, 2014). The government also launched the three-year Cassava Mechanisation and Agro-processing Project (CAMAP) in late 2012 as the pilot of a major public-private partnership initiative aimed at enhancing the contribution of cassava production and processing technologies by upgrading and expanding traditional planting, harvesting and processing techniques (AATF, 2012). The government also initiated a Presidential Committee on Cassava for Export Promotion in 2004 to ensure increased production, processing, packaging and export of cassava and cassava products to satisfy both domestic and export markets (Asante-Pok, 2013).

At various times, all of these drives to improve the cassava sector also had the financial backing of the FAO, USAID, GTZ, UNDP, IFAD, IITA and other donors (Nweke, 2004; Eke-okoro & Njoku, 2012; Asante-Pok, 2013). Additionally, the dissemination of information on cassava to the grassroots level was promoted through the production of various research manuals, extension guidebooks and targeted farmer trainings that have mainly been provided by IITA and NRCRI over the past four decades.

## 3 Methodology

### *3.1 Data sources*

The primary data used for this study were drawn from three regions of Delta State, Nigeria, which is situated in the southern (Niger Delta) part of the country: the North, Central and South Delta regions, which have different agro-ecological characteristics. The major foods grown in Delta State are cassava (leading producer), yam, plantain, maize, and vegetables (MANR, 2006). Delta state was selected as the case study area for this research because it has the ideal climatic and soil conditions for the cultivation of cassava, which is a very important staple crop for the state.

Farm sampling was based on the cell structure developed by the Agricultural Developmental Programme. First, nine local government areas (LGAs) of a total of 25 LGAs in the state were randomly selected. Next, 35 cassava growers from each LGA were selected using a stratified random sampling procedure with the size of the cassava farm operation as the strata. The cut-off points for farm size followed nationally defined categories (Apata *et al.*, 2011): marginal farms up to 1.00 ha, small farms from 1.01 to 2.00 ha, medium farms from 2.01 to 10.00 ha and large farms >10.01 ha. This provided a total of 315 cassava farmers as the sample size for the study. In addition, a survey of cassava marketers (i.e., wholesalers/retailers) was also conducted within the same three sampling regions. First, 35 marketers of cassava and cassava products were randomly selected from each region (i.e., 10–12 marketers from each of the nine LGAs). This provided a total sample size of 105 marketers (39 marketers from the Delta Central, 40 from the Delta South and 26 from the Delta North regions) spread across 20 markets in the three regions. The criteria used for selecting markets were that (a) markets must trade cassava and/or cassava products, and (b) markets must operate at least once a week. The average frequency of the market day was estimated to be 4 days (i.e., every 5$^{th}$ day is a market day with a range of 1–7 days).

To collect the primary data, two sets of structured questionnaires were administered containing both open and closed type questions, one for the farmers and the other for the marketers. A team of two research assistants was trained by one of the authors, and all three members were involved in collecting primary data through face-to-face interviews. The farmer survey included information on demographic and socio-economic characteristics, the size of the cassava farm operation, the inputs used and outputs received, the amount of cassava processed, and the constraints related to farming and processing cassava. The marketer survey included information on demographic and socio-economic characteristics, the quantities of cassava and its products that are purchased and marketed, the purchase and sale prices of each product, the cost of marketing, and the constraints on marketing. The surveys were conducted from September to December 2008.

The principal data for the trend analysis of the Nigerian agricultural sector were taken from the FAOSTAT database. The data included area cultivated (ha), total production (t), yield (kg ha$^{-1}$) and current prices (Naira/metric ton) of five major crops (i.e., cassava, groundnut, maize, millet and yam) covering the 52-year period (1961–2013). The information also included the share of agriculture in the GDP and the total government expenditure on agriculture for the period of 1970–2009. All the price data were then converted to constant 2010 prices using a GDP deflator for Nigeria from the Index Mundi (Index Mundi, 2015) to reflect real price changes over time.

Time-series data covering such a long period (e.g., 1960–2013) are likely to be unreliable for developing economies, and Nigeria is no exception. For example, Ammani *et al.* (2010) conducted an experiment comparing published maize production data from Nigeria for the period of 1992–2007 and an expected production based on twice the developing country average growth rate for maize production and noted that the official published data were significantly overstated. Jerven (2014) noted that the agricultural production data series for developing countries (i.e., Nigeria, Malawi, and India) are weak because they are subject to political pressure, particularly when governments subsidise agricultural inputs. Nevertheless, in the absence of suitable alternatives, we utilised FAOSTAT because it is widely used despite its limitations, but the interpretation of the results should account for the issue of data quality.

### *3.2 Analytical methods*

A range of methods were applied, including trend analysis, the estimation of growth rates and the decomposition of the sources of the growth in the production of cassava and the other competing crops under consideration. The purpose was to provide a comprehensive picture of cassava performance over time in relation to the other crops to judge its potential.

Average annual compound growth rates were computed to determine the rate of change of the variable of interest using a semi-logarithmic trend function: $\ln Y = \alpha + \beta T$, where $Y$ is the target variable; $T$ is time; ln is natural logarithm, and $\beta$ is the growth rate. To control for the cyclical nature of the time-series data, we used 3-year moving averages to compute the average annual compound growth rates (e.g., Gupta *et al.*, 2009), which is an improvement over previous studies (e.g., Nweke, 2004; Echebiri & Edaba, 2008; Ojiako *et al.*, 2007; Liverpool-Tasie *et al.*, 2011).

We have analysed the changes in the area planted, total production, productivity and the real prices of five major crops (i.e., cassava, groundnut, maize, millet, and yam) at the national level for the 52-year period (1961–2013). All of the analyses were classified by the SAP stages to examine whether this major policy instrument has any influence on the performance of cassava and the other crops under consideration. The 1961–1985 period depicts the pre-SAP stage; the 1986–1993 period depicts the main SAP implementation stage, and the 1994–2013 period depicts the post-SAP stage. This is because, the three main elements of SAP are: "(a) pursue macroeconomic stability by controlling inflation and reducing fiscal deficits; (b) open economies to the rest of the world through trade and capital account liberalisation, and (c) liberalise domestic product and factor markets through privatisation and deregulation" ("Washington Consensus" as summarized by Gore, 2000, p.789–790). The main argument in favour of SAP is that openness, trade liberalisation and free market operation will boost production in agriculture. Since Nigeria also adopted SAP in 1986 and subsequently removed subsidies, deregulated and reduced state control in agriculture to a large extent, it is important to examine whether SAP has any discernible impact, an approach also adopted by Ugwu & Kanu (2012) and Ojiako *et al.* (2007).

To identify the sources of the observed growth in production of all of the crops, we applied a decomposition method to examine the relative contribution of the growth in the cultivated area and the growth in the yield of the individual crops. The basic method of decomposing sources of the growth in production depends on the identity:

$$P\,(\text{Output}) = A\,(\text{area planted}) * Y\,(\text{yield}) \qquad (1)$$

Using the subscripts 0 and t for the base year and terminal year, respectively, the relationship between the two production levels can be expressed as

$$P_t/P_0 = (A_tY_t)/(A_0Y_0) = (A_t/A_0) * (Y_t/Y_0) \qquad (2)$$

Eq. (2) is known as the Venegas-Ruttan (V-R) method, which demonstrates that the total output can be explained in terms of changes in the area planted and productivity (Alauddin & Tisdell, 1991). A few extensions of this basic method have been provided by others (see Alauddin & Tisdell, 1991 for details), but we select the method proposed by Wennergren *et al.* (1984), hereafter named the Wennergren, Antholt and Whitaker (W-A-W) method. This is because the W-A-W method decomposes the growth of individual crop output into four components as follows:

$$\begin{aligned}&\text{Change in total output}\,(P_t - P_0) = \\ &\text{Area effect}\,(V) + \text{Yield effect}\,(S) + \\ &\text{Cropping pattern}\,(R) + \text{Interactions}\,(U)\end{aligned} \qquad (3)$$

where

Area effect $(V) = Y_0[A_t(1 + C_0 - C_t) - A_0]$

Yield effect $(S) = [A_t(1 + C_0 - C_t) * (Y_t - Y_0)]$

Cropping patternt $(R) = [A_t * Y_0(C_t - C_0)]$

Interactions $(U) = [A_t(Y_t - Y_0) * (C_t - C_0)]$

and $C$ = the proportion of the gross cropped area under the crop, and the other variables were defined above.

The advantage of the W-A-W method is that it includes the influence of the change in cropping pattern as well as the interaction between the change in cropping pattern and individual crop yield, which was not available in the V-R method. The implicit assumption of this method is constant returns to scale in terms of the output, which is reasonable.

## 4 Results

### 4.1 *Trends in area, production, productivity and the real price of cassava and its competitors*

Table 1 presents the mean values of area, production, productivity and the real prices of five major crops: cassava, maize, millet, groundnut and yam covering the period 1961–2013 and classified as the pre-SAP, SAP and post-SAP periods. The average area allocated for cassava was quite low during the pre-SAP period, and it only picked up during the post-SAP period, although millet area remained highest throughout. The total crop production increased due to a combination of increases in area as well as increases in yield, although the exact

**Table 1:** *Area cultivated, total production, productivity and real prices of cassava and other crops in Nigeria*

| Panel | Variables | Mean values of 3-year moving averages | | | |
|---|---|---|---|---|---|
| | | Pre-SAP (1961–1985) | SAP (1986–1993) | Post-SAP (1994–2013) | Overall (1961–2013) |
| | | *Area cultivated* ('000 ha) | | | |
| | Cassava | 1011.26 | 1761.84 | 3441.39 | 2071.86 |
| | Groundnut | 1387.08 | 819.38 | 2205.28 | 1608.81 |
| A | Maize | 1023.15 | 4013.29 | 4065.01 | 2703.95 |
| | Millet | 3683.14 | 4102.62 | 4686.04 | 4151.49 |
| | Yam | 786.83 | 1173.62 | 2741.80 | 1598.18 |
| | | *Total production* ('000 t) | | | |
| | Cassava | 9956.46 | 19079.05 | 39355.20 | 22715.55 |
| | Groundnut | 1153.31 | 1051.00 | 2913.15 | 1792.22 |
| B | Maize | 1016.26 | 5061.62 | 6505.60 | 3778.46 |
| | Millet | 2854.59 | 4334.14 | 5897.28 | 4240.22 |
| | Yam | 6563.83 | 11379.57 | 30363.46 | 16502.91 |
| | | *Productivity* (kg/ha) | | | |
| | Cassava | 9852.33 | 10939.99 | 11376.42 | 10584.59 |
| | Groundnut | 850.16 | 1291.12 | 1330.54 | 1093.01 |
| C | Maize | 1039.94 | 1289.31 | 1601.94 | 1285.64 |
| | Millet | 872.79 | 1069.90 | 1247.65 | 1040.66 |
| | Yam | 8593.03 | 9245.12 | 11035.71 | 9676.29 |
| | | *Prices* (Naira/t) | | | |
| | Cassava | 921.05 | 958.30 | 752.34 | 865.21 |
| | Groundnut | 1384.41 | 2889.32 | 1739.99 | 1796.70 |
| D | Maize | 1023.07 | 1498.10 | 1512.13 | 1293.43 |
| | Millet | 985.37 | 1475.27 | 1356.69 | 1214.68 |
| | Yam | 1432.74 | 2022.29 | 1990.15 | 1749.83 |
| | | *Government outlays* | | | |
| E | Agriculture's share of GDP (%) | 25.91 | 28.16 | 39.16 | 31.66 |
| | Agriculture's share of government expenditure (%) | 4.28 | 2.40 | 3.11 | 3.44 |

Note: Price data are from 1967–2008; Agriculture's share of GDP and government expenditure is from 1970–2009.
Source: Computed using the FAOSTAT database (FAOSTAT, 2014)

contribution of these sources of growth cannot be identified here. The analysis of the sources of the growth in production is relegated to the next section, but the average real prices of these crops were highly variable throughout the period. The growth in the overall yield level of cassava was very low, increasing from 9.8 t $ha^{-1}$ to 10.6 t $ha^{-1}$ over the 52-year period.

Figures 1 to 4 present the trends in the 3-year moving averages of the area indices, production indices, yields and real prices of the five major crops: cassava, maize, millet, groundnut and yam for the period 1961–2013. It is clear from the figures that the trends are highly variable with respect to all of the indicators for all of the crops during the period under consideration. Cassava area steadily increased from 1986 onward (i.e., from the beginning of the SAP period) although the rate of increase is lower than for yam area, which is another major staple in Nigeria (Figure 1). Similarly, total cassava production also increased steadily during the same period but lagged behind yam (Figure 2). In terms of productivity, cassava yield was largely stable, showing very little improvement compared to the other major crops, which experienced high levels of fluctuation (Figure 3). The trend in the yield increase in cassava is only apparent during the past few years, which may be due to renewed interest in boosting cassava yield through various projects (e.g., Cassava Transformation Project) and the influence of the TMS variety developments noted by Eke-okoro & Njoku (2012).

The trend in real prices (i.e., at constant 2010 prices) of these major crops presents an interesting contrast (Figure 4). With the exception of cassava, the prices for all major crops rose sharply during the SAP period, then fell sharply during the post-SAP period and kept falling. However, the level of fluctuation in cassava price is re-latively low among the five crops under consideration (Figure 4).

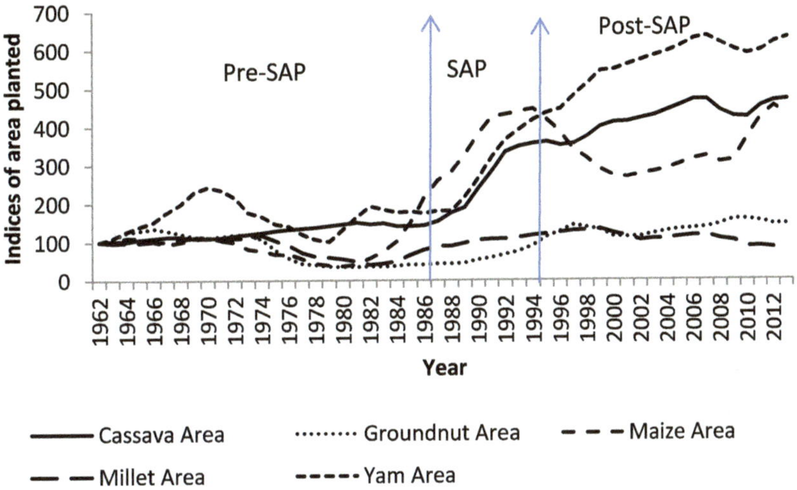


Source: Computed using the FAOSTAT database (FAOSTAT, 2014)

**Fig. 1:** *Three-year moving average indices of the area planted under major crops.*

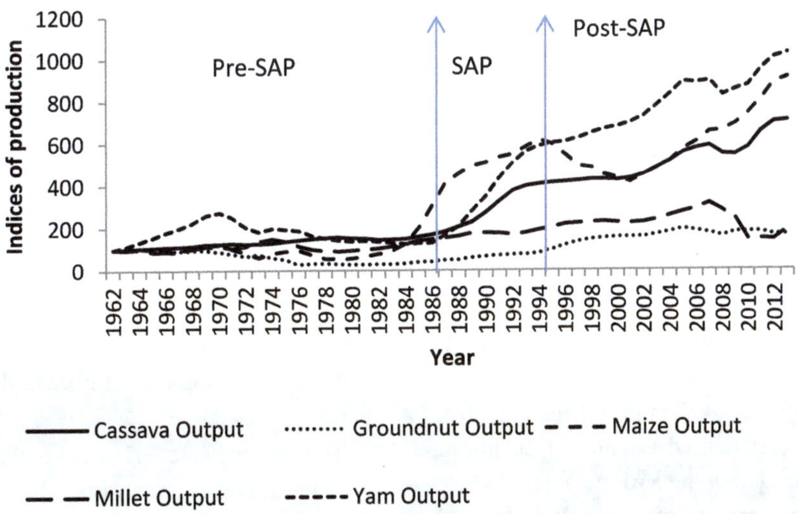


Source: Computed using the FAOSTAT database (FAOSTAT, 2014)

**Fig. 2:** *Three-year moving average indices of the total production of major crops.*

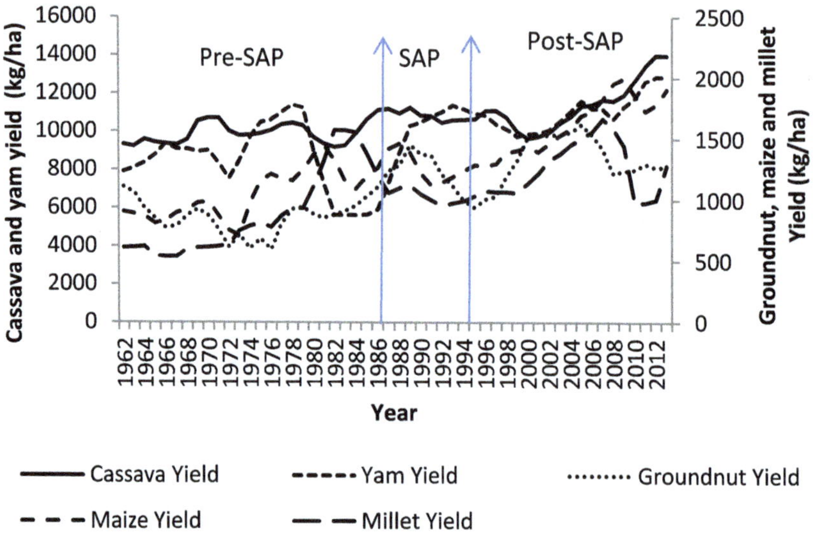


Source: Computed using the FAOSTAT database (FAOSTAT, 2014)

**Fig. 3:** *Three-year moving average of major crop yields (kg/ha).*

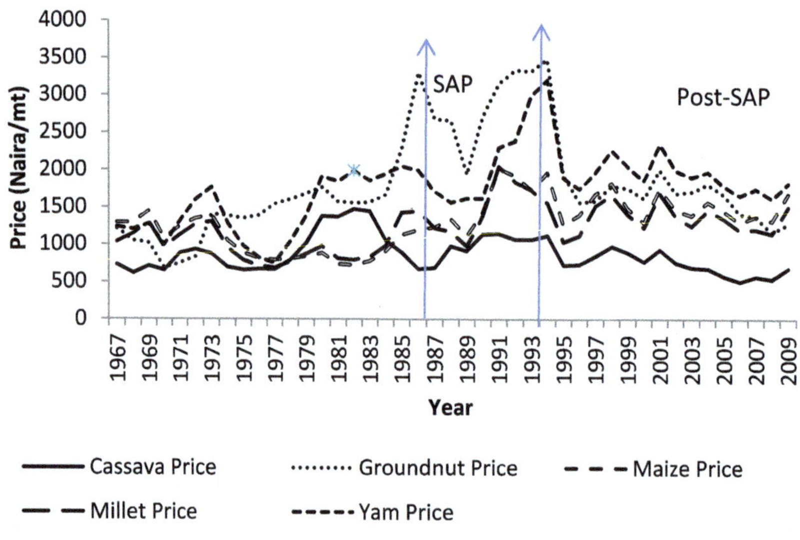


Source: Computed using the FAOSTAT database (FAOSTAT, 2014)

**Fig. 4:** *Three-year moving average prices of the major crops (Naira/t) in Nigeria.*

The annual compound growth rates from the 3-year moving averages of these data series are presented in Table 2, which displays interesting contrasts. For example, during the pre-SAP period, the area under cassava increased at 1.8 % p.a. while the area under maize, millet and groundnut experienced a significant decline at highly variable rates. During the SAP period, however, all of the crop areas increased at a very high rate led by cassava at 14.5 % p.a. Finally, during the post-SAP period, the growth rates fell sharply with millet area even experiencing a decline at 2.1 % p.a. The observed growth in total cassava production during the pre-SAP and SAP periods is due to increases in area, and the productivity of cassava and the other crops, except millet, experienced a low rate of growth during the post-SAP period.

The growth in the real prices of these major crops also presents interesting contrasts. The real prices of cassava grew significantly during the pre-SAP and SAP periods but then significantly declined at 3.5 % during the post-SAP period, resulting in no growth in real prices over the 52-year period. In contrast, despite the high level

**Table 2:** *Growth rate estimations from 3-year moving average data for Nigeria.*

| *Panel* | *Variables* | *Mean values of 3-year moving averages* | | | |
|---|---|---|---|---|---|
| | | *Pre-SAP* (1961–1985) | *SAP* (1986–1993) | *Post-SAP* (1994–2013) | *Overall* (1961–2013) |
| | | *Area cultivated* | | | |
| A | Cassava | 0.018 *** | 0.147 *** | 0.015 *** | 0.037 *** |
| A | Groundnut | –0.070 *** | 0.085 *** | 0.017 *** | 0.010 ** |
| A | Maize | –0.028 ** | 0.122 *** | 0.005 | 0.039 *** |
| A | Millet | –0.039 *** | 0.053 *** | –0.021 *** | 0.003 |
| A | Yam | 0.004 | 0.135 *** | 0.019 *** | 0.037 *** |
| | | *Total production* | | | |
| B | Cassava | 0.020 *** | 0.137 *** | 0.030 *** | 0.042 *** |
| B | Groundnut | –0.071 *** | 0.100 *** | 0.029 *** | 0.023 *** |
| B | Maize | –0.007 | 0.085 *** | 0.028 *** | 0.055 *** |
| B | Millet | 0.007 | 0.028 ** | –0.010 | 0.020 *** |
| B | Yam | –0.009 | 0.235 *** | 0.029 *** | 0.044 *** |
| | | *Productivity* | | | |
| C | Cassava | 0.002 | –0.009 ** | 0.015 *** | 0.005 *** |
| C | Groundnut | –0.001 | 0.019 | 0.011 * | 0.013 *** |
| C | Maize | 0.021 *** | –0.034 * | 0.024 *** | 0.015 *** |
| C | Millet | 0.048 *** | –0.029 ** | 0.008 | 0.016 *** |
| C | Yam | –0.011 * | 0.100 *** | 0.010 *** | 0.007 *** |
| | | *Prices* | | | |
| D | Cassava | 0.036 *** | 0.073 *** | –0.035 *** | –0.004 |
| D | Groundnut | 0.044 *** | 0.024 | -0.036 *** | 0.011 *** |
| D | Maize | –0.030 *** | 0.075 *** | –0.008 | 0.012 *** |
| D | Millet | –0.012 *** | 0.063 * | –0.003 | 0.011 *** |
| D | Yam | 0.029 ** | 0.068 ** | -0.021 ** | 0.014 *** |
| | | *Government outlays* | | | |
| E | Agriculture's share of GDP (%) | –0.051 *** | 0.048 ** | 0.011 *** | 0.014 *** |
| E | Agriculture's share of government expenditure (%) | 0.109 *** | –0.174 ** | 0.039 | –0.003 |

Note: *** significant at the 1 % level (p<0.01); ** significant at the 5 % level (p<0.05); * significant at the 10 % level (p<0.10)
Source: Computed using the FAOSTAT database (FAOSTAT, 2014)

of fluctuations during the SAP period, the real prices of the remaining four crops increased significantly at 1.1 %–1.4 % over the 52-year period under consideration. The lack of overall growth in the real price of cassava may act as a disincentive to the producers to boost cassava performance.

The last two rows of Tables 1 and 2 present the contribution of agriculture to the GDP and government expenditure on agriculture. These two sets of figures exhibit a clear mismatch between the contribution of the sector to national income and the corresponding support by the government to maintain the sector. While agriculture contributed approximately 30 % of the GDP on average, government expenditure was only 3.5 %, and Table 2 also shows that, even though the contribution of agriculture to the GDP continued to grow during the SAP and post-SAP periods, the corresponding government expenditure, which was already very low, showed no sign of growth.

### *4.2 Sources of the growth in production*

In this section, we examine the sources of the growth in the production of individual crops using the W-A-W method, as in Eq. (3). The results are presented in Table 3, and it is encouraging to note that the yield effect is the dominant source of the growth in cassava production, contributing 76.4 % of the total growth followed by an area effect that contributes 28.2 %. The contribution of the change in cropping pattern is negligible. Groundnut also experienced a similar pattern whereas the growth in the production of maize, millet and yam is mainly due to the area effect and the change in cropping pattern with a negative contribution from the yield effect. The implication is that, compared with the other four crops, cassava has the potential to support the growth of agriculture in Nigeria.

### *4.3 Constraints in the cassava sector*

A number of constraints on cassava production and productivity have been identified by many authors. For example, Addy *et al.* (2004) and Nweke (2004), among others, have argued that the efficient use of inputs, improvements in technology (including irrigation, pesticides, cuttings and storage methods), the provision of market information, and the determination and dissemination of the adequate numbers of cassava planters, harvesters, peelers, hydraulic presses and dryers would all improve efficiency in cassava production. The Cassava Master Plan noted that the primary challenge faced by the cassava sector is low productivity due to Nigeria's subsistence cassava farming culture with its large but rudimentary and underdeveloped industry. The plan also noted that Nigeria has the potential to earn revenue

**Table 3:** *Sources of the growth in production of major crops in Nigeria (1961–2013).*

| *Crops* | *Sources of growth using the W-A-W method* | | | | |
|---|---|---|---|---|---|
| | *Area effect* | *Yield effect* | *Cropping pattern* | *Interactions* | *Total* |
| Cassava | 28.20 | 76.39 | –4.71 | 0.12 | 100.00 |
| Groundnut | 38.14 | 64.53 | –3.18 | 0.52 | 100.00 |
| Maize | 128.82 | –50.47 | 21.12 | 0.53 | 100.00 |
| Millet | 129.56 | –58.92 | 27.15 | 2.21 | 100.00 |
| Yam | 144.94 | –59.18 | 17.58 | –3.34 | 100.00 |

Source: Computed using the FAOSTAT database (FAOSTAT, 2014)

**Table 4:** *Infrastructure constraints affecting the cassava sector.*

| *Constraints* | *Average value of farmers' responses on a 5-point Likert scale* | | | | |
|---|---|---|---|---|---|
| | *Delta Central* | *Delta South* | *Delta North* | *All Regions* | *Weighted Ranking* |
| Water Provision | 3.49 | 3.82 | 2.51 | 3.27 | 1 |
| Processing Facilities | 3.70 | 3.54 | 2,41 | 3.22 | 2 |
| Electricity Provision | 4.05 | 2.38 | 2.83 | 3.09 | 3 |
| Marketing Facilities | 3.79 | 2.52 | 2.90 | 3.07 | 4 |
| Credit Facilities | 3.69 | 2.37 | 2.42 | 2.83 | 5 |
| Road Network | 3.82 | 1.88 | 2.70 | 2.80 | 6 |
| Extension Services | 3.39 | 1.94 | 2.72 | 2.69 | 7 |
| Information Provision | 3.24 | 1.85 | 2.66 | 2.58 | 8 |

Likert scale ranking: 1: Strongly Disagree, 2: Disagree, 3: Neutral, 4: Agreed, 5: Strongly Agree
Source: Field survey in Delta State, Nigeria, 2008.

of USD 5 billion from cassava products if the existing bottlenecks along the cassava value chain are resolved (CMP, 2006).

We present a detailed analysis of the constraints to the cassava sector in Nigeria according to the responses of 315 cassava farmers/processors and 105 retailers/marketers collected from three regions of Delta State, Nigeria, in December 2008. Approximately 68.3 % of the 315 farmers/processors are marginal or small farmers with farm sizes <2.01 ha, and the average farm size of the entire sample is only 2.05 ha. Based on a Likert scale analysis of the ranks of individual responses for each of the eight infrastructure elements evaluated, the provision of water, which is mainly meant for irrigation and the washing and processing of cassava into gari, was ranked as the greatest constraint affecting the cassava sector followed by adequate processing facilities and electricity provision (Table 4). Ayoade & Adeola (2009), in their study investigating the constraints to domestic industrialization of cassava in Osun State in southwest Nigeria, also stated that inadequate processing equipment, high processing costs and ineffective linkages between farmers and processors were the most important factors constraining cassava production. Naziri *et al.* (2014) noted that the physical loss of cassava in southwest Nigeria is estimated at 481,258 ton per year accounting for 6.7 % of total production and 82 % of the physical loss takes place during processing stage alone. The main reason is delay in processing which in turn is due to shortage in peeling capacity and mechanical peeling. In fact, the economic loss of cassava production in southwest Nigeria is estimated at more than USD 20 million per year (Naziri *et al.*, 2014).

Farmers were also asked about the constraints to adding value to cassava through processing, and they identified a lack of transportation and adequate information as the top two constraints (Table 5). Approximately 91.5 % of the processors agreed that transportation of cassava root tubers from the farm/market to the processing site is costly as the average distance from the farmers/processors to the nearest marketplace is estimated to be 2.93 km (±3.13 km) with a maximum distance of 15 km. In fact, distance is a major factor that adversely affects the cost and efficiency of processing, which was also supported by Akinnagbe (2010) and Tonukari (2004).

Similarly, when asked about constraints to marketing cassava and cassava products, all of the marketers (i.e., retailers and wholesalers) noted inaccessible markets, unstable prices and the high costs of marketing as the main obstacles (Table 6). Rahman & Awerije (2014) noted that increases in the purchase price of cassava and cassava products as well as marketing costs per unit significantly reduce profit margins.

**Table 6:** *Marketing constraints to the cassava sector.*

| *Constraints* | *% of marketers/retailers responding* |
|---|---|
| Market Accessibility | 100 |
| Unstable Prices | 100 |
| High Cost of Marketing | 100 |
| Lack of Market Infrastructure | 98.1 |
| Storage Problems | 65.7 |
| Lack of Information | 61.9 |

Source: Field survey in Delta State, Nigeria, 2008.

**Table 5:** *Constraints to adding value in cassava through processing.*

| *Constraints* | *% of farmers responding* | *Rank* |
|---|---|---|
| Transportation Difficulties | 91.5 | 1 |
| Lack of Adequate Information | 91.4 | 2 |
| Too Many Buyers for Limited Raw Materials | 76.6 | 3 |
| Lack of Processing Equipment | 76.2 | 4 |
| High Cost of Raw Materials/Processing Equipment | 72.4 | 5 |
| Lack of Adequate Infrastructure | 70.5 | 6 |
| Others | 23.8 | 7 |

Source: Field survey in Delta State, Nigeria, 2008.

## 5 Conclusions and policy implications

The principal aim of this study is to explore the potential of cassava to support agricultural growth in Nigeria based on a review of the development of cassava; perform a trend analysis of the area planted, production, productivity and the real prices of cassava and major competing crops (i.e., yam, maize, millet and groundnut) covering a 52-year period (1961–2013); identify sources of the growth in production; and examine key constraints in the cassava sector at the local level based on a survey of 315 farmers/processors and 105 marketers/retailers from three regions of Delta State, Nigeria, conducted in December 2008.

The results revealed that although the average cassava yield only increased from 9.8 t ha$^{-1}$ to 10.6 t ha$^{-1}$ over a 52-year period, cassava productivity grew by 1.5 % during the post-SAP period, which may be a reflection of the use of improved TMS varieties developed by IITA and NRCRI during the pre-emptive–CMD cassava development period. A decomposition of the sources of the growth in cassava production confirmed that the yield effect is the dominant source, contributing 76.4 % of the total growth, followed by the area effect (28.2 %). The lack of growth in the real price of cassava over the study period and a significant decline of 3.5 % p.a. during the post-SAP period is a major obstacle to driving the cassava sector forward. It is important to note that the cassava sector was relatively more stable in terms of area, production, productivity and real prices compared to the other crops during the period under consideration. Therefore, considering all of these factors, it can be concluded that the cassava sector has the potential to support the growth of Nigerian agriculture if managed properly. This is because all of the other crops experienced significant fluctuations in all the indicators during the same period, thereby providing less confidence in their potential to support agricultural growth. Naziri *et al.* (2014) noted that innovations to extend shelf-life of fresh root cassava through technologies, such as waxing, paraffin coating and high humidity storage can be tried as such measures were successful in other economies. Also, use of mechanical peeling could substantially improve processing capacity.

Another point to note is the neglect of the agricultural sector by the government, as oil has become the main source of income in the Nigerian economy. This is evident from the fact that, although the contribution of the agricultural sector to national income grew consistently, the government did not provide proportional, corresponding support to maintain the sector; there is a clear mismatch that is largely responsible for the poor growth performance of cassava and other major crops over time. Ojiako *et al.* (2007) also noted a lack of expenditure support for the growth of roots and tubers in Nigeria. The Comprehensive African Agricultural Development Program (CAADP), founded on a declaration by African Head of States at Maputo in 2003, set the target to devote 10 % of their national budget to the agricultural sector by 2008, which very few countries actually achieved (Poulton *et al.*, 2014). However, as mentioned earlier, time-series data covering a long period are likely to be unreliable for the developing countries including data from Nigeria. Therefore, interpretation of the findings of this study should take into account such limitation although we have used data supplied by FAOSTAT, which is the most widely used source of such data.

A host of constraints affect the cassava sector at the local level, of which poor market and marketing infrastructure; the lack of irrigation/water provision, processing facilities, transportation and information on market prices; and unstable prices are the dominant factors as identified and prioritised by the farmers, processors and marketers surveyed from Delta State.

A number of policy implications can be drawn from the results of this study. The first is the need for investment to improve market and marketing infrastructure because although the price for cassava in Nigeria is determined by market forces, high fluctuations in price indicate that the market is not functioning properly (Rahman & Awerije, 2014), which is also reflected in Table 6. Improvements in market and marketing infrastructure will address these issues. Second, the improved TMS varieties developed by research stations with high potential yields need to be diffused widely to increase cassava productivity at the farm level. Asante-Pok (2013) noted that the favourable policy environment established by the government encouraged cassava development, leading to a new orientation in the research-extension-farmer linkage (e.g., Cassava Multiplication Programme and the Roots and Tubers Expansion Programme), which are positive steps that should be enhanced further. Third is the need for investment in improving cassava processing facilities and utilities that can also contribute to reduced fluctuations in prices. A key driver of prices in the Nigerian cassava economy is the relative price of gari; Rahman & Awerije (2014) noted that the price of gari is 2.81 times higher, estimated at Naira 79.84 per kg, compared with the price of raw cassava (Naira 28.41 per kg) in Delta State. This serves as an incentive to process cassava into gari provided that the processing costs do not outweigh the additional revenues. The Raw Material Research and Development Council funds research projects on the fabrication of cassava processing equipment (Asante-Pok, 2013), and such measures need to be enhanced further to improve processing technologies, which in turn will reduce processing costs. Fourth is the need for investment in irrigation infrastructure and/or water provision, which was identified as a major constraint by local farmers/processors (Table 4). Finally, improvements in the dissemination of information ranging from technological expertise to the production, processing and marketing of cassava and cassava products through mass media, agricultural extension services and digital technologies.

Although meeting all of these policy options is formidable, the effective implementation of these broader policy measures will drive the cassava sector forward and support agricultural growth in Nigeria.

**Acknowledgements**

The paper was partially developed from the co-author's PhD thesis completed at the School of Geography, Earth and Environmental Sciences, University of Plymouth, UK in 2014. The primary data required for this project was generously funded by the Seale-Hayne Educational Trust, UK. The authors gratefully acknowledge the critical comments of the two referees and the editor-in-chief which have substantially improved the paper. However, all caveats remain with the authors.

## References

AATF (2012). Public-private partnership project on cassava mechanisation, agro-processing launched. African Agricultural Technology Foundation (AATF). URL `http://www.aatf-africa.org/node/451` (last accessed: 26.08.2015).

Addy, P. S., Kashaija, I. N., Moyo, M. T., Quynh, N. K., Singh, S. & Awalekhwa, P. N. (2004). *Constraints and Opportunities for Small and Medium Scale Processing of Cassava in the Ashati and Brong Ahafoa Regions of Ghana*. International Centre for Research in Agriculture (ICRA), Netherlands.

Adubi, A. A. & Okunmadewa, F. (1999). *Price, Exchange Rate Volatility and Nigeria's Agricultural Trade Flows: A Dynamic Analysis*. African Economic Research Consortium, Nairobi, Kenya.

Akinnagbe, O. (2010). Constraints and strategies towards improving cassava production and processing in Enugu north agricultural zone of Enugu State, Nigeria. *Bangladesh Journal of Agricultural Research*, 35, 387–394.

Akoroda, M. O., Gebremeshel, I. & Oyinlola, A. E. (1985). *Impart of IITA Cassava varieties in Oyo State of Nigeria*. IITA, Ibadan, Nigeria. p. 105

Akpan, S. B. (2012). Analysis of food crop output volatility in agricultural policy programme regimes in Nigeria. *Developing Country Studies*, 2, 28–35.

Alauddin, M. & Tisdell, C. (1991). *The Green Revolution and Economic Development: The Process and Its Impact in Bangladesh*. Macmillan, London.

Ammani, A., Sani, B. & Alamu, J. (2010). Challenges to the reliability of officially published statistics for extension work. *Journal of Agricultural Extension*, 14, 48—52.

Apata, T. G., Folayan, A., Apata, O. M. & Akinlua, J. (2011). The economic role of Nigeria's subsistence agriculture in the transition process: implications for Rural Development. Paper Presented at the 85th Agricultural Economics, UK Conference at Warwick University During 18–20 April 2011.

Asante-Pok, A. (2013). Analysis of Incentives and Disincentives for Cassava in Nigeria. Monitoring African Food and Agricultural Policies project (MAFAP), FAO, Rome, Italy. URL `http://www.fao.org/3/a-at582e.pdf` (last accessed: 26.08.2015).

Awoyinka, Y. A. (2009). Cassava Marketing: Option for Sustainable Agricultural Development in Nigeria. *Ozean Journal of Applied Science*, 2 (2), 175–183.

Ayoade, A. R. & Adeola, R. G. (2009). Constraints to domestic industrialization of cassava in Osun State Nigeria. *Ozean Journal of Social Sciences*, 2 (2), 1–5.

CMP (2006). Cassava master plan 2006 A Strategic Action Plan for the Development of the Nigerian Cassava Industry. UNIDO, Geneva.

CTA (2011). Action plan for a cassava transformation in Nigeria (draft). URL `http://unaab.edu.ng/attachments/Cassava%20Report%20Final.pdf` (last accessed: 26.08.2015).

Echebiri, R. N. & Edaba, M. E. I. (2008). Production and utilization of cassava in Nigeria: prospect for food security and infant nutrition. *Production Agriculture and Technology*, 4, 38—52.

Eke-okoro, O. N. & Njoku, D. N. (2012). A review of cassava development in Nigeria from 1940–2010. *ARPN Journal of Agricultural and Biological Science*, 7 (1), 59–65.

FAO (2005). A review of cassava in Africa with country cases studies on Nigeria, Ghana, united Republic of

Tanzania, Uganda and Benin. Proceedings of the Validation Forum on the Global Cassava Development Strategy, vol. 2. International Fund for Agricultural Development and Food and Agricultural Organisation of the United Nations, Rome, Italy.

FAOSTAT (2014). FAOSTAT database on Nigeria. URL http://faostat3.fao.org/home/E (last accessed: 15.07.2015).

Gore, C. (2000). The Rise and Fall of the Washington Consensus as a Paradigm for Developing Countries. *World Development*, 28 (5), 789–804.

Gupta, D., Sahu, P. K. & Banerjee, R. (2009). Forecasting jute production in major contributing countries in the world. *Journal of Natural Fibres*, 6, 127–137.

IITA (2009). Processing potential for cassava production growth in sub-Saharan Africa. COSCA working paper No 11, Collaborative Study of Cassava in Africa, Nigeria.

Index Mundi (2015). GDP deflator of Nigeria 1960–2013. URL http://www.indexmundi.com/facts/nigeria/gdp-deflator (last accessed: 25.08.2015).

Jerven, M. (2014). The political economy of agricultural statistics and input subsidies: evidence from India, Nigeria and Malawi. *Journal of Agrarian Change*, 14, 129–145.

Knipscheer, H., Ezedinma, C., Kormawa, P., Asumugha, G., Mankinde, K., Okechukwu, R. & Dixon, A. (2007). *Opportunities in The Industrial Cassava Market in Nigeria*. International Institute for Tropical Agriculture (IITA), Surrey, UK.

Liverpool-Tasie, L. S., Kuku, O. & Ajibola, A. (2011). Nigeria Strategy Support Program. Working Paper # 21. International Food Policy Research Institute, Washington, D.C.

MANR (2006). Agricultural development policy report. Ministry of Agriculture and Natural Resources (MANR), Delta State, Nigeria.

Nang'ayo, F., Odera, G., Muchiri, N., Ali, Z. & Werehire, P. (2007). A strategy for industrialization of cassava in Africa. Proceedings of a Small Group Meeting, 14–28 November 2005, African Agricultural Technology Foundation, Ibadan, Nigeria. URL http://aatf-africa.org/userfiles/CassavaProceedings.pdf (last accessed: 15.11.2014).

Natsa, R. T. (2014). Turning the cassava dream into reality. Leadership Newspaper, April 2014. URL http://leadership.ng/business/368405/turning-cassava-dream-reality (last accessed: 25.08.2014).

Naziri, D., Quaye, W., Siwoku, B., Wanlapatit, S., Phu, T. V. & Bennett, B. (2014). Diversity of post harvest losses in cassava value chains in selected developing countries. *Journal of Agriculture and Rural Development in the Tropics and Subtropics*, 115 (2), 111–123.

Nkonya, E., Pender, J., Kato, E., Omobowale, O., Phillip, D. & Ehui, S. (2010). Enhancing Agricultural Productivity and Profitability in Nigeria. Nigeria Strategy Support Program, Brief # 19. International Food Policy Research Institute, Washington, D.C.

Nweke, F. (2004). *New Challenges in the Cassava Transformation in Nigeria and Ghana*. International Food Policy Research Institute (IFPRI), Washington, D.C.

OECD (2010). *Economic importance of agricultural for sustainable development and development and poverty reduction: findings from a Case study of Indonesia*. Organisation for Economic Co-operation and Development (OECD), Paris, France.

Ojiako, I. A., Asumugha, G. N., Ezedima, C. & Uzokwe, N. E. (2007). Analysis of production trends in the major root and tuber crops in Nigeria, 1961–2005. *Research on Crops*, 6, 371–380.

Okoh, S. O. & Dominic, O. (2004). Economics of cassava production in a developing economy: a case study of Akwa Ibom State of Nigeria. *Nigerian South-East Journal of Agricultural Economics and Extension*, 6, 22–25.

Olowu, A. U. (2011). *Agricultural financing and performance in Nigeria: A case study of the Agricultural Credit Guarantee Scheme (ACGS)*. Master's thesis, Stellenbosch University, South Africa.

Poulton, C., Berhanu, K., Chinsinga, B., Cooksey, B., Golooba-Mutebi, F. & Loada, A. (2014). The Comprehensive Africa Agriculture Development Programme (CAADP): Political Incentives, Value Added and Ways Forward. Futures Agricultures Working Paper # 077. URL http://www.future-agricultures.org/publications/research-and-analysis/1833-the-comprehensive-africa-agriculture-development-programme-caadp-political-incentives-value/file (last accessed: 08.12.2015).

Rahman, S. & Awerije, B. O. (2014). Marketing efficiency of cassava products in Delta state, Nigeria: a stochastic profit frontier approach. *International Journal of Agricultural Management*, 4, 28–37.

Tonukari, N. J. (2004). Cassava and the future of starch. *Electronic Journal of Biotechnology*, 7, 5–8.

Ugwu, D. S. & Kanu, I. O. (2012). Effects of agricultural reforms on the agricultural sector in Nigeria. *Journal of African Studies and Development*, 4, 51–59.

Walkenhorst, P. (2007). *Distortions to agricultural incentives in Nigeria. World Bank Agricultural Distortions Working Papers # 37*. World Bank, Washington, D.C.

Wennergren, E. B., Antholt, C. H. & Whitaker, M. D. (1984). *Agricultural Development in Bangladesh*. Westview Press, Boulder, Colorado.

# 15

# Explaining technical inefficiency and the variation in income from apple adoption in highland Ethiopia: The role of unequal endowments and knowledge asymmetries

Sintayehu Hailu Alemu [a,b,*], Luuk van Kempen [a], Ruerd Ruben [a,c]

[a] *Department of Cultural Anthropology and Development Studies, Radboud University, Nijmegen, The Netherlands*
[b] *Department of Economics, Hawassa University, SNNPRS, Hawassa, Ethiopia*
[c] *Agricultural Research Institute (LEI) - Wageningen University*

---

**Abstract**

This paper considers the performance and quality of apple fruits and seedlings production in Chencha district of southern Ethiopia. The estimated, three-factor (labour, land and capital) production frontier reveals that the technical inefficiency is 60 % and 48 % for fruits and seedlings production, respectively. Endowments in land, apple plantation and manure are important production factors for both fruits and seedlings, while labour is significant only for seedlings production. We could not reject constant returns to scale, neither for apple fruits nor for seedlings. Apart from capital and labour endowments, producer knowledge on apple cultivation is a positive and significant determinant of the level of output, product quality, and income generated from apples. The insignificance of the education variable for fruits and seedlings production suggests that what matters in the apple business is specific knowledge of apple-growing technology rather than formal education, at least beyond primary education.

*Keywords:* performance, knowledge, quality, apple fruits, apple seedlings, elasticity, returns to scale

---

## 1 Introduction

Apple (*Malus domestica*), a temperate climate zone fruit tree, is not an indigenous crop to Ethiopia and can thus be considered relatively 'new' for Ethiopian farmers (Ashebir *et al.*, 2010). According to Hayesso (2008) and Girmay *et al.* (2014) it was introduced to the country some fifty years ago by missionaries from Europe and North America. However, it is only after active promotion by NGOs, which led to the establishment of a marketing cooperative in 1998, that the product became widely known to a large number of producers and taken up as a means of income by those living in the highland agro-climatic zones of Chencha. Chencha, located in the Gamo Gofa administrative zone of the Southern Nations, Nationalities and Peoples' Regional State (SNNPRS), is the leading district in the country to adapt this fruit to the Ethiopian context and distribute the seedlings (plantlets)[1] to other parts of the country (Dagnew *et al.*, 2009). Nowadays, a number of other highland regions as well are becoming competent producers of different varieties of apple fruits and seedlings (Hruy *et al.*, 2012), assisted by government policies that promote apple as a strategic crop.

The farming systems of the Chencha district are mixed systems based on crop production and livestock rearing. According to the local Agricultural and Rural

* Corresponding author
P. O. Box 1278, Hawassa, Ethiopia
Email: sintayehu2007@gmail.com or sintayehuhl@yahoo.com

[1] The correct horticultural term for a grafted rootstock would be 'plantlet' rather than 'seedling', but for ease of comprehension we will refer to 'seedling' only, i.e., to distinguish this economic activity from fruit production in a consistent manner.

Development Office (personal communication with experts; Fetena et al., 2014), the district has a total area of 37,650 ha, situated at an altitude ranging from 1,600–3,200 m asl with an average annual rainfall of 1,100–1,300 mm. The mean annual temperature is around 17°C (min. 11°C to max. 23°C). The cereals most commonly grown in the area are wheat and barley, while beans and peas are the most popular leguminous crops. Potato and enset (false banana) are the dominant root crops in the area. In terms of livestock ownership, cattle, sheep, horses, mules and chickens feature most prominently. Being the most important cash crop, apple orchards are estimated to cover 728 ha in total. The commonly used rootstalk varieties are MM106, MM111, M26, M27, M9, M4 and what is locally known as CH6. Especially MM106 is a widely used variety (Fetena *et al.*, 2014). The common varieties used as scion wood are Ana, BR, CP29, Crispin, Princesa, Dorset, Red delicious, Jonica, Red Jonagold, and Galla must, among others. Some farmers produce their own rootstalk and scion, and sell the leftovers on the market.

Several researchers have studied the impacts of apple fruit[2] adoption in the highlands around Chencha (Hayesso, 2008; Dagnew *et al.*, 2009; Freeman, 2013; Fetena *et al.*, 2014; Girmay *et al.*, 2014). While using different methods, ranging from large-scale surveys to ethnographic fieldwork, the studies agree on the fact that the income generated from apple varies dramatically among those households who have adopted the crop. Dana Freeman, an anthropologist who has done extensive fieldwork in the locality of Masho, observed that the introduction of apple has led to a 'major increase in inequality' (Freeman, 2013). This is borne out in the study by Girmay *et al.* (2014), who calculated from their survey among 141 households who commercialise apple in eight different *kebeles* in and around Chencha, that the top 5 % of producers accounted for no less than 75 % of total income generated from apple, while the bottom 60 % fail to generate even 5 % of the total surplus.

In explaining this extreme heterogeneity in performance, Girmay *et al.* (2014) stressed the unequal distribution in production factors, not only in terms of available land and the number of mother trees acquired, but also for instance in the possession of livestock for manure or capital to purchase modern grafting and pruning equipment. At the same time, however, the research alluded to the existence of knowledge gaps. For instance, the finding that roughly three out of four producers interviewed failed to prune and graft their seedlings properly, is explained by 'the lack of technical know-how', in addition to the non-availability of equipment (ibid.). Fetena *et al.* (2014), who conducted a study among 181 apple farmers in the same area, confirmed the variation in knowledge levels. For instance, they observed that there is 'no standard for manuring and weeding of apple orchards in the study areas' (p.13), which signals that not all farmers have converged to best practice. This is also illustrated by the fact that sizeable portions of the respondents could neither identify insect pests and crop diseases nor know what control strategies are available. Unfortunately, both studies only presented selective descriptive statistics on both factor constraints and knowledge asymmetries and did not perform a statistical test to relate these directly to income differentials, so that the relative importance of endowments and knowledge remained unclear. This, however, is of interest from a policy point of view, as unequal access to endowments like capital, labour and land likely requires different types of interventions than differentiated access to, and mastery of, knowledge about apple cultivation. This paper purports to provide such an empirical test based on survey data from 380 apple-producing households from four selected localities in Chencha district.

In order to distinguish between input constraints and knowledge asymmetries, we first estimate the stochastic frontier production functions for apple fruits and apple seedlings in the area. Subsequently, we use the production function to estimate the output for each household, given their endowments, and then compare the 'fitted' output levels to the actual ones reported by the households. The degree of *technical efficiency* will be obtained by taking the ratio of the actual to fitted frontier output levels. This degree of efficiency, as well as the deviations between original and predicted values are then regressed on a knowledge index. We also link the knowledge index to quality differences in output, taking the shares of first and second grade apples in total production as dependent variables. Finally, we present an integrated regression that shows the relative importance of resource constraints and knowledge differentials. It is important to note that the current paper does not explore the various types of market and governance failures that may underlie both the inequality in endowments and the differentiated access to relevant knowledge. In this respect we should point out that limited resources devoted to apple production do not necessarily represent struc-

[2] While we will use the term 'apple fruits' throughout the paper, it should be pointed out that this also includes different varieties of pears. Local producers treat pears as a variety of apple, so we also ignore the distinction here. Apple is by far the dominant crop in terms of output and value, however.

tural inequalities. Households who are engaged in other types of agricultural production or in non-farm activities may deliberately choose to make limited investments in apple cultivation, even if their endowments are abundant.

## 2 Technical efficiency as performance measure

There is a wide array of literature that relates the use of various agricultural technologies to performance (Squires & Tabor, 1991; Bravo-Ureta & Pinheiro, 1993; Bravo-Ureta & Evenson, 1994; Kalirajan & Shand, 2001; Haji, 2007). A powerful measure of performance is the widely used notion of economic efficiency, which in turn can be decomposed into technical and allocative efficiency (Farrell, 1957). Allocative (or price) efficiency measures the firm's success in choosing the optimal input combination, i.e., where the ratio of marginal products for each pair of inputs is equal to the ratio of their market prices (Bravo-Ureta & Pinheiro, 1993). By contrast, a farmer's technical efficiency measures a farmer's ability to produce the maximum possible level of output from a given input and production technology (Squires & Tabor, 1991). Hence, technical inefficiency refers to the failure of attaining this maximum level. Since our dataset does not have complete information about prices and costs, we confine our analysis to technical efficiency only.

Technical efficiency is a concept that compares each farmer's production performance with the input-output relationship that is deemed 'best practice' (Squires & Tabor, 1991). The best practice can consist in the performance of a researcher in a farmer's field or the practice of the most efficient farmers in a comparable environment. Technical efficiency is then measured as the deviation of individual farmers from this best practice frontier. In our case the practice of the best producers has been taken as a benchmark.

Stochastic frontier analysis and data envelopment analysis (DEA) are two alternative methods for estimating frontier functions and thereby measuring efficiency of production (Coelli *et al.*, 1998). DEA involves the use of linear programming, whereas stochastic frontiers involve the use of econometric methods. The stochastic frontier production model incorporates a composite error structure with a two-sided symmetric error and a one-sided component as proposed by Aigner *et al.* (1977) and Meeusen & van den Broeck (1977). The one-sided component reflects inefficiency ($u$), while the two-sided error ($v$) captures the random effects outside the control of the production unit, including measurement error and other random factors that can affect the relationship (Bravo-Ureta & Pinheiro, 1993; Coelli *et al.*, 1998). The estimation of the stochastic frontier function can be accomplished using maximum likelihood estimation or corrected ordinary least squares (COLS) (Richmond, 1974), depending on whether an explicit distribution for the error term of the efficiency component is assumed or not (Bravo-Ureta & Pinheiro, 1993). When assumptions are made explicit, exponential, gamma or half-normal distributions are commonly used.

Several agricultural efficiency analyses have been carried out in the Ethiopian context, of which the studies by Gebreegziabher *et al.* (2004), Seyoum *et al.* (1998) and Haji (2007) stand out. Seyoum *et al.* (1998) estimated the stochastic frontier production function for maize farmers in eastern Ethiopia, distinguishing between participants and non-participants in a project that promotes high-input maize technologies (Sasakawa-Global 2000 project). They established that project farmers are technically more efficient than those who remained outside the project. The study by Haji (2007) estimated technical, allocative and economic efficiency levels for mixed farmers in eastern Ethiopia who are predominantly engaged in growing vegetables. Using non-parametric data envelopment analysis they revealed technical, allocative and economic efficiency of 91, 60 and 56 %, respectively. Finally, using a similar methodology, Gebreegziabher *et al.* (2004) found 80 % technical efficiency among farmers in northern Ethiopia in producing commonly grown crops in the region. They used the value of overall crop output (in birr) as a dependent variable.

## 3 Materials and methods

### *3.1 Sampling, data and measurement*

The sampling strategy for the survey was based on random selection of four out of six *kebeles* in the district, all of which have experience in apple production. In the selected villages all apple-producing farmers were surveyed. Hence, the survey was in fact a census, ensuring representativeness for these four localities. In total 380 households were interviewed.

In order to estimate the average and stochastic frontier Cobb-Douglas production function, the dependent variables are 1) apple fruits output and 2) number of seedlings produced. The explanatory variables include 1) plot size allocated to fruits and seedlings production, 2) labour used in the production of fruits and seedlings in man-days, 3) reported value of apple plantation owned, and 4) amount of manure applied. Together with apple plantation value, manure is considered to represent a capital input. Since there is no market for manure in Chencha, livestock wealth is the dominant constraint on the application of manure. All variables are transformed into logarithms.

The general performance in apple production among the surveyed households is proxied by the technical efficiency measure and individual performance is assessed by the ratio $\frac{(Y-\hat{Y})}{\hat{Y}}$, where $Y$ is the original observation of output and $\hat{Y}$ is the predicted value of output after fitting the logarithmic transformation of the Cobb-Douglas production function. This ratio has been generated for fruits and seedlings separately. The same procedure has been applied to income generated from apple and seedlings, which is the performance indicator that is of prime interest to explain the observed disparity in welfare outcomes from apple production.

The level of a producer's knowledge on apple 'technology' is measured by scoring this producer on a knowledge index. This index is constructed based on weighting the individual's knowledge on the 'best practices' in the various stages of cultivation of the crop. These stages were categorised into six; variety identification, grafting, tree management, pest control, post-harvest handling, and marketing. For instance, farmers who lack knowledge on variety identification may not properly graft their seedlings with the appropriate variety. Likewise, those who do not know the benefit of pruning might be unable to manage the size of their tree and thereby jeopardize the quality of the fruits. Knowledge pertaining to each stage of production has been scored on a five-point Likert scale, based on self-reports. The overall index scale for knowledge has been generated using principal components analysis (PCA). Three principal components with eigen values higher than unity were taken from the six knowledge level variables.

Apple fruits and seedlings quality is measured by the respective ratios of first grade output to total production, which are used as dependent variables in the quality regressions. In addition to the knowledge index, the models control for other explanatory variables hypothesised to affect performance, including a number of demographic and socioeconomic characteristics of farmers and their households, market availability, membership in a cooperative, and village dummies.

### 3.2 The model

The stochastic frontier Cobb-Douglas production function with the following form is assumed:

$$Y = A\, f(K^{\alpha}, L^{\beta}, N^{\gamma})\, e^{\varepsilon_i}, \tag{1}$$

where $Y$ = apple fruits output or the number of seedlings produced, $K$, $L$ and $N$ are capital, labour, and land that have been used in the production of apple fruits and seedlings, $\varepsilon_i = v_i - u_i$ is the composite error term containing the random error ($v$) and the technical efficiency component ($u$), A is production technology, and $\alpha$, $\beta$ and $\gamma$ are the elasticity coefficients of output with respect to capital, labour and land, respectively. All the inputs are assumed variable. Land for a given individual in a season is fixed but in cross-sectional data it varies across observations.

Equation (1) can be expressed in linear form by using the following logarithmic transformation:

$$\ln Y = \beta_0 + \alpha \ln K_i + \beta \ln L_i + \gamma \ln N_i + v_i - u_i \tag{2}$$

Therefore, to estimate elasticity parameters of each input, we run OLS regression on equation (2). These elasticity coefficients from the OLS estimation indicate the percentage change in output for a unit percentage change in the respective inputs. The sum of the elasticity coefficients is an indicator of the returns to scale in production, i.e., $\alpha + \beta + \gamma = 1$ indicates constant returns to scale and a sum $< 1$ ($> 1$) decreasing (increasing) returns to scale.

Based on the estimation of the stochastic frontier model of equation (2) using maximum likelihood estimation and the half-normal distribution assumption for the error term, the technical efficiency coefficients for each individual producer can be calculated by dividing the actual production figures by the predicted values of the frontier function $TE = \frac{Y}{\hat{Y}}$, and range between 0 and 1. The level of inefficiency $(1 - TE)$ is subsequently used in the regression.

The other output performance indicator, $\frac{(Y-\hat{Y})}{\hat{Y}}$, is modeled as follows:

$$Z = \sigma + \lambda X + e, \tag{3}$$

where $Z = \frac{(Y-\hat{Y})}{\hat{Y}}$, $X$ is a vector of explanatory variables, $\sigma$ the constant term, $\lambda$ is a vector of parameters to be estimated, and $e$ is the error term.

The ratio of first grade to total production, denoted by $\frac{Q_1}{Q_T}$, has been taken as a dependent variable to explore factors affecting product quality, where $Q_1$ is first grade quantity and $Q_T$ is total quantity of production. This ratio is calculated for fruits and seedlings quantities separately. A similar model as in equation (3) has been fitted for quality performance. For the inefficiency and quality analyses a Tobit model has been applied, while the other analyses are run using OLS regression techniques.

## 4 Results

### 4.1 *Descriptive statistics*

The descriptive statistics of the surveyed households are shown in Table A1 in the Appendix. The data indicate that the average household size is 6.2 and the maximum number of household members registered equals 14. The age of respondents ranges between 16 and 95 years with an average age of 44.2. The respondent with the best education profile reports having attended school for 16 years, but on average educational careers last for 7.8 years only. Regarding apple and seedlings production, the extent of experience varies widely. One respondent has been producing apple for 34 years, whereas others started as recently as one year ago. On average, respondents have 10.2 years of experience in apple cultivation. Average landholdings are small with 0.9 ha, ranging from a minimum of 0.005 to a maximum of 7.65 ha. Producers on average allot 0.08 and 0.03 ha of land for fruits and seedlings, respectively. The data reveal a strong preference to cultivate apple fruits and seedlings in their own backyard rather than on plots located at some distance from their residence. This might be due to fear of loss from theft [3], ease of management or soil fertility reasons.

The average income of the producers from apple fruits production was 5,555 Birr ($285) [4] in the 2012 production year. The maximum amount earned by a producer was 59,713 Birr ($3,062). Seedlings production generated an income of 17,400 Birr ($892) on average, up to 244,240 Birr ($12,525) in the best of cases in the same production season. First-grade fruits and seedlings fetched a market price that was on average 38 % and 68 % higher than the market price obtained for second-grade fruits and seedlings, respectively. Concerning quality of the produce, 65 % of the total production of fruits and seedlings of the average producer is first grade. Hence, more than a third of production tends to be classified as second grade.

### 4.2 *Production functions*

Table 1 reports the result from estimating equation (2) to arrive at the average production function and stochastic frontier for apple fruits and seedlings. As shown in column 2, the output of apple fruits is positively and significantly affected by both capital-intensive inputs and land. The column reports elasticity coefficients of apple fruits output to each factor input. Plantation asset has an elasticity of 0.30 and manure of 0.22. It follows that the elasticity of fruits output to capital is 0.52, which is the sum of the two elasticity coefficients. Therefore, a unit percentage increase in capital input yields a 0.52 % increase in fruits output.

The elasticity of land equals that of capital. Increasing the amount of land in hectares by a unit percentage point raises fruits output by 0.52 percentage points. The elasticity with regard to labour is negative but insignificant. As shown in the fourth column, all input factors make a significant positive contribution to seedlings production. The elasticity coefficients are 0.6, 0.13 and 0.28 for capital, labour and land, respectively.

The Likelihood Ratio test indicates technical inefficiency in the production of both apple fruits and seedlings. The inefficiency was 60 % for fruits and 48 % for seedlings production (Table 1). Table A3 in the annex displays the Tobit regressions of these levels of inefficiency on a set of exogenous explanatory variables. Inefficiency in fruits production is explained by a lack of cultivation experience and by the particular village in which the producer resides. By contrast, inefficiency in seedlings production can be linked to a lack of knowledge on the technology and not being able or willing to join a marketing cooperative. Moreover, older and female farmers tend to be less efficient.

### 4.3 *Productivity*

Table A2 of the annex shows the productivity of capital and labour per unit of land. Output per hectare was used as the dependent variable and labour per hectare, manure per hectare, and plantation asset per hectare, were treated as explanatory variables. We find that seedlings production is more intensive in manure than fruits production, while plantation asset is more productive for fruits than for seedlings. Also, the productivity analysis confirms that labour input is important for seedlings, but not significant in fruits production. As a robustness check, we ran an adapted version of the Cobb-Douglas production function in equation 2, in which labour and manure are included as logarithmic transformations while the other variables, including additional explanatory variables, are in levels. The results (available from authors on request) are almost identical to the ones obtained from our original specification.

[3] This particularly applies to seedlings. According to Freeman's (2009) observations in Masho village: "The problem [of theft] has become so intense that nowadays no-one plants apple saplings in regular agricultural plots, which are unfenced and unguarded. Instead people have started to plant the apple saplings within their homesteads." pp.251.

[4] At the time of data collection the average official exchange rate of USD to Ethiopian Birr was 1$(USD) = 19.5 Birr.

**Table 1:** *Cobb-Douglas frontier production functions for apple fruits and seedlings*

| *Variables* | *Fruits output (in kg)* | | *# of seedlings produced* | |
|---|---|---|---|---|
| | *Average function* | *Stochastic frontier* | *Average function* | *Stochastic frontier* |
| Quantity of manure applied (in kg) | 0.22 (0.07)** | 0.18 (0.06)** | 0.39 (0.05)** | 0.36 (0.05)** |
| Value of apple plantation asset (in Birr) | 0.30 (0.06)** | 0.28 (0.06)** | 0.21 (0.05)** | 0.20 (0.04)** |
| Land allocated (in ha) | 0.52 (0.06)** | 0.54 (0.05)** | 0.28 (0.05)** | 0.28 (0.04)** |
| Labour used (in man-days) | –0.05 (0.08) | –0.06 (0.07) | 0.13 (0.05)** | 0.16 (0.05)** |
| Constant | 2.88 (0.73)** | 4.80 (0.73)** | 3.00 (0.57)** | 3.73 (0.57)** |
| F-Statistic | 56.46 | – | 94.94 | – |
| Adjusted R-squared | 0.39 | – | 0.55 | – |
| Λ | – | 2.27 (0.21) | – | 1.63 (0.20) |
| $\sigma^2$ | – | 3.48 (0.42) | – | 1.48 (0.22) |
| Log Likelihood | – | –561 | – | –397.5 |
| $\chi^2$ for $u = 0$ | – | 22.5** | – | 7.87** |
| Average inefficiency, % | | 60 | | 48 |
| No. of obs. | 343 | 343 | 307 | 307 |

Note: $^+ p<0.1$; $^* p<0.05$; $^{**} p < 0.01$. The figures in parentheses represent the robust standard errors.
All variables are expressed in natural logarithms.
Source: Own survey 2013

### 4.4 *Variation in output performance*

Based on the Cobb-Douglas results presented in Table 1, output values have been predicted as well as the corresponding deviations of these predicted values from the actual production values, both for fruits and seedlings. The ratio of these deviations to the predicted values, defined as Z in equation (3), has been used as the dependent variable in the regressions shown in Table 2. We ran OLS for fruits (column 2) and seedlings (column 3) on the knowledge index and a number of exogenous explanatory variables.

The analysis reveals that knowledge positively and significantly contributes to seedlings production, but turns out to be an insignificant determinant of fruits output. Other significant determinants of output (for both fruits and seedlings) include the size of the cultivated area and the number of fruit trees owned by the producer. The age of trees plausibly affects only fruits output. Also, the cultivation experience (in years) of the producers improves output from fruits but not from seedlings. This indicates that new adopters are more inclined towards the production of seedlings, which is evidenced by the age of the producers as well. Age is a negative and significant determinant for seedlings output. Interestingly, market availability positively affects fruits output but is not significant for seedlings.

Another important predictor in the case of seedlings performance is the respondent's engagement in non-farm activities, which negatively affects output from seedlings. This indicates that those who achieve high output levels of seedlings tend to refrain from participation in non-farm activities. Another interesting result concerns (formal) education, which is insignificant in the model. Concerning gender, the sex of the producer only matters for seedlings production, where men outperform women.

Another relevant determinant is location, albeit for fruits production only. Fruits output in Doko Shaye subdistrict is significantly higher than in the reference village of Tolola. Finally, cooperative membership has a positive and significant contribution on both fruits and seedlings production. Cooperative members on average produce 2 % more fruits and 5.2 % more seedlings than non-members, other things being equal.

### 4.5 *Performance on Income*

In Table 3 output is replaced by the income that households report to have generated from fruits and seedlings as alternative performance indicator. Quality, to which we turn in the next section, is included as explanatory variable in this analysis. The results indicate that the size of the area cultivated, number of trees, product quality, and cooperative membership are significant and positive contributors towards higher income from the apple business. On the other hand, the knowledge of the producer and quantity of manure applied were found to be positive contributors to income from seedlings production only.

**Table 2:** *OLS regression for apple fruits and seedlings output performance*

| *Variables* | *Ratio for fruits output* | *Ratio for seedlings produced* |
|---|---|---|
| Apple cultivation experience in years | 0.29 (0.10)** | –0.08 (0.30) |
| Market availability (rated 1–5) | 1.68 (0.81)* | 1.12 (1.47) |
| Knowledge index on apple technology | –0.14 (0.28) | 2.96 (0.93)** |
| Area covered | 9.75 (5.88)+ | 63.54 (29.65)* |
| Number of trees | 0.05 (0.01)** | 0.13 (0.04)** |
| Average age of trees | 0.61 (0.15)** | –0.22 (0.41) |
| Quantity of manure applied | –0.00 (0.00) | 0.00 (0.00) |
| Age of respondent | –0.03 (0.04) | –0.19 (0.09)* |
| Sex of household head (male=0; female=1) | 1.54 (1.40) | –6.50 (2.11)** |
| Education, grade completed in years | –0.10 (0.13) | –0.37 (0.27) |
| Non-farm work participation dummy | 0.78 (1.22) | –6.18 (2.76)* |
| Asset index (value of livestock, equipment & house) | 0.24 (0.55) | 0.35 (1.97) |
| Cooperative membership | 2.03 (1.12)+ | 5.20 (2.06)* |
| Dummy for Doko Shaye village | 8.46 (1.66)** | –3.76 (3.00) |
| Dummy for Doko Losha village | –2.82 (1.74) | –0.85 (3.99) |
| Dummy for Chencha town | 2.24 (1.39) | 2.12 (3.45) |
| Constant | –15.37 (3.58)** | 25.23 (9.31)** |
| F-Statistic | 12.69 | 7.41 |
| Adjusted R-squared | 0.42 | 0.44 |
| No. of obs. | 296 | 310 |

Note: + $p<0.1$; * $p<0.05$; ** $p < 0.01$. The figures in parenthesis represent the robust standard error.
Source: Own survey 2013

**Table 3:** *Performance on fruits and seedlings income*

| *Variables* | *Ratio for fruits income* | *Ratio for seedlings income* |
|---|---|---|
| Area covered | 68.45 (32.9)* | 341.45 (181.14)+ |
| Number of trees | 0.33 (0.06)** | 0.67 (0.23)** |
| Average age of trees | 2.52 (0.55)** | –2.11 (2.52) |
| Quantity of manure applied | –0.01 (0.01) | 0.08 (0.03)** |
| Quality (first grade/total production) | 24.76 (8.82)** | 47.24 (27.78)+ |
| Sex of household head (male=0; female=1) | 5.88 (5.43) | –30.57 (9.31)** |
| Apple cultivation experience in years | 0.38 (0.39) | –1.33 (1.79) |
| Cooperative membership | 11.80 (2.86)** | 28.99 (10.95)** |
| Knowledge index on apple technology | –0.95 (0.92) | 16.90 (5.47)** |
| Constant | –46.21 (9.29)** | 27.58 (25.98) |
| F-Statistic | 22.61 | 8.89 |
| Adjusted R-squared | 0.47 | 0.42 |
| No. of obs. | 329 | 273 |

Note: + $p<0.1$; * $p<0.05$; ** $p < 0.01$. The figures in parenthesis represent the robust standard error.
Source: Own survey 2013

### *4.6 Quality*

The quality of fruits and seedlings, i.e., the share of first-grade produce, was regressed on a set of explanatory variables, including the level of knowledge of the producer. Table 4 shows the results from the Tobit regression. Knowledge on apple technology turns out to be a positive and significant determinant of both fruits and seedlings quality. Other relevant quality determinants of fruits include market availability, price incentives (first-grade to second-grade price ratio), cooperative membership and the frequency of visiting other producers' orchards. The number of visits paid to the orchards of others likely increases the production abilities of the visitor through sharing experiences and best practices of the producers visited. Location-wise the Doko Shaye village and Chencha town outperform the reference village of Tolola in fruits quality, whereas Doko Losha village produces lower-quality fruits than Tolola. The village dummies did not pick up any location-specific effects on the quality of seedlings.

## 5 Discussion

The result from the Cobb-Douglas stochastic frontier analysis generated the expected sign for all inputs, except labour, in the production of both fruits and seedlings. Capital is slightly more important for seedlings than for fruits production, but the real difference lies in land and labour endowments. Land availability is far less important in seedlings production, which results from the fact that seedlings can be planted in close range from each other, whereas fruit trees need to be properly spaced in an orchard. Moreover, labour is a significant input in seedlings production, which was not the case for fruits. Since the sum of the elasticities roughly equals unity, we can conclude that seedlings production exhibits constant returns to scale. Nor could we reject the constant returns to scale assumption for fruits, except for the negative sign for labour. While counterintuitive, it is not uncommon to find negative contributions from labour in this type of analysis (Sahota, 1968; Chowdhury *et al.*, 1975). It should be borne in mind that labour in fruits production mainly concerns land preparation and planting, which is a one-time activity in the first season and in consecutive years no additional land preparation is required except for relatively modest labour efforts in tree management activities, such as weeding, pruning, irrigation, and harvesting. Hence, compared to seedlings production, labour input in fruits production is low, except for the early years after planting. Even though it is plausible that labour is not a large contributing factor in apple production, we did not anticipate a (modest) negative effect. Sahota (1968) points to multicollinearity, measurement error or shortage of rain as potential

**Table 4:** *Tobit model estimation for fruits and seedlings quality*

| *Variables* | *Fruits quality* | *Seedlings quality* |
|---|---|---|
| Market availability (rated 1–5) | 0.05 (0.01)** | 0.00 (0.01) |
| Density (no. of trees/area) | −0.00 (0.00) | 0.00 (0.00) |
| Per-capita manure applied | −0.00 (0.00) | −0.10 (0.049)* |
| Price ratio (first grade/second grade) | 0.07 (0.03)* | −0.01 (0.01) |
| Knowledge index on apple technology | 0.02 (0.01)** | 0.02 (0.01)* |
| Sex of the respondent | 0.03 (0.03) | 0.02 (0.03) |
| Age of the respondent | −0.00 (0.00) | −0.00 (0.00) |
| Highest grade education in years | −0.00 (0.00) | 0.00 (0.00) |
| Cooperative membership dummy | −0.06 (0.04)+ | 0.00 (0.04) |
| Frequency of invitation by neighbours | 0.03 (0.01)** | 0.02 (0.01)** |
| Dummy for Doko Shaye village | 0.14 (0.03)** | 0.04 (0.03) |
| Dummy for Doko Losha village | −0.13 (0.04)** | −0.05 (0.04) |
| Dummy for Chencha town | 0.09 (0.03)** | 0.04 (0.03) |
| Constant | 0.30 (0.10)** | 0.54 (0.07)** |
| LR $\chi^2$ | (110.08)** | (29.09)** |
| Pseudo R- squared | −0.71 | −0.12 |
| No. of obs. | 239 | 199 |

Note: + $p<0.1$; * $p<0.05$; ** $p<0.01$. The figures in parenthesis represent the robust standard error.
Source: Own survey 2013

causes of a negative contribution from labour that resulted from his own work on Indian agriculture. We do not detect multicollinearity problems in our dataset, but cannot rule out measurement error [5].

The level of technical inefficiency is higher in fruits production relative to the seedlings case, which may suggest that people usually plant the trees (orchard) for trial and ignore the management for lack of know-how. Still, it will bear some fruits eventually. Whereas in the seedlings case, most people may not be strongly motivated to try the planting when lacking the necessary techniques and if they try, it will be with some understanding of how it works. Hence, seedlings production is more knowledge-intensive, requiring technical know-how on issues like variety identification, grafting and other intensive seedlings management activities. Therefore, there is more involvement in planting orchards relative to producing seedlings among those with limited information, resulting in higher levels of inefficiency in fruits production.

Performance in production is driven by a number of factors. For the fruits, producer's experience and the tree quality and quantity (i.e. age and number of trees) delivers a better outcome, but for the seedlings, technical know-how, age, and sex are important determinants. Young and male producers perform better. This might be due to better access to information relative to the other groups. Another important predictor in the case of seedlings performance is the respondent's engagement in non-farm activities, which negatively affects output from seedlings. This indicates that those who achieve high output levels of seedlings tend to refrain from participation in non-farm activities. Due to the relatively high labour-intensity of seedlings production, it competes for scarce labour with non-farm occupations.

The insignificance of education shows that what matters in apple production is the specific knowledge on the technology rather than the formal education attained by the producer. However, it should be noted that the average level of education of the producers is 7.8 years, so that having completed primary education could still be an important determinant of performance. The observed effect of education in the literature is mixed. Studies have encountered positive, negative and neutral effects. Kalirajan & Shand (1985) obtained non-significant effect of education on rice yield in Tamil Nadu, whereas Pudasaini (1983) found a positive effect of education in modern agriculture compared to the traditional one in Nepal [6]. Concerning the locations, the fieldwork did not bring forward an obvious explanation for this variation in performance across these specific localities, but variation in soil fertility or climatic conditions that affect dormancy may be responsible. No other location effects were detected.

Performance in income is driven by quality of produce and cooperative membership, among others. Since first-grade produce commands a much higher price than its second-grade counterpart, quality producers will benefit more. Cooperatives are the dominant market outlets in the area that increase the earning of the producers who supply to these channels, since they have better market access and reap a higher price compared to the other outlets. Market availability and price ratios of first and second grade produce were important predictors of fruits quality. This might be due to the fact that market demand for seedlings has so far been high, also as a result of government promotion, while demand for fruits is considered a constraint, given that most Ethiopian consumers are not yet very familiar with apples. It is expected that the tables might turn in the future, however. Demand for seedlings is likely to dwindle once other regions have expanded their apple cultivation, while demand for fruits is likely to grow steadily (Girmay *et al.*, 2014). The price gap between first and second grade output sometimes tends to be very large, such that it indirectly discourages production of second grade relative to premium quality, by which it contributes to quality fruits production. Manure application, unexpectedly, failed to show a significant effect on apple quality, while importantly affecting apple yields in volume terms.

## 6 Conclusion

With the objective to estimate technical efficiency using the Cobb-Douglas stochastic frontier production function and to identify how knowledge of the producer contributes to output and quality in fruits and seedlings production, we have found that; first, the stochastic frontier production function estimation has shown that there was 60 % and 48 % technical inefficiency in the production of apple fruits and seedlings in Chencha district, respectively. Table 5 provides an overview of the key drivers of this inefficiency, for both fruits and seedlings, beyond the 'usual suspects' of disparities in conventional input factors land, labour and capital. When comparing fruits and seedlings, the respective drivers show

[5] Guarding the seedlings 24/7 is an activity done parallel to other activities, especially during daytime, which has likely resulted in some double counting when eliciting a household's time allocation.

[6] See Weir & Knight (2004) and Weir (1999) for a detailed review on the effect of education on agricultural performance.

**Table 5:** *Summary of results from the regression analysis*

| *Variable* | *Fruits* | | | | *Seedlings* | | | |
|---|---|---|---|---|---|---|---|---|
| | *Output Performance* | *Income Performance* | *Technical Efficiency* | *Quality* | *Output Performance* | *Income Performance* | *Technical Efficiency* | *Quality* |
| Knowledge | (–) | (–) | (+) | (+)*** | (+)*** | (+)*** | (+)* | (+)** |
| Education | (–) | | (–) | (–) | (–) | | (–)* | (+) |
| Experience | (+)*** | (+) | (+)*** | | (–) | (–) | (–) | |
| Age | (–) | | (+) | (–) | (–)** | | (–)*** | (–) |
| Sex | (+) | (+) | (–) | (+) | (–)*** | (–)*** | (–)** | (+) |
| Cooperative membership | (+)* | (+)*** | (+) | (–)* | (+)** | (+)*** | (+)** | (+) |

Note: * $p<0.1$; ** $p<0.05$; *** $p<0.01$. The + and – signs in parentheses are the signs of the coefficients of the respective variables.
Source: Own survey 2013

considerable overlap. For example, cooperative membership is important in explaining the income obtained from both apples and seedlings, whereas formal education levels fail to emerge as an important determinant in either of these. At the same time, the overlap is only partial. While knowledge impacts positively on the quality of both apples and seedlings, it only boosts income from seedlings through higher output, which does not materialize in the case of fruits. Vice versa, experience in cultivation benefits fruits production, whereas it does not have an impact on the outcome variables for seedlings. The latter are, unlike those for apples, responsive to age and gender of producers.

Since the level of technical efficiency in fruits and seedlings production is quite low, there is ample room to take measures that push inefficient producers towards the frontier. A direct way to achieve this objective is to improve the knowledge of producers. The extension service and other stakeholders are expected to improve the knowledge of the producers on various aspects of the technology, but the potential of other sources of knowledge, such as social networks, training and experience, and their relative contribution to improve the technological competence of the producer, need further research. Second, market access could be optimised through strengthening weak cooperatives or linking farmers to strong cooperatives and other potential customers. Third, since capital is most significant, planting more orchard trees will improve the supply of planting materials for seedlings production and increase the volume of fruits output. Moreover, livestock production needs to be promoted in the absence of a market for manure.

Finally, as young producers were performing well, access to land and licensing of knowledgeable seedlings producers might be considered to exploit the maximum benefit from the technology and improve livelihood in highland areas like the Chencha district in Ethiopia.

## Acknowledgements

The authors acknowledge funding from the Netherlands organisation for international cooperation in higher education (Nuffic) within the framework of its NICHE programme ("Agribizz" project ETH019).

## References

Aigner, D., Lovell, C. K. & Schmidt, P. (1977). Formulation and estimation of stochastic frontier production function models. *Journal of Econometrics*, 6 (1), 21–37.

Ashebir, D., Deckers, T., Nyssen, J., Bihon, W., Tsegay, A., Tekie, H., Poesen, J., Haile, M., Wondumagegnehue, F., Raes, D., Behailu, M. & Deckers, J. (2010). Growing apple (*Malus domestica*) under tropical mountain climate conditions in northern Ethiopia. *Experimental Agriculture*, 46 (1), 53–65.

Bravo-Ureta, B. E. & Evenson, R. E. (1994). Efficiency in agricultural production: the case of peasant farmers in eastern Paraguay. *Agricultural Economics*, 10 (1), 27–37.

Bravo-Ureta, B. E. & Pinheiro, A. E. (1993). Efficiency analysis of developing country analysis: a review of the frontier function literature. *Agricultural and Resource Economics Review*, 22 (1), 88–101.

Chowdhury, S. R., Nagadevara, V. & Heady, E. O. (1975). A Bayesian application on Cobb-Douglas production function. *American Journal of Agricultural Economics*, 57 (2), 361–363.

Coelli, T., Rao, D. S. P. & Battese, G. E. (1998). *An Introduction to Efficiency and Productivity Analysis*. Kluwer Academic Publishers, New York.

Dagnew, T., Hayesso, T. & Visser, P. (2009). A Bright Future for Small Fruits Farmers in Southern Ethiopia. SNV Netherlands Development Cooperation Case Studies.

Farrell, M. J. (1957). The measurement of productive efficiency. *Journal of the Royal Statistical Society: Series A (General)*, 120 (3), 253–290.

Fetena, S., Shara, S., Anjulo, A., Gulie, G., Woldesenbet, F. & Yilma, B. (2014). Survey on apple production and variety identification in Chencha District of Gamo Gofa Zone, Southern Ethiopia. *Journal of Agriculture and Food Technology*, 4 (5), 7–15.

Freeman, D. (2009). Development and (un)happiness: a case study from rural Ethiopia. *In:* Ura, K. & Penjore, D. (eds.), *Gross National Happiness: Practice and Measurement*. Centre for Bhutan Studies, Thimpu.

Freeman, D. (2013). Value chains for development: an ethnography of pro-poor market interventions in Ethiopia. *Anthropology of This Century*, 3 (1).

Gebreegziabher, Z., Oskam, A. J. & Woldehanna, T. (2004). Technical efficiency of peasant farmers in northern Ethiopia: a stochastic frontier approach. Proceedings for the Second International Conference on the Ethiopian Economy, Addis Ababa, 3–5 June.

Girmay, G., Menza, M., Mada, M. & Abebe, T. (2014). Empirical study on apple production, marketing and its contribution to household income in Chencha district of southern Ethiopia. *Scholarly Journal of Agricultural Science*, 4 (3), 166–175.

Haji, J. (2007). Production efficiency of smallholders' vegetable-dominated mixed farming system in eastern Ethiopia: a non-parametric approach. *Journal of African Economies*, 16 (1), 1–27.

Hayesso, T. (2008). Hope in Apples. SNV Netherlands Development Cooperation Case Studies.

Hruy, G., Aregay, N., Teklebirhan, Y., Deckers, T., Bauer, H., Amare, K., Gebrehiwot, K., Deckers, J. & Keulemans, W. (2012). Apples in tropical highlands of northern Ethiopia: potentials and challenges. *Chronica Horticulturae*, 52 (3), 16–22.

Kalirajan, K. P. & Shand, R. T. (1985). Types of education and agricultural productivity: a quantitative analysis of Tamil Nadu rice farming. *Journal of Development Studies*, 21 (2), 232–243.

Kalirajan, K. P. & Shand, R. T. (2001). Technology and farm performance: paths of productive efficiencies over time. *Agricultural Economics*, 24 (3), 297–306.

Meeusen, W. & van den Broeck, J. (1977). Efficiency estimation from Cobb-Douglas production functions with composed error. *International Economic Review*, 18 (2), 435–444.

Pudasaini, S. P. (1983). The effects of education in agriculture: evidence from Nepal. *American Journal of Agricultural Economics*, 65 (3), 509–515.

Richmond, J. (1974). Estimating the efficiency of production. *International Economic Review*, 15 (2), 515–521.

Sahota, G. S. (1968). Efficiency of resource allocation in Indian agriculture. *American Journal of Agricultural Economics*, 50 (3), 584–605.

Seyoum, E. T., Battese, G. E. & Fleming, E. M. (1998). Technical efficiency and productivity of maize producers in eastern Ethiopia: a study of farmers within and outside the Sasakawa-Global 2000 project. *Agricultural Economics*, 19 (3), 341–348.

Squires, D. & Tabor, S. (1991). Technical efficiency and future production gains in Indonesian agriculture. *The Developing Economies*, 29 (3), 258–270.

Weir, S. (1999). The effects of education on farmer productivity in rural Ethiopia. Working Paper 99-7. Center for the Study of African Economies, Department of Economics, University of Oxford, UK.

Weir, S. & Knight, J. (2004). Externality effects of education: dynamics of the adoption and diffusion of an innovation in rural Ethiopia. *Economic Development and Cultural Change*, 53 (1), 93–113.

# Appendix

**Table A1:** *Descriptive statistics*

| *Variable* | *Obs* | *Mean* | *SD* | *Min.* | *Max.* |
|---|---|---|---|---|---|
| Family size | 380 | 6.20 | 2.51 | 1 | 14 |
| Sex of head (0=male; 1=female) | 380 | 0.12 | 0.32 | 0 | 1 |
| Age | 380 | 44.20 | 13.97 | 16 | 95 |
| Education in years | 380 | 7.79 | 4.64 | 0 | 16 |
| Cultivation experience in years | 380 | 10.52 | 5.47 | 1 | 34 |
| Knowledge index | 380 | 0.00 | 1.99 | –4.38 | 4.45 |
| Asset index | 380 | 0.00 | 1.19 | –1.06 | 9.89 |
| No. of apple trees | 380 | 47.93 | 69.55 | 2 | 718 |
| Age of trees | 380 | 7.32 | 3.60 | 1 | 27 |
| Manure amount applied for fruits production (in kg) | 380 | 107.4 | 230.64 | 1 | 2,320 |
| Manure amount applied for seedlings production (in kg) | 320 | 78.41 | 301.86 | 1 | 4,140 |
| Total plot size owned (ha) | 380 | 0.90 | 1.02 | 0.005 | 7.65 |
| Plot allotted for apple fruits (ha) | 380 | 0.08 | 0.12 | 0.001 | 1 |
| Plot allotted for seedlings (ha) | 314 | 0.03 | 0.09 | 0.0004 | 1 |
| Labour per fruit tree (man-days) | 380 | 9.18 | 15.07 | 0.11 | 146.4 |
| Labour per seedling (man-days) | 312 | 2.50 | 5.55 | 0.02 | 40.4 |
| Total fruits production in kg | 380 | 307.1 | 527 | 1 | 3923 |
| Total seedlings production in number | 314 | 711.8 | 1065.1 | 5 | 8150 |
| Income from fruits in Birr | 380 | 5,555.6 | 9,473.8 | 1 | 59,713 |
| Income from seedlings in Birr | 380 | 17,398 | 33,614 | 1 | 244,240 |
| Seedlings quality | 314 | 0.66 | 0.20 | 0 | 1 |
| Fruits quality | 354 | 0.64 | 0.22 | 0 | 1 |
| Fruits price ratio ($1^{st}/2^{nd}$) grade | 240 | 1.38 | 0.32 | 0.74 | 3 |
| Seedlings price ratio ($1^{st}/2^{nd}$) grade | 200 | 1.68 | 0.71 | 0.81 | 4.78 |
| Ratio fruits income deviation | 334 | 29.32 | 46.66 | –1 | 242.65 |
| Ratio seedlings income deviation | 277 | 86.48 | 145.87 | –1 | 954.26 |
| Ratio seedlings output deviation | 312 | 18.17 | 26.13 | –1 | 169.22 |
| Ratio fruits output deviation | 312 | 5.98 | 10.28 | –1 | 67.43 |
| Log fruits produced | 354 | 4.68 | 1.67 | 0 | 8.27 |
| Log seedlings produced | 314 | 5.77 | 1.36 | 0 | 9.00 |
| Log plantation asset | 380 | 9.79 | 1.38 | 6.21 | 13.82 |
| Log fruits land | 380 | –3.37 | 1.32 | –6.90 | 0 |
| Log seedlings land | 314 | –4.35 | 1.54 | –7.82 | 0 |
| Log labour fruits | 343 | 4.96 | 0.97 | 1.94 | 7.08 |
| Log labour seedlings | 310 | 5.47 | 1.14 | 1.80 | 7.48 |
| Log manure for fruits | 380 | 3.89 | 1.20 | 0 | 7.75 |
| Log manure for seedlings | 320 | 3.18 | 1.30 | 0 | 8.33 |
| Market availability for fruits | 363 | 3.60 | 1.05 | 1 | 5 |
| Market availability for seedlings | 364 | 3.59 | .97 | 1 | 5 |
| Cooperative membership | 380 | .81 | .40 | 0 | 1 |
| Non-farm work participation dummy | 380 | .72 | .45 | 0 | 1 |
| Frequency of invitation by neighbours for food or drinks (0 = None . . . 4 = >5 times) | 380 | 2.78 | 1.27 | 0 | 4 |

Source: Own survey 2013

**Table A2:** *Cobb-Douglas productivity function*

| *Variable* | *Log of fruits output per hectare* | *Log of seedlings produced per hectare* |
|---|---|---|
| *Log of manure applied in kilo grams per hectare* | 0.22 (0.06)** | 0.43 (0.05)** |
| *Log of apple plantation asset per hectare* | 0.303 (0.054)** | 0.07 (0.04)* |
| *Log of Labour used in man-days per hectare* | –0.12 (0.056)* | 0.192 (0.03)** |
| Constant | 3.37 (0.62)** | 4.065 (0.53)** |
| F- Statistic | 23.46 | 9110 |
| Adjusted R- squared | 0.16 | 0.47 |
| No. of obs. | 354 | 311 |

Note: + $p<0.1$; * $p<0.05$; ** $p<0.01$. The figures in parenthesis represent the robust standard error.
Source: Own survey 2013

**Table A3:** *Technical inefficiency (TIE) regressed on selected explanatory variables*

| *Variable* | *TIE for fruits* | *TIE for seedlings* |
|---|---|---|
| Gender of household head (0=male; 1=female) | 0.057 (0.051) | 0.140 (0.060)* |
| Education in years | 0.003 (.004) | 0.009 (.004)+ |
| Age in years | –0.000 (0.001) | 0.005 (0.001)** |
| Cooperative membership | –0.064 (.049) | –0.184 (0.072)* |
| Knowledge index | 0.004 (.011) | –0.024 (0.012)+ |
| Cultivation experience | –0.012 (.004)** | – |
| Doko Shaye village | –0.372 (.052)** | – |
| Doko Losha village | 0.158 (.054)** | – |
| Chencha Town | –0.126 (.048)* | – |
| Constant | 0.779 (.121)** | 0.198 (0.137) |

Note: + $p<0.1$; * $p<0.05$; ** $p<0.01$. The figures in parenthesis represent the robust standard error.
Source: Own survey 2013

# 16

# Crop-Specific EU Aid and Smallholder Food Security in Sierra Leone

Silvia L. Saravia-Matus [a,*], Sergio Gomez y Paloma [b], Sebastien Mary [c]

[a] *International Senior Consultant at European Commission – Joint Research Centre in Seville, Santiago, Chile*
[b] *European Commission – Joint Research Centre in Seville, Spain*
[c] *Du Paul University, Chicago, U.S.A.*

**Abstract**

The article analyses the viability of promoting crop-specific programs as a mean to improve smallholder net farm income and food security. The case study explores the relevance of European Union Stabilisation of Export Earnings (STABEX) funds in supporting Sierra Leone's agricultural development agenda. By analysing the drivers of food security for a number of targeted smallholders in the two most important agricultural zones of Sierra Leone, it is possible to compare the suitability of crop-specific support (in rice, cocoa and coffee) versus general aid programs (public infrastructure, on and off farm diversification opportunities, sustainable practices, access to productive assets, etc.). The results indicate that crop diversification strategies are widespread and closely related to risk minimisation and enhanced food security among smallholders. Similarly, crop-specific programs mainly focusing on commercialisation tend to overlook important constraints associated to self-consumption and productivity.

*Keywords:* EU aid policy, food security, smallholders, Sierra Leone, crop specialisation, crop diversification

## 1 Introduction

The Government of Sierra Leone requested in 2005 the use of European Union STABEX (Stabilisation of Export Earnings) funds (an instrument of the 8$^{th}$ European Development Fund) for the enhancement of national rice production and the rehabilitation of cocoa and coffee plantations (Government of Sierra Leone, 2007). The objective of the STABEX-funded projects was to achieve food security goals through the improvement of the agricultural sector (Gomez y Paloma *et al.*, 2012). Most of the support provided (4,378,000 EUR) was aimed at crop-specific commercial, organisational and technical assistance aimed at improving net farm income of smallholders [1].

In order to assess the relevance of EU aid policy in Sierra Leone, a 2009 survey designed by the Joint Research Centre of the European Commission is used (Gomez y Paloma *et al.*, 2012) [2]. Although there are no data reflecting interviewees' conditions prior to 2009, the cross-sectional survey provides an in depth picture of smallholders' farm income and crop portfolio. The

* Corresponding author
Email: silvia.saraviamatus@gmail.com

[1] Two separate implementing agencies were selected: Action Aid addressed rice cultivation in the Northern districts of Bombali and Tonkolili, while the The Deutsche Welthungerhilfe worked in the Eastern districts of Kenema, Kailahun and Kono supporting cocoa and coffee production. Each implementing agency operated through local farmer organisations (Gomez y Paloma *et al.*, 2012).

[2] The survey is based on primary data collected in 600 face-to-face interviews located in the main agricultural areas of Sierra Leone. Sample sizes throughout the different areas are representative of the agricultural population in the Northern and Eastern Regions, five districts, 11 chiefdoms, and 39 villages.

dataset offers valuable insights on the limitations and opportunities of crop-specific technical assistance in the context of rural, post-conflict settings as well as the key food security constraints faced by smallholders under tropical agricultural systems. From the experience in Sierra Leone it is possible to draw key lessons for future agricultural assistance programs or EU food aid and development policies. It can thus be argued that the paper is innovative as *a)* it is using farm-household level data which provides a detailed picture of smallholder's farm income and crop portfolio and *b)* it introduces a novel approach for analysing the capacity of the STABEX transfers aimed at increasing smallholder food security in Sierra Leone via crop-specific support. This examination is particularly relevant in the context of UN Sustainable Development Goals (SDG) 1 and 2 [3].

## 2 Method of analysis

Smallholders in low income countries under tropical settings often rely on their own production to secure (partially or entirely) their consumption, which is often far from reaching the nutritional balance required for a healthy life (Saravia-Matus *et al.*, 2012). In the present case study, the extent of food insecurity among smallholders is examined by contrasting agricultural income to a contextualised consumption level or basic needs basket. In other words, the ability of smallholders to become food self-sufficient is assessed on the basis of their agricultural performance or net farm income [4]. Frequently used thresholds include poverty lines (PL) and Food Consumption Scores (FCS). But for the present analysis FCS are preferred because the information on full income at farm-household level (i.e. farm as well as non-farm income sources) is not available. Thus, only the capacity to generate sufficient food intake from agricultural activities is captured in this analysis.

The key difference between PL and FCS is that FCS focus on the different levels and diversity of food consumption considering nutritional aspects that are country- and region-specific. FCS go beyond capturing the expenditure needed to attain the minimum daily nutritional requirement of 2700 calories per adult equivalent (usually employed in the extreme and full poverty lines) and include diet diversity, food frequency, and relative nutritional importance of different food groups [5]. Further rationale for using dietary diversity in the FCS construction is that it reflects the extent of adequate intake of essential nutrients. Dietary diversity is thus intended as a proxy of access to food (at household level), intake of energy and macronutrients and intake of micronutrients (FAO, 2008). Another main difference with respect to poverty lines is that other non-food expenditures such as shelter, access to safe water, education, health care are not covered in FCS expenditures. The latter is convenient given that only Net Farm Income (NFI) is contrasted against the FCS threshold. If full income information were available a poverty line threshold would be a more suitable choice.

The identification of smallholders who are food insecure on the basis of their agricultural performance is completed by an econometric assessment of the probability of their NFI to fall below a given threshold (Lovendahl & Knowles, 2005). In other words, the probability of being in a situation of food insecurity due to achieving an insufficient NFI level can be assessed through models of binary response variables (Baum, 2006). The resulting dependent variable "food poor" can be expressed as a binary variable taking the value of 1 if the smallholder reports a NFI below the identified threshold and 0 otherwise. For analytical purposes the "FCS poor level" threshold (equivalent to 0.30 $US per day per adult equivalent) is selected because it refers to the capacity of smallholders to generate NFI that could cover at least the minimum food consumption level for subsistence taking into consideration the rural context of Sierra Leone (World Food Program, 2008). Within this set up the binary response variable model can highlight the probability of a smallholder being categorised as "food poor" controlling for a series of factors. The model may be expressed as follows:

$$Prob\,(Food\,Poor_i = 1 \mid x_i) = F(x_i\beta)$$

[3] SDG 1: End poverty in all its forms everywhere. SDG2: End Hunger, Achieve Food Security and Improve Nutrition and Promote Sustainable Agriculture
`https://sustainabledevelopment.un.org/`

[4] For a detailed assessment of net farm income calculation procedures refer to Saravia-Matus & Gomez y Paloma (2014).

[5] The FCS procedure is to ask interviewees (in this case inhabitants of rural areas in Sierra Leone) about frequency of consumption (in days) over a recall period of the past 7 days. Food items are then grouped into 8 standard food groups with a maximum value of 7 days/week. The consumption frequency of each food group is multiplied by an assigned weight that is based on its nutrient content. Those values are summed obtaining the Food Consumption Scores which are then classified into adequate or poor levels (World Food Program, 2008). In Sierra Leone the FCS classified as Poor/Borderline varies between 0 to 35 and it is equivalent to 0.30 USD per day; while the Adequate FCS is above 35 which is closer to 0.40 USD per day. These expenditures are below the Full and Extreme Poverty Line of 2 USD and 1 USD per day. Note that in 2009 1 USD = 1700 SLL (Sierra Leonean Leone) approximately.

where $F$ is the logistic (or normal) cumulative distribution function (CDF) [6]. $x_i$ is a vector of values for the $i^{th}$ observation of the explanatory variables and $\beta$ is a vector of parameters. A constant was also included in the calculation. In this respect, the Sierra Leone dataset offers a series of theoretically relevant variables which are grouped into four categories: (i) *socio-demographic traits*, (ii) *productive asset ownership and access to infrastructure*, (iii) *livelihood diversification strategies* and (iv) *agricultural practices and crop portfolio* (Table 1).

It is expected that higher dependency ratios, higher educational achievements and enjoying a relevant position within the community may have a positive impact on NFI levels. Concerning gender, Ragasa (2012) argues that female farm-household heads experience more obstacles in the organisation of production and commercialisation activities. Similarly, women face constraints regarding their possibility to participate in innovation processes and access information, inputs (including cash) and extension services (FAO, 2010–2011) ultimately increasing their probability of being food insecure. Age may contribute either positively or negatively depending on whether it is correlated to higher dependency ratios (that is the farm-household head is far too old for productive work) or increased social capital that reduces transaction costs and increases NFI levels (which is associated to higher age as the individual has forged stronger long term relationships). Another relevant issue to consider is that there may be joint effects among the socio-demographic traits. For instance, female smallholders may report fewer years of education than their male counterparts and may not be as actively involved in community-level organisational structures as their male counter-parts.

In terms of productive assets, an increased access to land is expected to have a positive effect on NFI and consequently on food security. In the case of Sierra Leone, smallholders who are renting land are those who have limited access to this resource (Unruh & Turray, 2006). Consequently, if the smallholder relies on land rentals to secure land access, it is expected that this may have a negative impact on NFI levels and food security as there is an element of uncertainty besides increased cost (usually paid in kind i.e. rice bushels) [7]. In terms of storage capacity, if the smallholder farm-household is able to save part of their output for selling at different times throughout the years, it may indicate that higher selling prices may be achieved thus increasing NFI. Also seeds may be used for the next production cycle or for home consumption at times of scarcity or higher retail prices. Higher distances to markets imply higher transportation costs (and possibly transaction costs) and fewer sales or overall percentage of sold output and thus lower NFI levels. The ownership of a mobile phone is also expected to have a positive effect on NFI although alternative interpretations are also plausible. Jayne *et al.* (2011) argue that while a majority of sub-Saharan farmers now own or have access to a mobile phone, few feel that owning a mobile phone helps them find a better price for their farm output. Instead, the majority of farmers use their phones to notify a buyer that they have agricultural produce to sell, not to negotiate a price, or to search for price differences between buyers.

In terms of livelihood diversification indicators, the reception of gifts or remittances and having agriculture as the only income are considered. If gifts and remittances are used to invest in farm activities, NFI may increase; alternatively, on the basis of this inflow the smallholder may decide to reduce their farm labour efforts resulting in lower NFI levels. Relying on only agricultural income indicates that farm-households may be rather vulnerable to changes in agricultural markets and thus more likely to be food insecure if their NFI is reduced.

Agricultural practices and decisions around crop portfolio are also expected to have an effect on NFI and in the probability of achieving food security. In the case of fallowing (a practice expected to increase soil fertility, yields and NFI in the long term) a negative effect on NFI may be reported in the short term as land is taken out of cultivation (McCarthy *et al.*, 2011). In a study by Solis *et al.* (2008) which explored the connection between technical efficiency and environmental sustainability, smallholders that have a more diversified farm production plan (as well as off-farm work) reported both higher efficiency and sustainability. These findings largely coincide with Coelli & Fleming (2004) who argue that, in peasant economies, diversified production plans can lead to productivity gains that increase returns to land and labour, thus increasing food security status as supported by agricultural endeavours.

[6] For ease of interpretation, the analysis is made on the basis of the logistic CDF. This allows to calculate the log of the odds ratio which conveniently re-expresses the probability in terms of the odds of $y = 1$.

[7] In Sierra Leone (as in other African countries) land-owning families may decide to revoke the agreement at any moment and tenants are not allowed to plant trees as they may reduce the capacity of owners to claiming back their land. Meanwhile, trees are, inter alia, a symbol of land ownership (Unruh & Turray, 2006). Unruh & Turray (2006) describe that the prohibition against planting economic trees or making other long-term improvements to the land for people from outside the chiefdom has strong negative food security implications.

**Table 1:** *Explanatory variables of smallholder food insecurity*

| | |
|---|---|
| *Farm-household traits* | |
| Dependency ratio | Number of dependent household members over 18 years old / Total number of household members |
| Age of household head | Years |
| Gender of household head | 1 if Household head if Female; 0 if it is Male |
| Education of household head | Years of school attendance |
| Social capital of household head | 1 if Farm-household head holds a position in local government or communal organisation (councillor, section chief, village chief, village headman). 0 if ordinary citizen |
| *Access to productive assets & infrastructure* | |
| Acres per adult equivalent household member | Number |
| Land rental | 1 if farm-household rents land; 0 if not |
| Storage | Percentage of harvested crop which was stored |
| Mobile phone | 1 if Household head owns a mobile phone; 0 if not |
| Distance to market | Kilometres |
| *Livelihood diversification strategies* | |
| Only agricultural income | 1 if the farm proceed is the only reported income source of the household; 0 if additional income sources are reported: remittances, gifts, non-farm activities (i.e. petty trading, artisan etc.) |
| Remittances & gifts | 1 if remittances or gifts have been received by the farm-households; 0 if not |
| *Agricultural practices & crop portfolio* | |
| Fallow period | Number of years under fallow |
| Crop concentration/diversification | Shannon Index (0–1); Simpson Index (1 if $<0.5$ and 0 if $>0.5$) |

In the present case study, smallholders' decisions between crop specialisation and diversification are particularly relevant since STABEX-funded initiatives were aimed at promoting the production and commercialisation of selected crops. To measure inter-crop diversity, the Shannon index is used. This index expresses proportional abundance or evenness, accounting for the land shares allocated to each crop as well as the number of crops. The index gives less weight to rare species than common ones, but is more sensitive to differences to small degrees of relative abundances than the Simpson index which is another widely used evenness index measure of diversity. Both indices are used in the empirical analysis.

## 3 Empirical findings

As stated, the aim of the binary response variable model estimation is to analyse the probability of smallholders to achieve NFI levels above or below a predefined food security threshold. All of the explanatory variables presented in Table 1 were included in the regression with the exception of gender and age because the latter appeared to be strongly correlated to social capital.

Table 2 presents the binary response model estimation under logistic CDF and introduces average marginal effects. These average individual marginal effects can be interpreted as partial effects on the probability to be identified as "food poor". Among the statistically significant traits, the condition of holding a relevant position within the community (social capital) suggests that the smallholder will be less likely to fall under the food security threshold. That is 17 % lower probability of being identified as "food poor" than smallholders without this level of social capital as indicated in the marginal effects column (−0.1747). In terms of the dependency ratio, having a larger burden within the farm-household is positively related to being identified as "food poor" (16 % more likely to be identified as food insecure). Years of education which appeared as statistically insignificant may be associated to the general low educational levels observed in the rural areas of Sierra Leone.

Concerning livelihood diversification variables, only relying on agricultural income makes the smallholder more likely to be identified as food poor on the basis of the achieved NFI level. Similarly, farm-households receiving remittances and gifts appear to be those in most need; suggesting these resources may be used for subsistence rather than investment. Smallholders only relying on agricultural income and those receiving remittances and gifts have on average 13 % and 14 % higher probability, respectively, to be below the established food security threshold, *ceteris paribus*.

Regarding productive assets and infrastructure, only land availability per adult equivalent household member and the land rental dummy appear as statistically significant. As expected, the higher land availability, the less likely the smallholder is to be identified as food poor (−2.4 %), *ceteris paribus*. On the contrary, smallholders involved in land rental arrangements have on average 20 % higher probability to be "food poor". This may relate to the limitations and uncertainties associated to land tenants in Sierra Leone. For the case of storage and distance to markets, the statistical insignificance of these variables may be connected to the low market integration among the "food poor" smallholders.

**Table 2:** *Logit model estimation & average marginal effects after logit*

| | *Logit Dependent variable: FCS below poor* | | *Average marginal effects dependent variable: Pr (FCS below poor)* | |
|---|---|---|---|---|
| *Farm-household traits* | | | | |
| Education of household head | 0.0119 | (0.0249) | 0.0020 | (0.0038) |
| Social Capital | −0.826 *** | (0.2493) | −0.1747 *** | (0.0456) |
| Dependency ratio | 0.956 ** | (0.4315) | 0.1642 ** | (0.0656) |
| *Livelihood diversification* | | | | |
| Only agricultural income | 0.688 *** | (0.2370) | 0.1361 *** | (0.0384) |
| Remittances & gifts | 0.735 *** | (0.2240) | 0.1448 *** | (0.0356) |
| *Productive assets & infrastructure* | | | | |
| Acres per adult equivalent household member | −0.141 * | (0.0770) | −0.241 * | (0.0117) |
| Land rental | 1.070 *** | (0.3746) | 0.2012 *** | (0.0525) |
| Storage | 0.176 | (0.6355) | 0.0302 | (0.0980) |
| Distance to market | −0.0156 | (0.2140) | −0.0026 | (0.0033) |
| Mobile phone | 0.0956 | (0.2681) | 0.0199 | (0.0488) |
| *Agricultural practices* | | | | |
| Fallow period | −0.0904 *** | (0.0274) | −0.0155 *** | (0.0040) |
| Shannon index | −3.690 *** | (0.7456) | −0.6337 *** | (0.1063) |
| Constant | 2.089 *** | (0.4419) | | |
| Observations | 540 | | | |

Standard errors in parentheses; *** $p<0.01$, ** $p<0.05$, * $p<0.1$

The negative sign on the fallow practice suggests that smallholders who are able to keep land aside for improving soil fertility (and long term productivity) are less likely to be identified as food poor (−1.5 %) holding all other variables constant. Smallholders involved in fallowing are already capable of assuming the opportunity costs associated to this practice and maintain higher food security levels. Regarding the degree of crop diversification, captured through the Shannon Index, the negative sign suggests that smallholders maintaining a diversified crop portfolio (over 10 crops) are less likely to be 'food poor". In fact, the largest partial effect is associated to crop diversification practices reporting 63 % higher probability of *not* being identified as "food poor" if the smallholders have a highly diversified crop portfolio.

As evidenced in the logit regression results, the probability of being identified as food poor (under a very strict threshold of 0.30USD per person per day) was higher when smallholders had concentrated crop portfolios, could not undertake fallow practices, faced limited and/or uncertain land access, were only reliant on their agricultural activities for survival, had a larger number of dependent household members and reduced social capital within their community.

In order to explore the traits of the smallholders identified as "food poor" an agricultural typology is developed. The types of smallholders are defined on the basis of agro-ecological zones that may constrain the choice of crops cultivated and crop concentration/diversification indices. Each emerging type is then described in terms of food security, assets, market integration and output valuation. In the description of emerging types, the Richness Index is also introduced to illustrate how many and which kind of crops are produced. Table 3 provides an overview of the selected identification criteria, indicators and thresholds used to build the typology (Even *et al.*, 2016).

Table 4 presents the four basic emerging farm types. The "Crop Diversified" Types 1 & 3 representing both agro-ecologies and reporting on average 16 to 12 crops respectively account for 71.7 % of surveyed smallholders. Types 2 & 4, on the other hand represent the remaining percentage (28.3 %) and capture the "Crop Concentrated" smallholders reporting on average 6 and 4 crops in the sub-humid and humid zones respectively. Interestingly, in the humid or rainforest ecology there are very few crop concentrated smallholders (Type 4 represents only 5 % of surveyed smallholders in this agroecological region). In this zone, 95 % of surveyed smallholders prefer to combine cash tree crops with staple crops (Type 3). In the sub-humid zone, however, the presence of smallholders with both diversified (Type 1) and concentrated (Type 2) crop portfolios is more even (60 % to 40 % respectively). It is also relevant to point out that for all farm types upland rice has a dominant position in the crop portfolio with an average of 28 % of total cultivated area for Type 1, 50 % for Type 2; 20 % for Type 3 and 12 % for Type 4. The first reading from typology findings is that smallholders in Sierra Leone are inclined to manage a wide range of crops within their plots even if rice is well-recognized as the national staple crop. The latter is aligned to the widespread practice of inter-cropping in tropical and sub-tropical Sierra Leone.

In terms of social vulnerability among types, 88 % of smallholders belonging to Type 1 and 85 % of Type 2 smallholders are identified as "food poor" on the basis of the FCS threshold of 0.30 $ per person per day. Conversely, Types 3 and 4 from the rainforest ecology (or humid zone) present the lowest percentages of "food poor" smallholders (11 % and 9 % respectively). This is not surprising given the difference in average NFI per farm-household unit reported for Types 3 & 4 are roughly over 10 times higher than those reported under types 1 & 2. In order to draw accurate comparisons between types, the poverty gap index (PGI) is calculated. PGI is deemed suitable as it can be interpreted as the average percentage shortfall in income for the type population from the given threshold allowing for effective group comparisons. As illustrated, 76 % of the population of Type 2 farms are below the poor/borderline FCS expenditure threshold, making this the most vulnerable group among identified types. This is followed by 56 % for Type 1, 21 % for Type 4 and only 4 % for Type 3. Although the percentages are smaller for types in the humid zone, the findings suggest that smallholders with higher crop concentration are worse off than their counterparts in the same agroecological setting.

In terms of asset endowments Type 2 smallholders report the lowest land availability per adult equivalent household member, the lowest access to manual equipment at farm-household level and the lowest percentage of rice output sold. It is thus also not surprising that the overall output value per working unit is also the lowest for Type 2. Rice growers of the sub-humid zone (Type 2) are the poorest, rely mainly on self-consumption (for the case of rice) and have comparatively less access to productive assets. In other words, Type 2 smallholders coincide with Ellis & Freeman (2004) description of rural farm-households with low incomes which are associated with small land and livestock holdings, high reliance on food crop agriculture and low monetisation of the rural economy.

**Table 3:** *Identification of farm types based on two criteria*

| *Identification criteria, indicators & thresholds* | |
|---|---|
| Agro-ecological zone | Sub-Humid zone or Lowland & Upland ecologies suitable for rice cultivation (Sub-Humid zone = 0) *or* Humid zone or Rainforest ecology suitable for coffee and cocoa cultivation (Humid zone = 1) |
| Crop concentration / diversification | Crop concentrated = 1, if Simpson Index > 0.5 *or* Crop diversified = 0, if Simpson Index < 0.5 |

**Table 4:** *Emerging types & descriptive information*

| | *TYPE 1: Sub-Humid & DIVERSIFIED* | *TYPE 2: Sub-Humid & CONCENTRATED* | *TYPE 3: Humid & DIVERSIFIED* | *TYPE 4: Humid & CONCENTRATED* |
|---|---|---|---|---|
| Total # of obs | 240 | 159 | 191 | 11 |
| % of sample | 39.9% | 26.5% | 31.8% | 1.8% |
| % within ecology | 60.2% | 39.8% | 94.6% | 5.4% |
| Richness Index | 16 | 6 | 12 | 4 |
| Avg % rice cultivated area | 28% | 50% | 20% | 12% |
| Main crop in avg % area | Oil Palm: 30% | Rice: 50% | Cocoa: 33% Coffee: 22% | Cocoa: 39% Coffee: 21% |
| *Food security & poverty* | | | | |
| % of smallholders below the "poor/borderline" FCS expenditure | 88 | 85 | 11 | 9 |
| Avg NFI per hhunit (SLL) | 100,100 | 59,216 | 926,663 | 1,055,236 |
| Poverty gap index | 59 | 76 | 4 | 21 |
| *Asset endowment, market participation & output value* | | | | |
| Avg acre per adult equivalent household member | 1.36 | 0.95 | 2.42 | 1.55 |
| Avg manual equipment (# of hand tools) available at farm-household level | 15.3 | 14.8 | 28.8 | 18.3 |
| Avg fallow period | 5.7 | 6.8 | 10.1 | 4.8 |
| Avg % of rice sold | 30 | 25 | 65 | 64 |
| Avg output value per working unit (SLL) | 883,961 | 601,049 | 1,495,987 | 732,000 |
| Avg output value obtained per staple-cultivated Acre (SLL) | 169,743 | 179,557 | 259,468 | 300,000 |
| Avg output value obtained per cash tree crop-cultivated acre (SLL) | 53,252 | 48,134 | 668,823 | 729,838 |

Note: the distribution of observations between Type 3 and Type 4 in the rainforest / humid ecology is sustained even at a threshold of HH < 0.3 suggesting that crop portfolios for smallholders able to grow cocoa and coffee are highly diversified denoting a preference for integrated staple and cash crop systems.

Type 1 farm-households in the sub-humid zone who manage diversified crop portfolios appear to be better off than Type 2 farm-households. Regarding Types 3 & 4 from the rainforest ecology, it may be argued that their involvement with cash tree crops (cocoa and/or coffee) is a strong determinant in their achievement of higher NFI levels. However, given the distribution of observations between these two types, it can be stated that the majority of smallholders in this zone tend to prefer diversified crop portfolios even if the agro-ecology is suitable for cocoa and coffee cultivation. In fact, the PGI for Type 3 is by far the lowest. Smallholders in this zone are also on average marketing a higher percentage of their cultivated rice output. The latter suggests that a combination of staple and cash tree crop cultivation yields on average higher output value per cultivated acre. This finding fits rather well with the positive synergies emerging from mixed cropping systems as observed by Govereh & Jayne (2003). According to these authors, participation in cash crop schemes (especially under conditions of credit and input market failures) may enable smallholders to acquire key inputs and skills which can be used to increase their overall producitivty; thus exploiting synergies between cash crops and food crops not only in terms of production but also commercialy.

This typological analysis complements the econometric findings, and the combined evidence suggests that the STABEX-funded crop-specific policies partially failed to acknowledge the food security purposes of diversified crop portfolio management among targeted smallholders. For instance, the majority of smallholders have diversified plots (i.e. Types 1 & 3 representing more than two thirds of the sample). Type 3 smallholders who combine food and cash tree crop cultivation systems in the humid zone, were substantially better off than Type 4 smallholders with highly concentrated plots of cocoa and coffee. In fact, Type 4 farms were the smallest type in the sample and reported higher land limitations than Type 3, reinforcing the idea that when feasible smallholders in the humid zone prefer the mixed staple-cash tree crop cultivation system. In the sub-humid zone, rice-specific support seemed suitable to Type 2 smallholders who have on average 50 % of their plots under rice cultivation. However, this was also the group who claimed that the support was insufficient to raise income levels (Gomez y Paloma *et al.*, 2012).

## 4 Policy discussion

The econometric and qualitative findings suggest that STABEX-funded crop specific programs did not fully consider the implications of shifting cultivation and diversified crops at plot level which tend to be rather characteristic of tropical and sub-tropical agriculture. In fact, promoting a specific crop, even as relevant as rice in Sierra Leone, did not fulfil the expectations of particularly the poorest types of targeted farms who were actually the most reliant on this crop for self-consumption. In their particular situation, production requires to be substantially increased in order to fulfil internal consumption needs before engaging in sales (i.e. Type 2 smallholders consumed on average up to 75 % of their rice output). This situation calls for a reflection on the type of support provided for this specific segment of farms. Access to output increasing technology (for instance, fertilisers) was not central in the STABEX-funded technical assistance. Likewise, there was no emphasis on introducing sustainable land and water management approaches or supporting mixed production systems that could foster a varied diet. Smallholders did not have access to yield-increasing inputs or enhanced natural resource management. Actually, STABEX-funded initiatives were limited to rice market information, consolidation of farmer groups and training rather than efforts to increase actual productivity. The subjective evaluation of Type 2 smallholders concerning these initiatives suggests that a more balanced approach (which could have promoted both production and commercialisation) would have likely been more welcome. Similarly, a focus on systems of production (with diversified crops) could have been more aligned to smallholder food security concerns.

The situation for smallholders in the rainforest ecology (mainly Type 3 farm-households who represent the majority in this zone) is illustrative of the benefits associated to mixed agricultural systems (i.e. cash and food cropping). There are important indirect effects of cash cropping on the productivity of food cropping. Govereh & Jayne (2003) have classified two potential pathways by which these benefits occur: *farm-household-level synergies*, in which a farmer's participation in a commercialised crop scheme enables the acquisition of resources that would otherwise not be accessible for use on other crops and *regional spill-over effects* which occur when a commercialisation scheme attracts new investments to a region thereby providing benefits to all farmers in that region, regardless of whether they engage in the commercialisation of the given cash crop. According to Govereh & Jayne (2003) cash generating crops can help farmers overcome capital constraints on the purchase of lumpy assets and inputs, which can be used to expand food crop as well as cash crop production. In this setting, the STABEX-funded initiatives aimed at improving grade and standards (as well as international regulation enforcement) had a positive impact on price levels of cocoa and coffee grains aimed at export markets.

Crop diversification can also be seen as a path to break the cycle of rural poverty where smallholders are characterised by declining land availability, food deficit from own production, low monetisation of the local economy and little cash circulation to multiply rural activities (Ellis & Freeman, 2004). It is also relevant to remember that although farming systems of most regions are usually described in terms of a small number of crops, the majority of farming families, however, grow a wide variety of crops. Inter-planting two, three or more crops is widespread. Crop diversification also ensures a more effective use of aerial space. Crops are sometimes partly complementary in nutrient requirements, it minimises the effects of pest and disease attacks, combined yields are usually higher than the yields of individual crops and the soil is covered for a longer period by the combination of crops. Lastly, crop diversification is considered an important initial step in the transition from subsistence to commercial agriculture (Losch *et al.*, 2012).

But, if the benefits of crop diversification are so varied and to a large extent suitable to the agricultural environment of Sierra Leone, why were STABEX-funded initiatives focused on crop-specific support? From the part of the EU, there is strong commitment to frame development cooperation in line with the principles of ownership and partnership (EC, 2011). Dialogue at country level determines exactly where and how the EU intervenes.

According to Rodenburg *et al.* (2006), Sierra Leone's agricultural development policy has focused since independence on the achievement of rice self-sufficiency. Rice provides more proteins than cassava, maize or sorghum, it is available all year round because of its long shelf life (provided adequate storage is in place), making it preferable to other crops for food security (Norman & Kebe, 2006). Rice production also offers an important source of employment during cultivation, post-harvest and commercialisation. In Sierra Leone, rice is accepted as a medium of exchange and it drives the barter economy, often being used to procure coffee and cocoa, lure labour and purchase farm inputs and wage goods (*ibid.*). One important downside is that national-level specialisation in a given crop puts higher pressure to maintain a stable policy environment, to engage in WTO negotiations which safeguard the country's competitive position as well as to improve contract law (Kelly *et al.*, 2003). Another constraint surrounding rice production is the limited scope for area expansion under the traditional bush fallow system. As a result, most of the productivity increases must come from yield improvements which require the adoption of new technologies by the smallholders who produce the bulk of agricultural output (Rodenburg *et al.*, 2006). In this respect, STABEX-funded initiatives included a strong training component that may have had stronger effects if accompanied by adequate physical access to yield increasing technology (mainly fertilisers and improved seeds or machinery).

The National Sustainable Agriculture Development Plan of Sierra Leone (NSADP, 2009) focuses on a move towards permanent cultivation of food crops (mainly rice and cassava) and promotion of export tree crops (i.e. cocoa and coffee). In the case of smallholders that due to agro-environmental conditions cannot grow traditional export trees such as cocoa and coffee, the diversification of staple crops at plot level appears to be a dominant practice in comparison to mono-cultivation; if sufficient productive resources are available (namely, both labour and land). The main challenges for the implementation of the NSADP are on one hand the transition period between shifting cultivation systems to permanent agriculture and on the other the sustainability of the newly adopted system (Saravia-Matus & Gomez y Paloma, 2015). In the first matter, effects on short-term food security and employment opportunities must be considered. In the second, market organisation is crucial (particularly for the supply of key inputs such as fertilisers, improved seeds, pesticides or machinery).

In summary, the government's request for STABEX funds were guided by NSADP principles and the long term objective of securing and maintaining rice self-sufficiency as well as the promotion of key agricultural exports (i.e. coffee and cocoa). The EU funds were thus aimed at basic training (in the cultivation of selected crops according to the targeted zones), the provision of market information and the consolidation of farmer groups. Clearly, these aspects are relevant for the transition from shifting cultivation to permanent cultivation. However, these efforts are insufficient to tackle poverty reduction and food insecurity of smallholders.

The literature on agricultural development policies in sub-Saharan Africa is extensive and a great variety of interventions can be identified, yet in most cases these are not related to specific crops or to choosing winning crops. The latter can actually be classified into two main groups which roughly relate to *failures of production* on one hand and *failures of exchange and response* on the other hand (Devereux, 2009):

(i) *Addressing failures of production*: Most of the initiatives to increase production are usually connected to the improvement of natural resources management. These include tasks to enhance soil fertility (with adequate training in the fertiliser use and access), water access and management (irrigation facilities and water control) or improved seeds and fertilifers (natural or chemical). According to Jama & Pizarro (2008) in Africa only a small proportion of farmers use fertilisers and the amounts used are often inadequate. On average each hectare receives less than 9 kg of nitrogen and 6 kg of phosphorus. Typical crop requirements per hectare are at least 60 kg of nitrogen and 30 kg of phosphorus.

Chemical fertiliser use per hectare of farmland in Africa is about 10 % of the world's average, by far the lowest. It is important to highlight that the provision of fertilisers was not among the key activities of the STABEX-funded programs and it was one of the most frequent complaints of the targeted smallholders (Gomez y Paloma *et al.*, 2012). The same applies for the case of access to improved seeds. According to Norman & Kebe (2006) the low agricultural productivity in West and East Africa is due to an important extent to high incidence of pests, weeds, and diseases, drought and poor water control, poor seed management, poor soil fertility management and the necessary labour requirements to conduct efficient sustainable management of natural resources.

(ii) *Addressing failures of exchange and response*: initiatives to enhance institutional coordination and align incentive structures are often connected to the improvement of market access (both in terms of output and inputs markets including credit), extension services (training, weather forecasts, technology, etc.) and rural infrastructure. According to Barrett (2008) smallholders face two types of market entry barriers. One entry barrier is found at the micro-level where smallholders face insufficient access to productive assets, financing and improved production technologies which may generate adequate marketable surplus and make market participation feasible. The other entry barrier takes place at the meso-level. In remote areas the high costs of commerce limit market access (both in terms of spatial price transmission and trader competition). The latter leads to thinner and more volatile markets limiting smallholders' incentives to increase productivity so as to generate marketable surpluses. Agricultural productivity growth depends on functioning input distribution systems and *vice versa* (Jama & Pizarro, 2008) as well as on reliable infrastructure (i.e. roads, storage facilities, electrification) (Kelly *et al.*, 2003). Similarly, farmers need to be aware of what technologies work best, know how to use them and generate effective demand for viable new technologies to provide signals to input distribution system to supply them (*ibid.*). The STABEX-funded initiatives were not focusing on securing input provision networks; instead attention was placed on the consolidation of farmers' associations in order for them to create a demand for productive inputs as well as to organise selling channels. However, the reduced yields and the internal consumption requirements served as barriers for these efforts to be entirely effective. The STABEX-funded initiatives were aimed at reducing the costs associated to accessing markets by providing price information but no particular efforts were targeted at increasing productivity beyond training. Similarly, the development of public infrastructure was not a priority under STABEX-funded interventions. As highlighted by Kelly *et al.* (2003) without commitment to providing basic public goods, large scale government input subsidies, credit access or distribution programs are unlikely to have any lasting impact on agricultural intensification, rural incomes, national food security and poverty reduction. Consequently, governments need to also focus on public goods as a pre-requisite that will encourage farmers to intensify agricultural production and encourage the private sector to expand commercial input supply.

## 5 Conclusions

The present case study has provided evidence of the various agronomic and socio-economic strategies that farmers can adopt in order to improve their food security, considering their agro-ecological settings and market constraints. In Sierra Leone, such strategies were mainly related to crop diversification and self-consumption orientation. However, EU-funded aid policy measures on food security in Sierra Leone followed a crop-specific support mechanism as a mean to enhance production and market participation. The main idea behind this crop-specific support is that enhanced market interactions would increase incentives to expand rice, cocoa and coffee production thus lifting NFI and procuring a positive effect on food security.

Although rice is without doubt a key staple crop, the majority of targeted smallholders held diversified crop portfolios. At the same time, the strong reliance on agricultural output to fulfil self-consumption needs partially diminished the potential positive effects of policies aimed at increasing output market interactions. In fact, the poorest smallholders were the ones reporting strong specialisation and self-consumption in rice. On the other hand, smallholders with mixed agricultural systems of both cash tree crops and staple crops seemed to achieve higher food security levels on the basis of their achieved NFI. The agricultural development literature suggests that a series of positive synergies emerge between export crops and domestic food crops which allow increasing both productivity and market integration.

One of the key lessons of the Sierra Leone case study for both national officials and international donors is that in order to achieve food security, failures of production, exchange and response, have to be jointly addressed in a balanced manner. In Sierra Leone, the government favoured a crop-specific approach aimed at improving commercialisation, thus targeting the exchange failure aspect. An in-depth analysis of the constrained environment in which smallholders operate reveals that without addressing the obstacles to production enhancement (which needs to first cover substantial self-consumption needs) effective market responses will not develop. Similarly, a private input distribution network will not emerge if farmers are first interested in fulfilling their self-consumption requirements rather than integrating into output markets. Also, aid programs may largely increase their impact by focussing on the role of public investment in key assets and infrastructures. Efforts to provide training for improving market understanding and access, organisation of farmers' groups and even crop specific techniques are insufficient if access to yield increasing technology is not secured along with adequate establishments for storage and suitable transportation (both in terms of rural roads and machinery).

The latter calls for a more in-depth analysis of the various policy interventions directed towards a given agricultural area or areas. It is necessary to first account for all existing programs and assess the potential interlinkages and set of incentives that are being transmitted to targeted farms. In the particular situation of Sierra Leone, donor interventions such as that of the EU Aid policy are well aligned to the National Sustainable Agriculture Development Plan (NSADP) which the Government of Sierra Leone launched in 2009. The NSADP which is currently under implementation (2010–2030), proposes the gradual eradication of shifting cultivation practices and the active promotion of vertically integrated processing and marketing chains for selected staples (mainly rice and cassava) and export crops (cocoa and coffee). According to Saravia-Matus & Gomez y Paloma (2015), the government will have to create substantial transaction benefits to promote crop-specific commercialisation in the main agricultural areas of the country. These benefits should be accompanied by an overall incentive package that addresses the entire rural economy in Sierra Leone, otherwise, smallholders highly dependent on diversified crop portfolios for self-consumption will not be adequately accounted for in the NSADP.

Slow growth in agricultural productivity and income translates into slow overall economic development and increased poverty. Although agricultural investment and support continues to be an indisputable measure to achieve food security, the main recommendation emerging from the Sierra Leone experience with EU aid resources is that interventions need to be balanced and adapted to the agricultural and institutional context in which smallholders operate. Similarly, it is important to further analyse the interlinkages across existing policy interventions and plans. Only focusing on production enhancement without adequate market integration will prove deficient, in the same way that the opposite approach will not be enough to create incentives for expanding agricultural output, particularly when poverty and food insecurity are as severe as in the case of rural Sierra Leone and its agricultural setting.

## References

Barrett, C. B. (2008). Smallholder market participation: Concepts and evidence from eastern and southern Africa. *Food Policy*, 33, 299–317.

Baum, C. F. (2006). *An Introduction to Modern Econometrics Using Stata*. Stata Press Publication, Texas, USA.

Coelli, T. J. & Fleming, E. (2004). Diversification economies and specialisation efficiencies in a mixed food and coffee smallholder farming system in Papua New Guinea. *Agricultural Economics*, 31 (2–3), 229–239.

Devereux, S. (2009). Why does famine persist in Africa? *Food Security*, 1, 25–35.

EC (2011). Communication from the Commission to the European Parliament, the Council, the European Economic and Social Committee and the Committee of the Regions – Increasing the impact of EU Development Policy: an Agenda for Change. European Commission (EC), Brussels, 13 October 2011.

Ellis, F. & Freeman, H. A. (2004). Rural livelihoods and poverty reduction strategies in four African countries. *The Journal of Development Studies*, 40 (4), 1–30.

Even, M. A., Saravia-Matus, S., Giovannetti, J. F. & Groppo, P. (2016). The WAW International Typology as a Mechanism to Monitor Agricultural and Rural Transformations. *Development*. doi:10.1057/s41301-016-0001-9.

FAO (2008). Interagency Workshop Report WFP – FAO Measures of Food Consumption - Harmonizing Methodologies. FAO, Rome, 9 - 10 April 2008.

FAO (2010–2011). *The State of Food and Agriculture (SOFA): Women in agriculture: closing the gender gap for development*. FAO, Rome, Italy.

Gomez y Paloma, S., Acs, S., Saravia-Matus, S., Lakoh, A., Michel, B., Hites, G. & Sammeth, F. (2012). Rural poverty reduction and food security: The case of smallholders in Sierra Leone. EC-JRC-IPTS Technical Report EUR Number: 25264 EN. Seville, Spain.

Govereh, J. & Jayne, T. (2003). Cash cropping and food crop productivity: synergies or trade-offs? *Agricultural Economics*, 28, 39–50.

Government of Sierra Leone (2007). Prospects for the Poor in the Face of National Agriculture Reforms: A Poverty & Social Impact Analysis of Changes to Agricultural Policy in Sierra Leone. Economic Policy and Research Unit, Ministry of Finance. May 2007.

Jama, B. & Pizarro, G. (2008). *Agriculture in Africa: Strategies to improve and sustain smallholder production systems*. UNDP. New York Academy of Science, N.Y.

Jayne, T. S., Haggblade, S., Minot, N. & Rashid, S. (2011). Agricultural Commercialisation, Rural Transformation and Poverty Reduction: What have We Learned about How to Achieve This? Synthesis report prepared for the African Agricultural Markets Programme Policy Symposium, Alliance for Commodity Trade in Eastern and Southern Africa April 20–22, 2011, Kigali, Rwanda.

Kelly, V., Adesina, A. A. & Gordon, A. (2003). Expanding access to agricultural inputs in Africa: a review of recent market development experience. *Food Policy*, 28, 379–404.

Losch, B., Fréguin-Gresh, S. & White, E. T. (2012). *Structural Transformation and Rural Change Revisited: Challenges for Late Developing Countries in a Globalizing World*. World Bank, Washington D.C. 277 pp

Lovendahl, R. C. & Knowles, M. (2005). *Tomorrow's Hunger: A Framework for Analysing Vulnerability to Food Insecurity*. FAO-ESA Working Paper No. 05-07. Rome, Italy.

McCarthy, N., Lipper, L. & Branca, G. (2011). *Climate-Smart Agriculture: Smallholder Adoption and Implications for Climate Change Adaptation and Mitigation*. Mitigation of Climate Change in Agriculture Series 4. Food and Agriculture Organization of the United Nations (FAO), Rome, Italy.

Norman, J. C. & Kebe, B. (2006). African smallholder farmers: rice production and sustainable livelihoods. Regional Perspectives. *International Rice Commission Newsletter – Regional Perspectives*, 55, 33–42.

NSADP (2009). *National Sustainable Agriculture Development Plan 2010–2030*. The Republic of Sierra Leone.

Ragasa, C. (2012). Gender and Institutional Dimensions of Agricultural Technology Adoption: A Review of Literature and Synthesis of 35 Case Studies. International Association of Agricultural Economists (IAAE) Triennial Conference, Foz do Iguaçu, Brazil 18-24 August, 2012.

Rodenburg, J., Diagne, A., Oikeh, S., Futakuchi, K., Kormawa, P. M., Semon, M., Akintayo, I., Cissé, B., Sié, M., Narteh, L., Nwilene, F., Diatta, S., Sere, Y., Ndjiondjop, M. N., Youm, O. & Keya, S. O. (2006). Achievements and impact of NERICA on sustainable rice production in sub-Saharan Africa. *International Rice Commission Newsletter – Regional Perspectives*, 55, 45–58. FAO, Rome

Saravia-Matus, S. & Gomez y Paloma, S. (2014). Farm Viability of (Semi) Subsistence Smallholders in Sierra Leone. *The African Journal of Agricultural and Resource Economics (AfJARE)*, 9 (3), 165–182.

Saravia-Matus, S. & Gomez y Paloma, S. (2015). Implementation challenges to the National Sustainable Agriculture Development Plan (NSADP) for (semi)subsistence farmers in Sierra Leone. *Cahiers Agricultures*, 24 (4), 240–245. doi: 10.1684/agr.2015.0757.

Saravia-Matus, S., Gomez y Paloma, S. & Mary, S. (2012). Economics of Food Security. *Bio Based and Applied Economics Journal*, 1 (1), 65–80.

Solis, D., Bravo-Ureta, B. E. & Quiroga, R. E. (2008). Technical Efficiency among Peasant Farmers Participating in Natural Resource Management Programmes in Central America. *Journal of Agricultural Economics*, 60 (1), 202–219.

Unruh, J. D. & Turray, H. (2006). Land tenure, food security and investment in post-war Sierra Leone. LSP Working Paper 22, Access to Natural Resources Sub-Programme, FAO.

World Food Program (2008). *Sierra Leone: Household Food Security Survey in Rural Areas*. World Food Programme, Vulnerability Analysis and Mapping Branch (ODAV). Data collected in May 2007

# 17

# Assessing the impact of social grant-dependency on participation of KwaZulu-Natal rural households in farming: Application of the generalised propensity score method

Sikhulumile Sinyolo[a,*], Maxwell Mudhara[b], Edilegnaw Wale[b]

[a]*Human Sciences Research Council (HSRC), Pretoria, 0001, South Africa*

[b]*Discipline of Agricultural Economics, University of KwaZulu-Natal, Scottsville, 3209, South Africa*

**Abstract**

Social grants are an important instrument of social protection in South Africa, reaching millions of the poor each month. Although social grants have been found to reduce poverty and promote human development, considerable uncertainty remains about some of their incentive effects. This study uses a sample of 984 rural households selected from KwaZulu-Natal, South Africa, to investigate the potential incentive/dis-incentive effects of social grant-dependency on rural households' participation in farming activities. The data are analysed using the generalised propensity score (GPS) matching method and ordinary least squares. The results showed that the effect of social-grant dependency on households' farm participation levels varies at different dependency levels. While social grants had a negative effect on the households' farming participation levels when social grants income contribute 20–60 %, they had a positive effect at lower (< 20 %) and higher (> 60 %) dependency levels. The positive effect of social grants at the lower and higher levels supports the hypothesis that social grant beneficiaries use part of the grant income to alleviate financial constraints in agricultural production. However, the negative effect at the 20–60 % dependency levels is consistent with the dis-incentive hypothesis, suggesting that social grants may generate dis-incentives to farm production. The study identified several policy variables that affect the participation of rural households in smallholder farming activities, highlighting the importance of expectations of farming success as a key motivator.

*Keywords:* continuous treatment, farm labour, incentives to farm, social grants

## 1 Introduction

Social grants are an important instrument of social protection in South Africa, benefiting over 16 million South Africans each month in 2014 (SASSA, 2014). There are seven different grants: the old age grant, child support grant, disability grant, war veterans' grant, foster care grant, care dependency grant, and grant-in-aid (ibid). Eligibility for social grants is dependent on an income and asset-based means test, which varies according to the grant, the marital status of the beneficiary and other characteristics. Despite these means tests, the majority of South Africans are eligible to receive these grants (Abel, 2013). The child support grants have the largest number of beneficiaries, reaching over 11.5 million beneficiaries in 2014. However, it pays the least amount to individual beneficiaries, R 320 per month in 2014 per beneficiary (equivalent to about US$ 30 using the 2014 exchange rate) (SASSA, 2014).

The old age grant, which benefitted over 3 million people in 2014, pays the largest amount (R 1350 per beneficiary, equivalent to about US$ 120). The remaining grant types benefit fewer beneficiaries than the above two, and they pay amounts equal to or ranging between the two grant types. Although South Africa's social

* Corresponding author
Email: sksinyolo@gmail.com

grants have been found to address the poor's immediate basic needs, reduce poverty and promote human development (Neves et al., 2009; Samson, 2009), considerable uncertainty remains about some of their incentive effects (van der Berg et al., 2010). Conventional economic theory suggests that social grants may undermine the incentives of the poor to participate in economic activities, such as farming, by reducing the opportunity cost of failing to do so (Samson et al., 2004). Social grants may affect people's social and economic behaviour negatively and entrench a culture of dependency and entitlement (Gomersall, 2013). This is especially the case if the beneficiaries depend on social grants for many years.

Even though social grants in South Africa are targeted at specific vulnerable groups, such as the old, young, orphans and the sick, the potential for spill-over effects among household members has been noted by several researchers in the past (Klasen & Woolard, 2008; Abel, 2013; Devereux, 2013). According to Devereux (2013), the spill-over effects result in many unemployed or underpaid adults depending on these grants, i.e., becoming dependent on social grants beneficiaries. The concern, therefore, is that access to social grants by one or two household members may result in prime-aged, able-bodied household members choosing not to participate in economic activities, or reducing their work hours.

The debate in literature on the potential disincentive effects of the social grants on labour supply has been inconclusive. On one hand, some studies (e.g., Posel et al., 2006; Williams, 2007; Ardington et al., 2009, 2013) have concluded that additional income from social grants has a positive impact on employment by easing the constraints associated with job search. On the other hand, others (e.g., Bertrand et al., 2003; Abel, 2013) have concluded that social grants reduce incentives to work, as the additional income from social grants causes household members to work less and take additional leisure. In general, the previous studies have focused on the incentives of household members to engage in formal job activity, ignoring the incentives to engage in informal activities such as smallholder farming.

Given South Africa's high unemployment rates, which has been over 26 % in the past 10 years (Herrington et al., 2015), limited prospects for labour absorption in the off-farm sector (Aliber & Hall, 2012), and that the government has prioritised the expansion of the smallholder farming sector as part of its broader job creation strategy (DED, 2011), the incentives to work cannot be fully captured if the smallholder farming sector is ignored. Most of the rural people are self-employed in informal activities such as smallholder farming (Gollin, 2014), implying that studies on incentives to work should also consider this fact. Also, most of the recent social grant impact studies in South Africa, with the exception of Agüero et al. (2007), have focused on evaluating causal effects of a binary treatment, i.e., a dummy variable showing whether or not a household has access to grants. A weakness of the binary treatment approach is that it does not account for the heterogeneous effects of social grants, as it classifies all social grants recipient households in the same way, irrespective of the level of social grant support (Agüero et al., 2007; Bia & Mattei, 2012).

Therefore, this study aimed to contribute to the literature on social grants impacts by investigating the extent to which social grant-dependency affect rural households' incentives to participate in farming activities in the KwaZulu-Natal province. Social grant-dependency was defined as the extent to which social grants contribute to household income. The study's contribution to the literature is in three ways. Firstly, the study focuses on the smallholder farming sector, which has not been adequately investigated by the social grant impact studies. Secondly, the study uses the continuous treatment approach, instead of the limited binary treatment approaches. Lastly, the study uses the proportion of all the social grants to household income, thus capturing the importance of all the social grants to household income. Previous social grant studies (e.g., Posel et al., 2006; Agüero et al., 2007; Abel, 2013) have mostly focused on the impact of, mostly, one or a few of the social grants. This study argues that it is the relative importance of the social grant that influences household behaviour as it better captures the level of social grant-dependency compared to the absolute amount of social grants income.

## 2 Materials and methods

### *2.1 Data*

The multi-stage random sampling technique was adopted in this study to draw a sample of 984 farming households from four districts in the KwaZulu-Natal (KZN) province of South Africa. Firstly, four districts were purposively chosen out of the 11 districts in KZN. The four districts that were selected were Harry Gwala, Umzinyathi, Umkhanyakude and Uthukela. These districts have a significant number of rural communities engaged in smallholder farming activities. Secondly, a total of 984 households were randomly selected from the four districts. The lists of farmers were obtained from the respective local offices of KZN's Department of Agriculture. The total sample comprised of 239 households from Umzinyathi, 191 from Uthukela, 143 from Umkhanyakude and 411 from Harry Gwala districts. The number of households sampled was proportional to the number of farming households in the selected districts. The data were collected using a pre-tested structured questionnaire. The questionnaire captured the farmers' main information used to generate variables.

### 2.2 *Dependent and independent variables*

The dependent variable, capturing households participation in farming activities, was proxied by the total number of man-day equivalents the household members were involved in crop farming activities in the previous 60 days, following Abdulai et al. (2005). The 60-day period was considered short enough for the households to recall easily and give relatively accurate and reliable responses. The two months under study, October and November, represent the peak periods of labour demand for land preparation, cultivation and planting the summer crops. A man-day of work was defined as the amount of farm work that can be carried out by an adult male in an 8-hour work period (ILCA, 1990). The conversion factors (weights) applied to males and females in different age groups and carrying out different farming tasks to express labour time in terms of man-days were those presented in Panin (1986).

To capture social grant-dependency, the treatment variable, the study used the proportion of household income from social grants. The fact that social grants are given to individuals in households, and that different households differ in terms of the number of social grant beneficiaries leads to a high degree of heterogeneity in contribution levels received by households. Treating a household that receives just 10 % of its income from social grants the same as a household that receives more than 50 % of its income from grants seems likely to understate the potential effect of social grants. The proportion variable used in this study captures this variation.

The econometric models included other variables that were hypothesised to influence households' decision-making processes and incentives to participate in farming. Since the labour variable focused on the participation of family labour on crop farming activities, livestock farming was controlled for by including livestock size as one of the explanatory variables in the model. Off-farm commitments, captured as off-farm employment and off-farm business ownership, were also controlled for by entering these variables as explanatory variables.

### 2.3 *Estimation methods*

The generalised propensity score (GPS) method was used to estimate the impact of social grant-dependency on household members' participation in farming activities. The use of experimental or randomised designs is not applicable when studying social grants in South Africa because the social grants were not implemented with an experimental design (Agüero et al., 2007; Patel et al., 2013). The GPS is the conditional probability of receiving a particular dosage level given a set of observable variables (Imbens, 2000; Hirano & Imbens, 2004). The GPS method was applied under the un-confoundedness assumption that adjusting for differences in a set of observed pre-treatment variables removes all biases in comparisons by different levels of social grant-dependency. Compared to estimates based on full samples, the impact estimates based on matched samples are less biased and more reliable (Rubin & Thomas, 2000).

The GPS matching method was estimated following Hirano & Imbens (2004), using a Stata ado file developed by Bia & Mattei (2008). The estimation procedure consisted of three main steps. The first step involved estimation of the propensity score. To estimate the conditional distribution of the level of dependency on social grants, it was assumed that the level of dependency on social grants follows a normal distribution, conditional on the covariates:

$$g(T_i)|X_i \sim N\left[h(\gamma, X_i), \sigma^2\right], \qquad (1)$$

where $T_i$ is the level of dependency on social grants (the treatment variable), $g(T_i)$ is a suitable transformation of the level of dependency on social grants, $h(\gamma, X_i)$ is a function of covariates ($X_i$) with linear and higher-order terms, which depends on a vector of parameters, $\gamma$, and $\sigma$ is the standard deviation. The higher-order terms were included to obtain an estimate of the GPS that satisfies the balancing property (Bia & Mattei, 2008). The tests for normality and the balancing property were done to ensure that these assumptions were met before estimating the GPS. An important assumption of the GPS method is that adjusting for differences in a set of observed variables removes all biases in comparisons by different level of dependency on social grants (unconfoundedness). To strengthen the plausibility of this strong unconfoundedness assumption, a rich number of covariates was introduced in the estimation of the propensity score.

The GPS was estimated as follows:

$$\hat{R}_i = \frac{1}{\sqrt{2\pi\hat{\sigma}^2}} \exp\left[-\frac{1}{2\hat{\sigma}^2} g(T_i) - h(\hat{\gamma}, X_i)\right], \quad (2)$$

where $\hat{R}_i$ is the estimated score, and other variables and parameters are as defined before. $\hat{\sigma}$ and $\hat{\gamma}$ are the parameters estimated in Eq. (1).

The second main step of the GPS matching technique involved estimating the conditional expectation of the outcome Yi (household's participation in farming activities in man-day equivalents), given the level of dependency on social grants ($T_i$) and the GPS ($R_i$). The conditional expectation of the outcome was estimated as a function of the two scalar variables, $T_i$ and $R_i$, as follows:

$$\varphi\,[E(Y_i|T_i, R_i)] = \psi(T_i, R_i, \alpha)\,, \quad (3)$$

where $\varphi(.)$ is a link function that relates the predictor, $\psi(T_i, R_i, \alpha)$ to the conditional expectation, $\alpha$ are the parameters to be estimated using the polynomial approximations, and other variables are as defined before. As suggested by Bia & Mattei (2008), the polynomial approximations of order higher than three were not used.

The final step involved estimating the dose-response function. The estimated regression function was averaged over the score function and evaluated at the desired level of the treatment. The average potential outcome (dose-response function) for each level of dependency on social grants was estimated as follows:

$$\begin{aligned} E[\widehat{Y(t)}] &= \frac{1}{N}\sum_{i=1}^{N}\hat{\beta}\,[t, \hat{r}(t, X_i)] \\ &= \frac{1}{N}\sum_{i=1}^{N}\varphi^{-1}\left[\hat{\psi}\,(t, \hat{r}(t, X_i); \hat{\alpha})\right], \end{aligned} \quad (4)$$

where $\hat{\alpha}$ is the vector of the estimated parameters in the second stage and other variables and parameters are as described above. The analysis here used variation in the extent of level of dependency on social grants (the treatment) to identify the causal impact.

For robustness checks as well as to investigate other factors that determine the participation levels of rural households in farming activities, ordinary least squares (OLS) was used to estimate the following equation:

$$Y_i = \beta x_i + \delta T_i + \varepsilon_i\,, \quad (5)$$

where $Y_i$ is the number of man-days the household members were engaged in farming activities, $T_i$ is the proportion of household income from social grants, $x_i$ is a vector of household characteristics, $\beta$'s and $\delta$ are parameters to be estimated and $\varepsilon_i$ is the residual term.

## 3 Results

### 3.1 Descriptive statistics

Table 1 presents the demographic and socio-economic characteristics of the 984 sampled households. The table shows that the age of the rural farming households' heads was 56 years, on average, and that 47 % of households were male-headed. The sampled household heads attained low levels of education, and only 20 % were employed. On average, 0.97 prime-aged, able-bodied household members, representing just over 25 % of the total 3.67 prime-aged, able-bodied household members, were employed in the off-farm sector, highlighting the lack of economic opportunities in the rural areas. This underscores the importance of smallholder farming as a livelihood option among the rural households. The households spent an average of 36 man-days on farming activities in the previous 60 days.

The households had access to small land sizes (2 ha), and practiced both crop farming and livestock rearing. Maize was grown by most of the farmers (78 %) the season prior to the survey. Further crops cultivated included beans (56 %), spinach (53 %), onions (46 %), tomatoes (43 %), cabbages (38 %), potatoes (26 %), butternut (18 %) and beetroot (14 %). A significant proportion of the households practiced some form of irrigation. Some were members of smallholder irrigation schemes, while others watered their crops (such as spinach, tomatoes, etc.) using cans and hosepipes. The households owned moderate livestock sizes, with cattle (mean = 3), goats (mean = 5) and chickens (mean = 10) being the main animals kept. Cattle and goats are driven to communal grazing areas every morning to feed and collected at sunset, while the chickens fend for food around the yard.

The majority (84 %) of the households had access to social grants. On average, each household had about three social grants beneficiaries, showing the important role of social grants among rural households, in view of household size of seven. Social grants contributed significantly (38 %) to household income, which is bigger than the contribution of farming (13 %). The households reported poor access to markets and support services such as training, extension and credit. Few households hired people to help with their farming activities.

**Table 1:** *Household descriptive variables and their means (n = 984).*

| *Variable name* | *Variable description* | *Mean* | *SD* |
|---|---|---|---|
| Age | Household head age (Years) | 56 | 13 |
| Gender | Household head gender (1=Male) | 0.47 | – |
| Marital status | Household head marital status (1=Married) | 0.46 | – |
| Education level | Household head education level (Years of schooling) | 4.67 | 4.17 |
| Household size | Household size (Numbers) | 7.04 | 3.60 |
| Employment status of HH head | Household head off-farm employment (1=Yes) | 0.20 | – |
| Total family labour | Able-bodied, prime-aged household members | 3.61 | 2.30 |
| Number of employed HH members | Able-bodied household members employed off-farm | 0.97 | 1.32 |
| Farm labour | Household participation in farming activities in the last 60 days (Man-day equivalents) | 36.37 | 21.84 |
| Access to grants | Access to social grants (1=Yes) | 0.84 | – |
| Grants beneficiaries | Number of social grant beneficiaries per household | 3.18 | 1.81 |
| Land size | Land size household has access to (ha) | 1.90 | 4.47 |
| Livestock size | Livestock size per household (TLUs) | 3.53 | 17.40 |
| Assets | Value of assets (Rands†) | 82 105 | 38 937 |
| Total income | Total annual household income (Rands†) | 46 757 | 32 707 |
| Income from grants | Annual income from grants (Rands†) | 16 916 | 15 877 |
| Income from farming | Annual income from farm activities (Rands†) | 6553 | 12 438 |
| Social grant income proportion | Proportion of income from social grants | 0.38 | 0.26 |
| Farm income proportion | Proportion of income from farming activities | 0.13 | 0.14 |
| Farming experience | Household head farming experience (Years) | 18.70 | 13.28 |
| Hire labour | Hiring in farm labour (1=Yes) | 0.37 | – |
| Rainfall | Perceived rainfall (1=Good) | 0.67 | – |
| Soil quality | Perceived soil quality (1=Good) | 0.55 | – |
| Tenure | Secured land tenure (1=Yes) | 0.37 | – |
| Tillage access | Tillage access (1=Yes) | 0.45 | – |
| Market access | Market access (1=Yes) | 0.20 | – |
| Group membership | Farmer association member (1=Yes) | 0.42 | – |
| Credit access | Access to credit (1=Yes) | 0.36 | – |
| Extension access | Access to extension (1=Yes) | 0.46 | – |
| Training | Access to agricultural training (1=Yes) | 0.41 | – |
| Off-farm business | Small off-farm business ownership (1=Yes) | 0.08 | – |
| Irrigation access | Access to water for watering crops (1=Yes) | 0.46 | – |
| Distance to road | Distance to the nearest all-weather road (km) | 17.75 | 39.93 |
| Harry Gwala | Harry Gwala district (1=Harry Gwala) | 0.42 | – |
| Umzinyathi | Umzinyathi district (1=Umzinyathi) | 0.24 | – |
| Uthukela | Uthukela district (1=Uthukela) | 0.19 | – |
| Umkhanyakude | Umkhanyakude (1=Umkhanyakude) | 0.15 | – |

Notes: † Exchange rate: R 11.28 = US$ 1; TLU: Tropical Livestock Unit, HH: household

Most of the farmers perceived their soils to be fertile and rainfall to be good.

The results also showed low levels of off-farm entrepreneurship among the interviewed households, as only a small proportion of households owned some micro-business ventures such as weaving, handicrafts and tuck shops. The roads were generally poor and inaccessible in the study areas, with many rural households found far from the nearest all-weather road. These roads to the households are generally inaccessible by car, with households having to use wheelbarrows or bicycles to get to the all-weather roads.

### 3.2 The impact of social grant-dependency on farming participation levels: The generalised propensity score (GPS) method results

Before presenting the GPS results, a mean comparison of households' farm participation levels at three levels of social grant-dependency was done and results are presented in Table 2. The table shows that only 155 of the total sampled households were not recipients of any type of social grant, representing 16 % of the sample. A significant proportion (36 %) of the total sampled husehold derived more than 50 % of their income from social grants, while 48 % received less than 50 % of their income from social grants. The results indicate progressive decline in participation levels as the contribution of social grants to household income increases. While this suggest that households participate less in farming acitivities as they become more dependent on social grants, this result should not be accepted yet as the confounding factors were not controlled for.

The GPS approach was implemented following Bia & Mattei (2008) and the results are presented in Fig. 1. The tests for normality and balancing property indicated that these assumptions were satisfied. Figure 1 shows the average dose-response and treatment effect functions and the 95 % confidence bands for farm participation levels. The confidence bands were based on 100 bootstrap replications to account for the uncertainty associated with the estimation of the GPS and the parameters, as suggested by previous studies (Bia & Mattei, 2008, 2012).

**Table 2:** *Mean comparisons of household farm participation according to social grant-dependency.*

| *Variables* | *No income (0 %) from grants (n = 155)* | | *Low income (< 50 %) from grants (n = 478)* | | *High income (> 50 %) from grants (n = 351)* | | *F test* |
|---|---|---|---|---|---|---|---|
| | *Mean* | *SD* | *Mean* | *SD* | *Mean* | *SD* | |
| Man days over the last 60 days | 39.55 | 23.64 | 37.08 | 21.92 | 34.04 | 20.72 | 3.86** |

** means significant at 5 % significance levels.

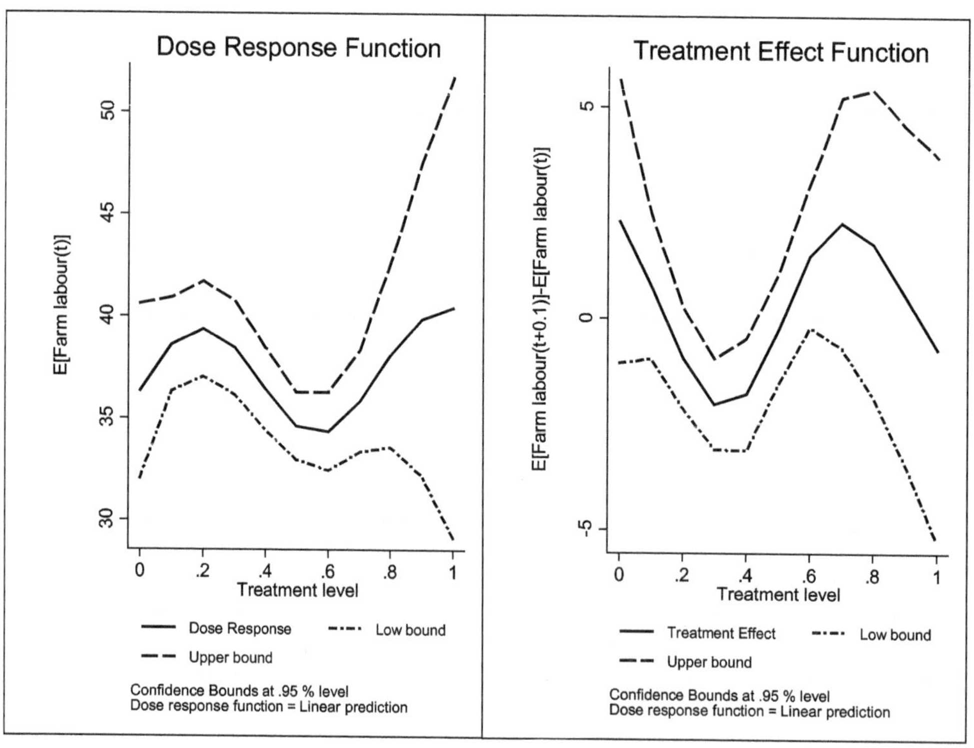


**Fig. 1:** *The average dose-response and treatment functions and 95 % confidence bands for the number of man-days the household members were engaged in farming activities during the last 60 days.*

Notes: E[Farm labour(t)] means the average number of man-days household members were engaged in farming activities at treatment level t, while treatment level means the contribution level of social grants income to total household income. The E[Farm labour(t+0.1)] − E[Farm labour(t)] shows the average effect of a social grant contribution increase of 10 % to household income on the household's farm participation level.

The confidence bands are narrow in the 0–80 % range, implying that the results are reliable up to 80 %. The wide 95 % confidence bands imply a high level of uncertainty. As such, the shape of the graph after dosages greater than 80 % should be considered with caution, and was not interpreted in this study.

The results showed that the effect of social grant-dependency is not uniform at different treatment levels. Figure 1 shows that increasing treatment (i.e., increasing dependency on social grants) is associated with increasing participation at lower (0–20 %) and higher (60–80 %) treatment levels. The implication is that the additional income from social grants, at lower and higher levels of social grant-dependency, plays a positive role in motivating households to participate more in farming. At lower levels, the social grant income is not significant enough to create a dependency syndrome. On the other hand, at higher levels, the households are poorer and have fewer other income sources, such that households must look for economic activities to augment their inadequate income. However, Fig. 1 indicate that additional income from social grants results in decreased incentives to supply more family labour to farming at dosages between 20 and 60 %.

### 3.3 *The impact of social grant-dependency on farming participation levels: Ordinary least squares (OLS) results*

Table 3 presents the OLS results estimated for robustness checks on the GPS matching results as well as to investigate the determinants of households' farm participation levels. The highly significant F statistic indicates that, collectively, the variables are significant determinants of farm participation, suggesting good model fit. The tests for collinearity were done using variance inflation factors (VIFs), and they indicated that there was little evidence of severe multi-collinearity among other variables. Heteroscedasticity was remedied by reporting robust standard errors. The Hausman test indicated that the level of dependency on social grants was not endogenous in the model. The results indicate a negative relationship between dependency on social grants and household participation in farming activities. Even though the GPS results indicated that this negative relationship applies only when households receive between 20 and 60 % of their income from social grants, the OLS results indicated that this relationship dominates on average.

The results showed that the relationship between the age of the household head and farming participation levels is non-linearly. This means that households are less likely to decrease their farm participation up to a certain age of the household head. However, after the household head reaches a particular age, more household members become involved in farming activities. This can be explained in terms of the earning opportunities of the household head. At younger ages, additional years mean more experience and connections, such that the household head accumulates opportunities. Such opportunities result in the household becoming less reliant on smallholder farming, hence less commitment to farming. However, at older years, the household head

**Table 3:** *The impact of social grant-dependency on household farm participation, OLS results (n = 984).*

| *Variables* | *Coef.* | *Std. Err.* |
|---|---|---|
| Social grant income proportion | −3.985 ** | 1.914 |
| Age | −1.090 *** | 0.346 |
| $Age^2$ | 0.009 *** | 0.003 |
| Gender | 1.484 | 1.495 |
| Education level | −0.094 | 0.189 |
| Marital status | 2.351 | 1.493 |
| Total family labour | 2.167 *** | 0.203 |
| Hired labour | −1.112 *** | 0.246 |
| Employment status of HH head | −3.035 * | 1.768 |
| Number of employed HH members | −1.627 *** | 0.542 |
| Off-farm business | 5.647 ** | 2.860 |
| Farming experience | 0.044 | 0.053 |
| Land size | 0.900 | 0.626 |
| Livestock size | −0.034 ** | 0.019 |
| Assets | 0.298 | 0.857 |
| Rainfall | −1.044 | 1.696 |
| Soil quality | 1.112 *** | 0.364 |
| Tillage access | −3.109 *** | 1.320 |
| Market access | 2.206 * | 1.339 |
| Group membership | 5.145 *** | 1.687 |
| Credit access | 1.125 *** | 0.372 |
| Extension access | 3.909 *** | 1.365 |
| Training | 0.958 | 1.564 |
| Distance to road | 0.033 | 0.019 |
| Irrigation access | 3.699 *** | 1.345 |
| Umzinyathi | 0.010 | 1.961 |
| Uthukela | −6.901 *** | 1.831 |
| Umkhanyakude | 2.879 | 2.600 |
| _Constant | 2.638 | 2.914 |
| *N* | 984 | |
| *F* | 9.46 *** | |
| $R^2$ | 0.43 | |
| Mean VIF | 4.50 | |
| Hausman test: $F = 1.43$, $p = 0.23$ | | |

Notes: *** means significant at 1 %; ** means significant at 5 %; * means significant at 10 % significance levels

is retiring or about to retire. The likely loss of income due to the head retiring forces households to be more involved in farming activities.

The results showed a positive relationship between total family labour and farm participation. This is expected, since higher number of able-bodied, prime-aged household members imply increased labour availability, which results in an increase in the number of man-days supplied to farming. The results also demonstrate that those households who hired in labour supplied less household labour to farming activities. This is because hired labour substitute for family labour, such that increasing hired labour leads to decreased farm participation by household members.

The employment status of the household head as well as the number of employed family members had negative estimated coefficients. This indicates that family members engage less in farming when the household head or a greater number of household members are employed off-farm. The explanation here is that households where members are employed in the off-farm sector have less remaining labour time to supply to farming activities. Moreover, these households are less dependent on farming, and therefore less committed to farming. Ownership of a off-farm small business had a positive estimated coefficient, suggesting a positive relationship between off-farm businesses and farming.

The results showed a negative relationship between livestock size and household farm participation levels. A bigger livestock size implies a wealthier household, and increasing wealth is associated with decreasing dependency on crop farming only. As such, household members become less committed to crop farming as the household becomes wealthier. As expected, access to tillage services reduces the participation of household members involved in farming. Access to tractors or animals for draught power reduces the work burden and/or drudgery. Households with access to tillage services such as tractors require fewer people to perform farming activities such as land preparation compared to households that rely on manual labour for the same tasks.

The households who felt that their soils were of good quality supplied more labour to farming activities than those who felt otherwise. The reason is that better soils increase the prospects of better yields, such that the households with good land quality, expecting better yields, would participate more in farming than those with poor soils. The results also indicate that access to the market motivates rural households to participate more in farming activities. Members of farmer groups supplied more family labour to farming than those who are not members. This is because association membership may help the individual farmers through pooling of resources and sharing of knowledge and experiences. The significant credit access estimate highlights the importance of credit support to the success of smallholder producers.

Access to extension is a motivator for households to become more involved in farming, hence the positive estimated coefficient. The same explanation applies to the positive estimated coefficient of irrigation access. Access to irrigation implies reduced chances of crop failure, higher productivity as well as higher expected revenues, hence more family members are likely to be involved in farming. Compared to Harry Gwala, the results indicate that households in the Uthukela district supplied less family labour to farming. The result may be indicative of the differences in economic opportunities between the two districts, suggesting that households in Uthukela district are less dependent on farming than those in Harry Gwala district.

## 4 Discussion

This paper examines the extent to which social grant-dependency has incentive / dis-incentive impacts on rural households' incentives to participate in farming activities. The paper addresses this issue by focusing on the causal effect of receiving different amounts of social grants to households and the subsequent relative importance of social grants income to the households. The GPS results indicate that the effect of social-grant dependency on households' farm participation levels differs at different dependency levels. This indicates the importance of moving beyond the dummy variable showing whether or not a household had access to social grants, as this would have failed to show these heterogeneities at different social grant-dependency levels.

The GPS matching approach indicates that while social grants have a negative effect on farming participation when social grants income contribute 20–60 %, they had a positive effect at lower and higher dependency levels. This suggests that increasing income from social grants motivates households to participate more in farming at lower and higher levels of social grant-dependency. Whereas households may be forced to participate at higher levels of dependency, as this implies poorer households, the study findings suggest that social grants may play a positive role if their contribution levels are kept below 20 % of household income. The positive effect of social grants at the lower and higher treatment levels is consistent with the presence of credit constraints that limit poor rural households' ability to engage in economic activities, supporting the hypothesis that social grant beneficiary households use some of the grant income to alleviate financial constraints in agricultural production.

This result is consistent with several studies (e.g., Devereux, 2002; Lund, 2002; Woolard, 2003; Diao et al., 2012; Mabugu et al., 2014; Proctor, 2014), that have reported that social grants can promote livelihoods and enhance economic activities by easing the financial constraints facing the poor - the so-called 'irrigation function' of social security. This suggests that social grants have the potential to complement livelihood activities such as farming as options for rural livelihoods in Africa. This is supported by Sinyolo et al. (2016a) who found that the households with access to social grants adopt more modern technologies such as inorganic fertilisers compared to households without access to social grants. However, findings from Sinyolo et al. (2016b) suggest that the increase in technology use due to access to social grants is not associated with increase in land under cultivation, as the study found no significant impact of social grants on the proportion of land cultivated. Similarly, the results of this current study suggests that the increase in household members participating in farming activities does not necessarily result in increasing proportion of land cultivated by households.

The GPS results also suggest that social grants have a dis-incentive effect when they contribute between 20 and 60 % to household income. This result supports other studies (e.g., Bertrand et al., 2003; Abel, 2013), which reported that an increase in social grants income increases the reservation wage and lowers labour force participation. This implies that, at least some of the social grant income that is, in theory, targeted towards the elderly, young or sick, ends up being redistributed (as cash or food, etc.) towards the working-age members of the household. The result of this intra-family redistribution is a significant reduction of the number of man-days that the household members engage in smallholder farming activities when social grants contribute between 20 and 60 % to household income.

The study identified key variables that affect the participation levels of rural households in smallholder farming. In general, the study highlighted the importance of expectations of success in motivating households to commit more family labour on farming activities. For example, households with access to irrigation had more members participating because of reduced risk of crop failure under irrigation. Moreover, access to institutional and/or organisational support such as extension, markets, credit or farmer groups were positively related to farming participation as these also increase chances of farming success. Agricultural extension officers remain the main sources of information with regards to new technologies or markets among the rural households. In South Africa, access to agricultural extension officers also means higher chances of accessing government support. Since farming is more likely to succeed where households have access to government support, household members are more likely to engage in farming where success is expected. Further analysis of the data indicated that households with access to extension had significantly higher farm income than those without extension access.

The result that households with better access to markets supplied more labour to farming than those with less access to the market is consistent with other studies in South Africa (e.g., Kirsten & Sartorius, 2002; van der Heijden & Vink, 2013). This highlights the important role played by access to markets in the success of smallholder farming. Market access speaks of opportunities of making good profits out of farming activities and it is these prospects that encourage households to supply more labour to farming. This study also demonstrates the positive role of farmer organisation in smallholder farming success, in line with what has been reported by earlier studies (e.g., Hellin et al., 2009; Markelova et al., 2009; HLPE, 2013; Sinyolo et al., 2014).

The positive and significant credit access estimate highlights the importance of credit support to the success of smallholder producers, as has been reported by others (Louw, 2013; Rahman & Smolak, 2014). Access to credit reduces the liquidity problem that usually affects the farmers during the planting season and this enhances the use of agricultural inputs in production, by ensuring that farmers secure the inputs in time. This leads to improved agricultural productivity, resulting in increased farm revenues and incentives for the farmers to participate more in farming.

Interestingly, whereas one would have expected that households who own off-farm businesses would be less committed to farming, the results of this study imply the opposite (Table 3). This indicates that rural households diversify their livelihoods and engage on a number of income generating activities, as has been reported by several studies (Shackleton et al., 2007; Jacobs & Makaudze, 2012; Aliber & Mdoda, 2015). According to these studies, households diversify into off-farm small businesses to supplement farming, not to substitute it. This suggests that starting off-farm businesses may be motivated by the need to find ways and means to relax credit constraints in farming, or raise supplementary income to augment to what they get from farming, especially during lean seasons.

To conclude, the study indicates that the effect of social grants on incentives of households to participate in smallholder farming activities is dependent upon the contribution levels of the grants. Given the importance of social grants in poverty reduction, the study recommends that social grants should continue but policy makers should be particularly cognisant of their possible adverse consequences on smallholder farming at moderate dependency levels. To address the disincentive effects of social grants and ensure successful smallholder production activities in the rural areas, this study recommends a holistic approach that addresses both the social grants side as well as the smallholder farming side. Policy-makers should aim to find strategies of reducing social grant-dependency and dis-incentive effects, while simultaneously creating a conductive environment to improve the attractiveness, viability and success of smallholder farming. The study suggests that introducing and/or supporting existing irrigation schemes as well as increasing institutional and/or organisational support could encourage the rural people to participate more in smallholder farming. This is especially important in South Africa where the government's employment strategy puts agriculture at the core of its drive.

## References


Abdulai, A., Barrett, C. B. & Hoddinot, J. (2005). Does food aid really have disincentive effects? New evidence from sub-Saharan Africa. *World Development*, 33 (10), 1689–1704.

Abel, M. (2013). Unintended labour supply effects of cash transfer programmes: Evidence from South Africa's old age pension. SALDRU Working Paper 114. University of Cape Town, Cape Town, South Africa.

Agüero, J., Michael, C. & Ingrid, W. (2007). The impact of unconditional cash transfers on nutrition: The South African Child Support Grant. Working Paper number 39. International Poverty Centre, Brasilia, Brazil.

Aliber, M. & Hall, R. (2012). Support for smallholder farmers in South Africa: Challenges of scale and strategy. *Development Southern Africa*, 29 (4), 548–562.

Aliber, M. & Mdoda, L. (2015). The direct and indirect economic contribution of small-scale black agriculture in South Africa. *Agrekon*, 54 (2), 18–37.

Ardington, C., Bärnighausen, T., Case, A. & Menendez, A. (2013). Social protection and labour market outcomes of youth in South Africa. SALDRU Working Paper 96. University of Cape Town, Cape Town, South Africa.

Ardington, C., Case, A. & Hosegood, V. (2009). Labour supply responses to large social transfers: Longitudinal evidence from South Africa. *American Economic Journal: Applied Economics*, 1 (1), 22–24.

van der Berg, S., Siebrits, K. & Lekezwa, B. (2010). Efficiency and equity effects of social grants in South Africa. Stellenbosch Economic Working Papers 15/10. University of Stellenbosch, Cape Town.

Bertrand, M., Mullainathan, S. & Miller, D. (2003). Public policy and extended families: Evidence from pensions in South Africa. *World Bank Economic Review*, 17 (1), 27–50.

Bia, M. & Mattei, A. (2008). A Stata package for the estimation of the dose–response function through adjustment for the generalized propensity score. *The Stata Journal*, 8 (3), 354–373.

Bia, M. & Mattei, A. (2012). Assessing the effect of the amount of financial aids to Piedmont firms using the generalized propensity score. *Statistical Methods and Applications*, 21, 485–516.

DED (2011). The New Growth Path: Framework. DED (Department of Economic Development), Pretoria. Available at: http://www.economic.gov.za/communications/publications/new-growth-path-series (accessed on: 25 August 2015).

Devereux, S. (2002). Can social safety nets reduce chronic poverty? *Development Policy Review*, 20 (5), 657–675.

Devereux, S. (2013). Trajectories of social protection in Africa. *Development Southern Africa*, 30 (1), 13–23.

Diao, X., Thurlow, J., Benin, S. & Fan, S. (2012). Strategies and priorities for African agriculture: economy wide perspectives from country studies. International Food Policy Research Institute, Washington DC.

Gollin, D. (2014). Smallholder agriculture in Africa: An overview and implications for policy. IIED Working Paper. IIED, London. Available at: http://pubs.iied.org/14640IIED (accessed on: 5 March 2015).

Gomersall, J. (2013). The performance of the Child Support Grant: Review and research priorities. *Development Southern Africa*, 30 (4-05), 525–544.

van der Heijden, T. & Vink, N. (2013). Good for whom? Supermarkets and small farmers in South Africa: A critical review of current approaches to increasing access to modern markets. *Agrekon*, 52 (1), 68–86.

Hellin, J., Lundy, M. & Meijer, M. (2009). Farmer organization, collective action and market access in Meso-America. *Food Policy*, 34, 16–22.

Herrington, M., Kew, J. & Kew, P. (2015). 2014 GEM South Africa Report. South Africa: The crossroads – a goldmine or a time bomb? Global Entrepreneurship Monitor (GEM). Available at: http://www.gemconsortium.org/country-profile/108 (accessed on: 1 December 2015).

Hirano, K. & Imbens, G. W. (2004). The propensity score with continuous treatments. *In:* Gelman, A. & Meng, X.-L. (eds.), *Applied Bayesian modeling and causal inference from incomplete-data perspectives*. Ch. 7, Wiley, New York.

HLPE (2013). Investing in smallholder agriculture for food security. HLPE Report 6. A report by the High Level Panel of Experts on Food Security and Nutrition of the Committee on World Food Security. Food and Agricultural Organisation (FAO), Rome.

ILCA (1990). Livestock systems research manual. Working Paper 1, Vol. 1. ILCA (International Livestock Centre for Africa), Addis Ababa, Ethiopia.

Imbens, G. (2000). The role of the propensity score in estimating dose-response functions. *Biometrika*, 87 (3), 706–710.

Jacobs, P. & Makaudze, E. (2012). Understanding rural livelihoods in the West Coast District, South Africa. *Development Southern Africa*, 29 (4), 574–587.

Kirsten, J. & Sartorius, K. (2002). Linking agribusiness and small-scale farmers in developing countries: Is there a new role for contract farming. *Development Southern Africa*, 19 (4), 503–529.

Klasen, S. & Woolard, I. (2008). Surviving unemployment without state support: Unemployment and household formation in South Africa. *Journal of African Economies*, 18 (1), 1–51.

Louw, A. (2013). Sustainable policy support for smallholder agriculture in South Africa: Key issues and options for consideration. *In:* Greenberg, S. (ed.), *Smallholder and agro-food value chains in South Africa: Emerging practices, emerging challenges*. Institute of Poverty, Land and Agrarian Studies (PLAAS), Cape Town.

Lund, F. (2002). 'Crowding in' care, security, and micro-enterprise formation: Revisiting the role of the state in poverty reduction and development. *Journal of International Development*, 14, 681–694.

Mabugu, R., Chitiga, M., Fofana, I., Abidoye, B. & Mbanda, V. (2014). Assessing the general equilibrium effects of social grants in South Africa. Paper presented at the 17th Annual Conference on Global Economic Analysis "New Challenges in Food Policy, Trade and Economic Vulnerability", Dakar, Senegal.

Markelova, H., Meinzen-Dick, R., Hellin, J. & Dohrn, S. (2009). Collective action for smallholder market access. *Food Policy*, 34, 1–7.

Neves, D., Samson, M., van Niekerk, I., Hlatshwayo, S. & du Toit, A. (2009). The use and effectiveness of social grants in South Africa. Institute for Poverty, Land and Agrarian Studies (PLAAS) and Economic Policy Research Institute (EPRI), Cape Town.

Panin, A. (1986). *A comparative socio-economic analysis of hoe and bullock farming systems in northern Ghana*. Ph.D. thesis, University of Goettingen, Goettingen, Germany.

Patel, L., Hochfeld, T. & Moodley, J. (2013). Gender and child sensitive social protection in South Africa. *Development Southern Africa*, 30 (1), 69–83.

Posel, D., Fairburn, J. A. & Lund, F. (2006). Labour migration and households: A reconsideration of the effects of the social pension on labour supply in South Africa. *Economic Modelling*, 23 (5), 836–853.

Proctor, F. J. (2014). Rural economic diversification in sub-Saharan Africa. IIED Working Paper. IIED, London.

Rahman, A. & Smolak, J. (2014). Financing smallholder farmers in developing countries. *In:* Hazell, P. & Rahman, A. (eds.), *New directions of smallholder agriculture*. Oxford University Press, Oxford.

Rubin, D. & Thomas, D. (2000). Combining propensity score matching with additional adjustments for prognostic covariates. *Journal of American Statistical Association*, 95, 573–585.

Samson, M. (2009). Social cash transfers and employment: A note on empirical linkages in developing countries. The Organisation for Economic Cooperation and Development (OECD), Paris.

Samson, M., Lee, U., Ndlebe, A., Quene, K. M., van Niekerk, I., Gandhi, V., Harigaya, T. & Abrahams, C. (2004). The Social and Economic Impact of South Africa's Social Security System. Summary Report. Department of Social Development, Economic Policy Research Institute, Cape Town, South Africa.

SASSA (2014). Fact sheet: Issue no 12 of 2014. A statistical summary of social grants in South Africa. South Africa Social Security Agency (SASSA), Pretoria. Available at: http://www.sassa.gov.za/index.php/statistical-reports (accessed on: 28 June 2015).

Shackleton, C. M., Shackleton, S. E., Buiten, E. & Bird, N. (2007). The importance of dry woodlands and forests in rural livelihoods and poverty alleviation in South Africa. *Forest Policy and Economics*, 9, 558–577.

Sinyolo, S., Mudhara, M. & Wale, E. (2014). Water security and rural household food security: Empirical evidence from the Mzinyathi district in South Africa. *Food Security*, 6 (3), 483–499.

Sinyolo, S., Mudhara, M. & Wale, E. (2016a). The impact of social grants on the propensity and level of use of inorganic fertiliser among smallholders in KwaZulu-Natal, South Africa. *Agrekon*, 55 (4), 436–457.

Sinyolo, S., Mudhara, M. & Wale, E. (2016b). To what extent does dependence on social grants affect smallholder farmers' incentives to farm? Evidence from KwaZulu-Natal, South Africa. *African Journal of Agricultural and Resource Economics*, 11 (2), 154–165.

Williams, M. J. (2007). The social and economic impacts of South Africa's Child Support Grant. Working Paper #40. Economic Policy Research Institute, Cape Town.

Woolard, I. (2003). Social assistance grants, poverty and economic growth in South Africa. Paper presented at the TIPS/DPRU Forum 2003: The Challenge of Growth and Poverty: The South African Economy Since Democracy, Johannesburg.

# 18

# Determinants of participation in cavy marketing: Evidence from the Democratic Republic of Congo

Franklin Simtowe [a,*], Birthe K. Paul [b], Benjamin M. M. Wimba [c], Samy B. Bacigale [d,**], Wanjiku L. Chiuri [e,***], Brigitte L. Maass [b,****]

[a] *International Maize and Wheat Improvement Center (CIMMYT), Nairobi, Kenya*
[b] *International Center for Tropical Agriculture (CIAT), Nairobi, Kenya*
[c] *Institut National pour l'Etude et la Recherche Agronomiques (INERA)-Mulungu, Bukavu, Democratic Republic of Congo*
[d] *Université Evangélique en Afrique (UEA), Bukavu, Democratic Republic of Congo*
[e] *International Center for Tropical Agriculture (CIAT), Kigali, Rwanda*

**Abstract**

Smallholder supply chain participation remains low despite the potential welfare gains that would result from choosing a market-oriented production. Yet, studies on determinants of market participation for commodities with underdeveloped value chains are scanty. Employing a double-hurdle model, this paper examines factors determining households' participation in cavy marketing among cavy farmers from Sud-Kivu province in the Democratic Republic of Congo. We find that wealthier households participated less in cavy marketing while those producing more cavies were more likely to participate in their marketing. Moreover, smaller households tended to sell more cavies, while households with other livestock sold fewer cavies. The findings underscore the significance of increasing the participation in the cavy supply chains by farmers through the promotion of appropriate husbandry practices that enhance cavy productivity and production and that enable farmers to participate in markets as sellers.

*Keywords:* domestic cavy, marketing, double-hurdle, supply chain participation, guinea pig, *Cavia porcellus*

## 1 Introduction

The participation of small-scale producers in commodity supply chains is widely acknowledged in development literature as a basis for economic growth and poverty reduction. Yet, current smallholder supply chain participation remains low despite the potential welfare gains that would result from choosing a market-oriented production (Barret, 2008). Cross-country and multiple year studies consistently show that typically less than one-quarter of the smallholder farmers in Africa sell some of their cereals produced (Levinsohn *et al.*, 2007). Literature on causes of low market participation in the developing world is exhaustive, attributing it to a mix of institutional, infrastructural and policy failures, price risk and uncertainty, difficulties of contract enforcement, insufficient numbers of middlemen, cost of putting small dispersed quantities of produce together and the inability to meet standards. Indeed, in the absence of institutional arrangements that can link producers to markets, Alene *et al.* (2008) demonstrate that high transport and transaction costs undermine the process of exchange and result in limited markets with

* Corresponding author
International Maize and Wheat Improvement Center (CIMMYT)
ICRAF House, United Nations Avenue – Gigiri,
P.O. Box 1041-00621, Nairobi, Kenya; Email: f.simtowe@cgiar.org

** Current affiliation: International Institute of Tropical Agriculture (IITA), Bukavu, Democratic Republic of Congo

*** Current affiliation: Current affiliation: University of Laikipia, Nyahururu, Kenya

**** Current affiliation: Georg-August University of Göttingen, Göttingen, Germany

little rural-urban market linkages. Poor infrastructure and supply chain problems, compounded by weak contracting environments, make it more costly for farmers to access input and output markets as well as the benefits from technology adoption.

Yet, empirical studies to understand the market participation behaviour by small producers have concentrated on commodities with long[1] and well-developed supply chains and among producers accessing spatially differentiated markets. Such studies are based on the assumptions that market access is not uniform because households may face different transactions costs for market participation (Key *et al.*, 2000; Renkow *et al.*, 2004), In this paper we analyse the participation of farmers in cavy supply chains in the eastern region of the Democratic Republic of Congo (DRC).

Domestic cavy, also widely known as Guinea pig (*Cavia porcellus*) is considered one of the small livestock species with potential for commercialisation by smallholders or resource-poor farmers, due to its low requirements for capital, equipment, space and labour (Lammers *et al.*, 2009). Moreover, cavies provide high quality white meat (Kouakou *et al.*, 2013) and can be a good source of income and manure (Lammers *et al.*, 2009; Yiva *et al.*, 2014). Like for most small livestock, the information on the importance of cavies in the livelihoods of rural communities remains scanty. Cavies do not specifically appear in national or regional livestock census, which further reduces the likelihood of their inclusion in research and development initiatives, creating a vicious circle of neglect (Maass *et al.*, 2014).

It is estimated that more than two million cavies are kept in DRC, contributing significantly to nutrition security as well as income generation and empowerment of women and youth; the two Kivu (Nord and Sud) provinces of the DRC are said to be the largest cavy producing regions (Maass *et al.*, 2014). According to Cox (2012), aside from their comparatively low capital requirements, cavies are relatively secure to keep in times of unrest as large livestock have been depleted through looting by armed groups in the Kivu provinces. In a survey in Sud-Kivu, Maass *et al.* (2012) found that about 50 % of livestock keepers had cavies, while Mètre (2011) even suggested eight out of ten rural households in Sud-Kivu have cavies. Since the mid-1990s, cavies have become an integral part of people's diets in the region (Maass *et al.*, 2014). Farmers in the study region largely keep cavies for food, while a few keep them as an important source of income. However, unlike markets for conventional livestock, cavy markets are highly informal and underdeveloped, and little is known about the characteristics of suppliers in the cavy market as well as the demand for cavies in the region.

We use farm-level data collected from 250 cavy farmers from Sud-Kivu province of the DRC to assess determinants of household participation and the extent of participation in cavy supply chains. Literature on market participation models tends to describe marketing decisions as occurring in two steps: (1) whether to participate in the market and (2) what volume to be sold (Goetz, 1992; Key *et al.*, 2000; Bellemare *et al.*, 2006). We adopt the two-step framework and apply a double-hurdle market participation model proposed by Cragg (1971).

## 2 Theoretical and empirical framework

### 2.1 *Theory of market participation*

To explain market participation, we start with the standard economic choice problem of the agricultural household, making its production and consumption choices to maximize the utility of consumption, subject to some constraints (on available resources and technologies). We follow Key *et al.* (2000) to explore the role of transaction costs on cavy marketing. We adopt a static model that ignores a number of components of household decision making to reduce its complexity but also to more specifically capture the impact of transaction costs, liquidity, productivity and other wealth-related factors on the marketing of cavies. Market participation is specified as a choice variable. That is, in addition to deciding how much of each of the good $i$ to consume $c_i$, produce $q_i$ and use as an input $x_i$, the households also decide how much of each good to 'market' $m_i$ (*where m is positive when it is a sale and negative when it is a purchase*). In the absence of transaction costs, the households' problems could be to maximize the utility function (1) subject to equations (2) through (5):

$$u(c; z_u) \tag{1}$$

$$\sum_{i=0}^{n} p_i^m m_i + T >= 0 \tag{2}$$

$$q_i - x_1 + A_i - m_i - c_i = 0, \; i = 1, \ldots, N \tag{3}$$

$$G(q, x; z_q) = 0 \tag{4}$$

$$c_i, q_i, x_i \geq 0 \tag{5}$$

[1] More in general, a food supply chain can be defined as "short" when it is characterized by short distance or few intermediaries between producers and consumers.

where $p_i^m$ is the market price of good $i$, $A_i$ is an endowment in good $i$, $T$ is exogenous transfers and other incomes, $z_u$ and $z_q$ are exogenous shifters in utility and production, respectively, while $G$ represents the production technology. The cash constraint (Eq. 2) states that expenditures on all purchases must not exceed revenues from all sales and transfers. The resource balance (Eq. 3) states that, for each of the N goods, the amount consumed, used as inputs and sold is equal to what is produced and bought plus the endowment of the good. The production technology (Eq. 4) relates inputs to outputs. The empirical analysis focuses on how different socioeconomic and geospatial characteristics affect the participation of households in cavy marketing.

Following Stephens *et al.* (2011), we also test the extent to which liquidity constraints may affect household participation in commodity markets. Liquidity-constrained households may be forced to sell their cavies in order to meet their current income needs based on the intuition that wealthy households may have a wide range of alternatives through which to participate in commodity markets. Other aspects explored by Stephens *et al.* (2011), which we do not investigate in this paper but that are equally important in explaining market participation behaviour include seasonality of commodity prices and how credit access can affect household's commodity market participation decisions.

### 2.2 *Empirical estimation strategy*

The theoretical model discussed above leads us to some testable hypotheses regarding household liquidity and other socio-economic factors on cavy marketing. We expect households facing liquidity constraints to have a higher likelihood using cavies as an entry into commodity markets. A complete understanding of market participation decisions would require that we also explore determinants of household level purchases of cavies but we lack data on household purchases, hence, the focus on cavy market supplies. We are interested in understanding factors affecting two types of decisions: cavy market entry and quantity of cavy sales. We thus follow Holloway *et al.* (2002) and Cragg (1971) to address the empirical problem above by applying the double-hurdle framework.

The underlying assumption in the double-hurdle approach is that individuals make two decisions with regard to their participation in the marketing of a commodity. The first decision is whether they will sell some amount of the commodity at all. The second decision is about the quantity of that will be sold conditional on the first decision. In this study the two decisions are, therefore, whether to sell cavies and how many cavies to sell. The importance of treating the two decisions independently lies in the fact that factors that affect one's decision to sell may be different from those that affect the decision on how much to sell. The double-hurdle model allows for the possibility that these two decisions are affected by a different set of variables. The advantage with this approach is that it allows us to understand characteristics of a class of households that would never participate in cavy markets. Originally proposed by Cragg (1971), the double-hurdle model has been recently applied in a variety of areas.

The double-hurdle model assumes the decision to participate in a market and the intensity of participation are determined by two separate stochastic processes and, therefore, two equations. The first equation in the double-hurdle (Eq. 6) relates to the decision to participate in cavy markets, expressed as follows:

$$d_i^* = \alpha x_i + \varepsilon;\ \varepsilon \sim N(0, 1)\,, \tag{6}$$

$$\text{where } d_i = \begin{cases} 1 \text{ if } d_i > 0 \\ 0 \text{ otherwise} \end{cases}, \tag{7}$$

while the decision on how many cavies to sell can be described as

$$y^* = \beta z_i + \mu_i;\ \mu \sim N(0, \sigma^2)\,, \tag{8}$$

$$\text{where } y_i = \begin{cases} y^* \text{ if } y^* > 0 \text{ and } d_i = 1 \\ 0 \text{ otherwise} \end{cases}. \tag{9}$$

$d_i$ is a discrete variable measuring whether or not a household sold some cavies, while $d_i^*$ is the latent variable for $d_i$. $y_i$ refers to the observed number of cavies marketed, while $y^*$ is the latent variable for $y_i$. $x_i$ and $z_i$ are vectors of characteristics, allowed to overlap (household, market, regional) that influence market entry as well as the quantities of cavies to be marketed. $\beta$ and $\alpha$ are vectors of parameters, while $\varepsilon_i$ and $\mu_i$ are error terms. Given the expression in Eq. (9), the number of cavies sold is only observed when $d_i = 1$ and $y^* > 0$.

The model has an advantage over the standard univariate Tobit model in that it provides a more flexible framework to model the observed consumer's behaviour as a joint choice of two decisions.

### 2.3 *Data collection*

The data used in this analysis draws from a survey of cavy-keeping households conducted by the International Center for Tropical Agriculture (CIAT) in collaboration with Université Evangélique en Afrique (UEA), Bukavu and Institut National pour l'Etude et la Recherche Ag-

**Table 1:** *Household characteristics of cavy farmers in eastern DR Congo (2011–2012).*

| *Variable* | *Territoire* | | | *Overall mean* |
|---|---|---|---|---|
| | *Kalehe* (*n*=90) | *Kabare* (*n*=80) | *Walungu* (*n*=80) | (*n*=250) |
| Family size | 5.75 (2.6) | 5.75 (2.6) | 5.86 (2.5) | 5.79 (2.7) |
| Age of household head (years) | 39.2 (10.9) | 46.3 (15.0) | 44.2 (15.6) | 43.1 (14.1) |
| Farming experience (years) | 20.3 (10.7) | 26.2 (16.3) | 25.7 (15.2) | 23.9 (14.4) |
| Cavy farming experience (years) | 9.9 (8.6) | 11.7 (10.3) | 9.0 (8.3) | 10.2 (9.1) |
| Main motivation for keeping cavies (%) | | | | |
| Consumption | 71.1 | 57.5 | 63.7 | 64.4 |
| Cash income | 22.2 | 20.0 | 18.8 | 20.4 |
| Manure | 1.1 | 15.0 | 10.0 | 8.4 |
| Second motivation for keeping cavies (%) | | | | |
| Cash income | 30.0 | 18.7 | 40.0 | 29.6 |
| Consumption | 21.1 | 30.0 | 23.7 | 28.8 |
| Manure | 17.7 | 42.5 | 20.0 | 26.4 |
| Others | 11.1 | 6.25 | 12.5 | 10.0 |
| Female-household (%) | 33.3 | 46.3 | 47.5 | 42.0 |
| Average land holding (ha) | 2.16 (3.6) | 1.46 (2.4) | 1.84 (2.0) | 1.83 (2.8) |
| Can read and write (%) | 66.7 | 50.0 | 58.8 | 58.8 |
| Amount of off-farm income (CDF/year)* | 192,897 (973,361) | 63,241 (110,402) | 67,618 (152,930) | 111,318 (594,717) |
| Distance to market (km) | 3.1 | 3.7 | 2.5 | 3.1 |
| Cavies reared throughout the year (no./year) | 31.2 (18.3) | 28.8 (17.2) | 34.8 (27.2) | 31.6 (21.3) |
| Other livestock kept (no.) | | | | |
| Chicken | 2.2 | 2.2 | 2.4 | 2.2 |
| Goats | 1.3 | 1.5 | 1.7 | 1.5 |
| Pigs | 0.2 | 0.1 | 0.9 | 0.4 |
| Cattle | 0.5 | 0.2 | 0.3 | 0.3 |

Standard deviation in brackets.
* During the time of the survey in 2012, 1 USD was equivalent to about 900 Congolese Francs (CDF).

ronomiques (INERA). The data were collected from three *territoires* Kabare (81 households), Kalehe (90) and Walungu (79) by 10 trained enumerators between 26 June and 9 July 2012 in Sud-Kivu province in eastern DRC. These *territoires* and the 13 sampled villages were chosen after reconnaissance surveys and key informant interviews to represent different production systems and agro-ecologies, and if they were known to have high cavy concentrations. The survey data contains various aspects of cavy production, breeding, feeding, husbandry practices, marketing and consumption.

## 3 Results

### *3.1 Household characteristics*

Summary statistics for selected key household characteristics appear in Table 1. Sampled households were relatively large, averaging 6 persons per household. The average age of heads of households was 43 years with an average land holding size of 1.8 ha; about 42 % of them were from female-headed households. Cavy farming is not new in the area as the surveyed farmers have been keeping cavies for at least a decade. Farmers reared an average of 32 cavies per household in a year. The majority of farmers (64 %) reported the first reason for rearing cavies as consumption. About 20 % and 8 %, reported cash income and manure as the first reasons for rearing cavies, respectively. The findings underscore the significance of cavy farming as a food security commodity. Aside from rearing cavies, they also reared other livestock such as chicken, goats, pigs and cattle. There was significant variation in household off-farm incomes and in the ease of access to output markets across territories.

The distribution of cavies reared by land holding size category (Table 2) reveals that the farmers in the smaller land holding size category (0–0.5 ha) kept relatively less cavies (about 30) than farmers with larger land holdings (more than 2 hectares) who had an average of 35 to >41 cavies, a finding suggesting that smaller land holdings did not necessarily drive farmers into cavy intensification.

**Table 2:** *Scale of cavy production (% of cavy farmers) by landholding size in eastern DR Congo (2011–2012).*

| | *Landholding size category* | | | | | |
|---|---|---|---|---|---|---|
| *Number of cavies held (July 2011–July 2012)* | 0–0.5 ha (*n*=72) | 0.51–1 ha (*n*=52) | 1.1–2 ha (*n*=78) | 2.1–5 ha (*n*=32) | >5 ha (*n*=16) | *Overall mean* (*n*=250) |
| Average no. of cavies | 29.7 | 29.3 | 31.4 | 34.5 | 40.5 | 31.6 |
| Proportion of household by cavy and land holding categories (%) | | | | | | |
| 1–10 cavies | 38.9 | 34.6 | 29.5 | 15.6 | 31.3 | 31.6 |
| 11–15 cavies | 26.4 | 32.7 | 24.4 | 25.0 | 25.0 | 26.8 |
| 16–20 cavies | 20.8 | 9.6 | 21.8 | 25.0 | 18.8 | 19.2 |
| 21–25 cavies | 5.6 | 15.4 | 12.8 | 18.8 | 18.8 | 12.4 |
| >25 cavies | 8.3 | 7.7 | 11.5 | 15.6 | 6.3 | 10.0 |
| Total | 100.0 | 100.0 | 100.0 | 100.0 | 100.0 | 100.0 |

**Table 3:** *Characteristics of cavy keepers according to their market positions.*

| *Variable* | *Autarky* (*n*=107) | *Sellers only* (*n*=64) | *Buyers only* (*n*=52) | *Buyer and seller* (*n*=27) | *Overall mean* (*n*=250) |
|---|---|---|---|---|---|
| Family size (no. of HH members) | 5.7 | 6.2 | 5.2 | 6.2 | 5.7 |
| Age of household head (years) | 42.8 | 46.4 | 40.1 | 42.2 | 43.1 |
| Male-hh (%) | 57.8 | 61.6 | 58.6 | 50.0 | 58.0 |
| Farming experience of HH head (years) | 23.8 | 25.9 | 21.3 | 24.4 | 23.8 |
| Average land holding (ha) | 1.67 | 2.30 | 1.60 | 1.70 | 1.83 |
| Cavies reared throughout the year (no./year) | 29.2 | 41.6 | 24.3 | 32.3 | 31.5 |
| Cavy farming experience (years) | 9.5 | 10.8 | 9.2 | 13.3 | 10.2 |
| Cavies at the time of survey (no.) | 16.2 | 17.8 | 13.0 | 13.7 | 15.6 |
| Cavies sold (no.)* | 0 | 11.7 | 0 | 6.6 | 10.7 |
| Proportion of cavies sold (%) | n.a. | 26 | n.a. | 22 | 25 |
| Selling price of cavies (CDF)† | n.a. | 1323 | n.a. | 1559 | n.a. |
| Amount of off-farm income (CDF) | 176,338 | 67,184 | 69,034 | 39,693 | 111,318 |
| | (896,148) | (132,590) | (134,674) | (62,310) | (594,717) |
| Distance to the market (km) | 2.9 | 3.4 | 2.9 | 3.7 | 3.1 |
| Cavies consumed (no. per year) | 13.1 | 12.2 | 11.3 | 12.0 | 12.4 |

n.a.: not available; Standard deviation in brackets
* These figures are highly likely an underestimation considering the long recall period, and they do not include cavies that died due to diseases or other causes. † During the time of the survey in 2012, 1 USD was equivalent to about 900 Congolese Francs (CDF).

### 3.2 *Marketing participation*

#### 3.2.1 *Market positions and characteristics of cavy producers*

There are four categories of market positions of cavy producers (Table 3); the cavy buyers only that buy for both consumption and breeding, cavy sellers only, those that buy and sell, and the autarkic household (those that did not sell or buy). Subsistence production is common with 44 % of the cavy farmers being autarkic implying that they did not buy or sell any cavies in the year 2011–2012. A quarter of them participated as sellers only, while 20 % participated as buyers only. The category involving cavy producers that participated in the market as both buyers and sellers comprised of 11 % of the farmers. Our interest in this paper is on a category of farmers that had some marketed surplus; thus, a combination of those that only sold plus those that sold and purchased. The two groups accounted for about 35 % of the total farmers.

Characteristics of cavy producers by the category of their market positions are presented in Table 3. Sellers of cavies tended to have older heads of households (46 years) than those in autarky (43 years) and buyers only (40 years). Sellers only also tended to have more years of experience in farming (26 years) than buyers only (21 years). The sellers only also had larger land holdings (2.3 ha) than the rest of the categories whose land holdings were less than 2 hectares. Cavy sellers had larger flocks of cavies (42) than those in autarky (29), buyers only (24) and those that bought and sold (32). Cavy buyers only had fewer (1.7) chicken than all the other categories. Autarkic households had substantially higher off-farm incomes than the rest of the household categories. Those that participated in both selling and buying tended to be a bit further away from the markets (3.7 km) than the remainder.

### *3.2.2 Scale of cavy marketing*

About a third (35 %) of households reported selling cavies (Table 4) for several reasons. Distress selling to cover emergency expenditures was reported by 29 % of the cavy sellers. Moreover, payment of school fees ranked highly, as major reason for selling cavies and was reported by 36 % of the cavy farmers, while 26 % sold cavies to cover planned expenditures. There was considerable variation in cavy marketing across the territories. The highest proportion of cavy producers that participated in cavy markets as sellers were from Kabare (40 %), followed by Kalehe (36 %) and Walungu (30 %). Cavy producers stated to sell about 25 % of their annual production. There was no variation in the proportions of cavies sold across the *territoires*. They sold an average of 10 cavies per household and at an average price of 1400 CDF per cavy (min. 900 CDF, max. 2000 CDF), which is equivalent to 1.6 USD per cavy. About 47 % of cavy-selling households reported that marketing decisions were made by female members of the household, while 24 % were made by men, and 21 % jointly by women and men.

From the foregoing, the cavy supply chain is quite short with most of the sales involving the interaction of producers and final consumers/buyers. The short supply chain has several advantages to both buyers and sellers, as it helps in establishing and strengthening closer ties between producers and consumers. Moreover, it is also expected to facilitate fair pricing and value addition for the cavies being sold in a way that addresses consumer preference. Consistent with the preceding explanation, farmers expressed great preference for individual consumers for a number of preferred traits (Table 5).

Over 75 % of the sellers highly ranked individual consumers for the good price they offered, while 63 % of them also expressed preference for individual customers for their timely delivery of payment.

### *3.3 Econometric results on determinants of cavy marketing*

The econometric analysis was used to assess the factors that influence participation in cavy marketing. The factors tested included, transaction costs, household wealth and liquidity as well as other household socio-economic characteristics. The existence of transaction costs raises effective prices for buyers, while lowering effective prices for sellers, creating a price band within which some households find it unprofitable to either sell or buy. For lack of data on the actual proportional transaction costs (e.g. transport costs), we use distance from the homestead to output markets as a proxy for the proportional transaction costs while the ownership of transport facilities such as bicycle and information tools such as radio and mobile phone, were used as proxies of the magnitude of fixed costs. We expect households that own bicycles and radio to have a higher likelihood to participate in marketing, although being fixed costs; they should not affect the quantity traded, while we expect households close to the markets to face lower transaction costs and to sell larger numbers of cavies than those far from the market.

Regarding variables reflecting household liquidity and wealth such as the amount of household off-farm income, ownership of other livestock, we expect that household liquidity constraints will increase the propensity of cavy sales and the quantity of sales. This would imply a lesser likelihood of participation in cavy selling for households with access to liquidity through alternative income sources and wealth. This follows an expansive literature on market participation in developing countries (de Janvry *et al.*, 1991; Goetz, 1992; Key *et al.*, 2000) showing that, if people have no other means of addressing liquidity constraints, they might find it optimal to convert non-cash wealth in the form of grains or livestock products into cash by selling them. Thus, cavies should be seen as an emerging farm commodity in eastern DRC whose importance in the household economy is likely to increase as better markets develop and are accessed by producers. Other variables tested for their effect on market participation, but without any theoretically justifiable *a priori* expectations in terms of their impact on market participation included; age of the household-head, household size, gender of the farmer, education levels as well as variables that have a bearing on cavy demand such as land size and years of experience in farming.

Estimates of the market entry (decision to participate) and quantity (intensity of participation) equations, respectively, are displayed in Table 6. The coefficient for bicycle ownership was insignificant in the market entry equation but was negative and significant in the quantity equation suggesting that bicycle ownership did not influence the decision to sell cavies, but that bicycle owners significantly sold fewer cavies whenever they decided to sell some.

**Table 4:** *Scale of cavy marketing by region ('Territoire') in eastern DR Congo.*

| *Variable* | *Territoire* | | | *Overall mean* |
|---|---|---|---|---|
| | *Kalehe* (*n*=90) | *Kabare* (*n*=80) | *Walungu* (*n*=80) | (*n*=250) |
| Proportion of cavy-farmers selling cavies (%) | 35.5 | 40 | 30 | 35.2 |
| Proportion of households selling cavies through different channels (%) | | | | |
| Individual customers | 71.4 | 80.0 | 75.0 | 75.6 |
| Relatives | 39.3 | 70.0 | 50.0 | 53.5 |
| Cavy traders | 14.3 | 10.0 | 32.1 | 18.6 |
| Others | 25.0 | 3.3 | 7.1 | 11.6 |
| Butcher/slaughter | 10.7 | 10.0 | 10.7 | 10.5 |
| Cavies sold last year (average no.) | 9.5 | 9.5 | 11.6 | 10.1 |
| Proportion of cavies sold (%) | 24 | 24 | 28 | 25 |
| Average price received per cavy (CDF)* | 1488 | 1492 | 1137 | 1397 |
| Place of the market relative to the household | | | | |
| Within village | 92.6 | 96.9 | 96.4 | 95.4 |
| Within *territoire* | 7.4 | 3.1 | 3.6 | 4.6 |
| Who decides on market participation (%) | | | | |
| Women | 42.9 | 52.9 | 44.8 | 47.3 |
| Men | 21.4 | 26.5 | 24.1 | 24.2 |
| Jointly made | 35.7 | 11.8 | 17.2 | 20.9 |
| Children | 0 | 5.9 | 13.8 | 6.6 |
| Non-household members | 0 | 2.9 | 0 | 1.1 |
| Major motivation for selling cavies (%) | | | | |
| Paying school fees | 37.5 | 29.0 | 44.0 | 36.3 |
| Cover emergency household expenses | 20.8 | 38.7 | 24.0 | 28.8 |
| Cover planned household expenses | 37.5 | 19.4 | 24.0 | 26.3 |
| Unwanted behaviour | 0 | 3.2 | 4.0 | 2.5 |
| Culling due to disease | 0 | 0 | 4.0 | 1.3 |
| Culling due to performance | 0 | 3.2 | 0 | 1.3 |
| Reduce stock size | 0 | 3.2 | 0 | 1.3 |
| Other | 4.2 | 3.3 | 0 | 2.2 |
| Cavies consumed last year (no.) | 12.4 | 10.4 | 14.3 | 12.4 |

* During the time of the survey in 2012, 1 USD was equivalent to about 900 Congolese Francs (CDF).

**Table 5:** *Ranking by preferred traits of individual cavy consumers/buyers in eastern DR Congo (% of cavy farmers).*

| *Proportion (%) of cavy farmers ranking buyers by preferred buyer traits (n=250)* | *Individual buyer characteristic* | | | | | |
|---|---|---|---|---|---|---|
| | *Best prices* | *Buys large quantities* | *Reliability (buys throughout)* | *Timely payment* | *Road accessibility* | *Easy transport access* |
| Most preferred | 59.0 | 10.0 | 13.3 | 23.3 | 12.5 | 11.3 |
| Preferred | 26.2 | 33.3 | 31.7 | 40.0 | 28.6 | 15.1 |
| Neutral | 4.9 | 21.7 | 21.7 | 20.0 | 33.9 | 34.0 |
| Not preferred | 4.9 | 11.7 | 25.0 | 8.3 | 17.9 | 22.6 |
| Never preferred | 4.9 | 23.3 | 8.3 | 8.3 | 7.1 | 17.0 |

**Table 6:** *Double hurdle estimates of determinants of participation in cavy supply chains in eastern DR Congo.*

| *Variables* | *Market entry equation* | | *Quantity sold equation* | |
|---|---|---|---|---|
| | *Coefficient* | *SE* | *Coefficient* | *SE* |
| Latent Demand | | | | |
| Age of household head (years) | 1.221** | 0.558 | 0.198 | 0.165 |
| Gender of household head (1=M, 0=F) | −0.259 | 0.354 | 0.121 | 0.098 |
| Household size (no.) | 0.411 | 0.290 | −0.273*** | 0.072 |
| Can read and write (1=yes, 0=otherwise) | 0.224 | 0.366 | 0.106 | 0.098 |
| Latent supply | | | | |
| Cavies produced annually (no./year) | 1.156*** | 0.291 | 1.238*** | 0.081 |
| Cavy farming experience (years) | 0.106 | 0.152 | 0.025 | 0.047 |
| Land holding size (ha) | 0.024 | 0.121 | 0.030 | 0.036 |
| Wealth related variables | | | | |
| Chicken (no.) | −0.022 | 0.039 | −0.030*** | 0.010 |
| Goats (no.) | −0.071 | 0.076 | 0.009 | 0.020 |
| Pigs (no.) | 0.115 | 0.155 | −0.132* | 0.069 |
| Cows (no.) | 0.083 | 0.060 | −0.025 | 0.016 |
| Liquidity measures | | | | |
| Amount of off-farm income (CDF)† | −0.025 | 0.018 | −0.004 | 0.004 |
| Hiring farm labour (1=yes, 0=otherwise) | −0.313 | 0.341 | −0.213** | 0.086 |
| Transaction cost variables | | | | |
| Distance to the nearest market (proportional) | −0.059 | 0.059 | −0.018 | 0.014 |
| Ownership of radio (1=yes, 0=otherwise) (fixed) | 0.089 | 0.343 | 0.100 | 0.095 |
| Ownership of bicycle (1=yes, 0=otherwise) (fixed) | 0.037 | 0.671 | −0.519*** | 0.173 |
| Ownership of mobile (1=yes, 0=otherwise) (fixed) | 0.114 | 0.159 | 0.043 | 0.041 |
| Others | | | | |
| *Ex-ante* decision not sell cavies (1=yes,0=others) | −0.739** | 0.322 | −0.184** | 0.087 |
| Protestant (1=yes, 0=otherwise) | −0.036 | 0.659 | 0.423** | 0.184 |
| Catholic (1=yes, 0=otherwise) | −0.156 | 0.632 | 0.368* | 0.191 |
| *Teritoire* level fixed effects | | | | |
| Kabare (1=yes, 0=otherwise) | 0.792* | 0.412 | −0.158 | 0.108 |
| Kalehe (1=yes, 0=otherwise) | 0.827* | 0.430 | −0.075 | 0.129 |
| Constant | −9.363*** | 2.570 | −3.093*** | 0.774 |
| Observations | 250 | | 250 | |
| *P* | 0.0434 | | 0.0434 | |
| $\chi^2$ | 34.53 | | 34.53 | |
| Ll | −408.5 | | −408.5 | |

***, **, *: Significance at 1 %, 5 %, and 10 % level.
† During the time of the survey in 2012, 1 USD was equivalent to about 900 Congolese Francs (CDF).

The point estimates for other proxies of transactions costs (distance to the market, ownership of radio and mobile phones) were neither significant in the market entry decisions nor in the quantity sold.

The coefficient estimates for variable capturing liquidity constraints (amount of off-farm income) had an expected negative coefficient, but it was not significant in explaining neither market participation nor the intensity of participation. However, households with capacities to hire agricultural labourers tended to sell fewer cavies. The ownership of livestock is an indicator of the household's wealth status (Maass *et al.*, 2013) as well as a measure of the household's flexibility to enter the commodity market. Households with a diversified livestock portfolio may enter commodity markets using livestock other than cavies, hence, reducing the likelihood as well as quantities of cavies to sell. Results indicate that the ownership of chicken and pigs did not affect cavy market entry, but conditional on cavy market entry, household owning chicken and pigs tended to sell fewer cavies than those with fewer or no chicken and pigs.

The entry coefficient for the number of cavies kept has a significant expected positive sign in both the participation as well as the quantity equation, suggesting that households with more cavies are more likely to sell some, as well as transact in larger quantities of cavies.

Larger family size did not influence the decision to participate in cavy marketing, but once households decided to participate in markets, those with larger family sizes sold significantly fewer cavies than those with smaller family size.

Moreover, we found that *ex-ante* motivations of keeping cavies tended to influence the way cavies were utilised by the producers. Cavy farmers who indicated that their motivation to keep cavies was just for food were less likely to sell cavies and that, even when they decided to sell some cavies, they did so on a smaller scale than those whose principle motivation for keeping cavies was to sell. Older farmers with more years of farming experience tended to have a large propensity to enter cavy markets, but their age did not influence the quantities of cavies they sold. Interestingly, both Protestant and Catholic religious affiliation had a positive effect on market intensity, but not on market entry as opposed to other religions (i.e., Moslem, traditional African religion, Seventh day Adventists and Jehova Witnesses).

The fixed effects location variables are significant in explaining the variability in market participation. Results reveal that households in Kabare and Kalehe *territoires* were more likely to participate in cavy marketing compared to Walungu (reference *territoire*). The location dummy variables did not seem to influence the intensity of market participation as the coefficients were insignificant. These location dummy variables are probably capturing systematic differences in transaction costs and/or unobserved household characteristics, reflecting geographic concentration of market participants in *territoires* with better marketing infrastructure despite longer distance. The lack of differentiation might also be caused by high intra-*territoire* variability of infrastructure and market access. Point estimates for a few of the other variables were not significant. Such variables include literacy levels, land holding size and the gender of the head of household.

## 4 Discussion

The results show that cavy farmers are confronted by a limited and small size market, with most producers having direct contacts with the final consumers and selling within their villages. The combination of a limited market with the poor road infrastructure in the study area (Ulimwengu *et al.*, 2009), discourages traders from reaching some of the remote villages where cavies are produced. This in turn impacts negatively on the prices that farmers receive. To reduce the risk of market failure, a multi-stakeholder approach has been employed by the project in applying an integrated agricultural research for development (IAR4D) approach comparable to that by Chiuri *et al.* (2013), setting up of Innovation Platforms to encourage investments in marketing and infrastructure around the cavy producing region as to enhance access to markets, especially for women.

Currently cavies are sold as live animals without any value addition. There will be need to train producers and traders in making potential value-added products that can be created from cavies such as smoked cavies, making dumplings, minced meat, or mixing cavy meat with other meats that are all currently under investigation in Cameroon (F. Meutchieye, pers.comm.) and whose market acceptability is tested, for example, in southern Colombia (Argote *et al.*, 2009). A problem of acceptability of cavy meat by some consumers in some restaurants calls for concerted efforts to sensitize communities about the nutritional advantage of cavy and the delicacy of its meat.

Overall the findings from the econometric analysis show that some factors affecting the decision to participate in cavy marketing are different from those determining the intensity or extent of participation in cavy marketing. The significant negative effect of the ownership of bicycle on the intensity of market participation is inconsistent with *a priori* expectation that, if farmers owned bicycles and if they used them for transportations, they would pay less for transport and then sell more cavies. Hence the apparent negative effect could be suggestive of the fact that the ownership of a bicycle is rather an indicator of wealth; to the extent that wealthier households reared cavies for consumption and not for sale.

The lack of significance of distance to market, ownership of radio and mobile is consistent with expectation in that the short cavy supply chains with no or few intermediaries (middle men) have lower transaction costs as well as increase the share of value added that goes to the farmer. In this study, nearly all cavy sales were done within the village and around homesteads or road sides where transaction costs were low. The fact that most transactions were between relatives or people that knew each other also reduced information and search costs, hence, reducing transaction costs. Moreover, this finding is also consistent with those by Renkow *et al.* (2004) who found little relationship between the distance to the market and other related transaction costs for villagers in Kenya without access to motorized transport. The positive and significant coefficient of household with the capacity to hire labour makes intuitive sense if such households have enough liquidity from other sources to the extent that they do not have to rely on cavies to enter commodity markets. The positive effect of the ownership of pigs and chicken on cavy marketing is consistent with expectation especially if chicken and pig markets are more lucrative than cavy markets, such that chicken and pork income substitute income from cavy sales. Moreover, these farmers may already have climbed up some steps on the "livestock ladder" (Maass *et al.*, 2013) and, thus, they kept cavies as a "normal" livestock commodity that serves home consumption, while chicken and pig sales reflect a higher business orientation. The negative effect of family size on cavy marketing seems to suggest that larger families consume more of what is produced and have less for the market, a finding that is consistent with those reported by Benfica *et al.* (2006) for tobacco contract farmers in Mozambique and Key *et al.* (2000) for maize farmers in Mexico. The positive effect of the age of the farmer on market entry is plausible and might be explained by the fact that older and more experienced farmers have greater contacts, which might enhance mutual trust and allow trading opportunities to be undertaken at lower costs.

## 5 Conclusions

This paper investigates the determinants of smallholder cavy farmers' participation in cavy marketing. Our analysis shows that about a third of households participated in cavy selling. Households facing liquidity constraints tended to sell more cavies. Although older farmers and those with larger numbers of cavies were more likely to participate in cavy marketing, it was smaller households and those with less chicken and pigs that tended to sell more cavies, as they were probably still at the lower end of the "livestock ladder" (Maass *et al.*, 2013) and, consequently, they were still more dependent on micro livestock to generate income.Moreover, we found that *ex-ante* motivations of keeping cavies tended to influence the way cavies were utilised. Cavy farmers who indicated that they decided to keep cavies just for food were less likely to sell them and that, even when they decided to sell some cavies, they did so on a smaller scale than those whose principle motivation for keeping cavies was to sell. The findings suggest that there is potential for scaling out participation in cavy supply chains if cavy productivity is improved as this will impact strongly on how and in what way the cavy supply chain evolves. Nevertheless, there are costs involved in improving cavy husbandry for increased production, which has to be seen in relation to their contribution to the overall household income. On the other hand, common sales and purchases through the Innovation Platforms implemented within the integrated agricultural research for development (IAR4D) approach of the project (Chiuri *et al.*, 2013) has already helped to take advantage of economies of scale, resulting in sales increases in all three research locations. Improved road infrastructure would immediately provide cavy farmers with better access to markets, a necessary pull for the supply chain to meet the potential high demand in an environment that is currently facing agricultural growth. In general and for sustainability and increased welfare gain, it is also important to sensitize households about the nutritional value, delicacy of the meat and the convenience in raising cavies as they require less capital to raise relative to larger livestock. The promotion of appropriate husbandry practices that enhance cavy productivity and production should be encouraged as they can go a long way in enabling farmers participate in markets as sellers.

***Appendix**: The cavy supply chain in Sud-Kivu Province, eastern DR Congo.*

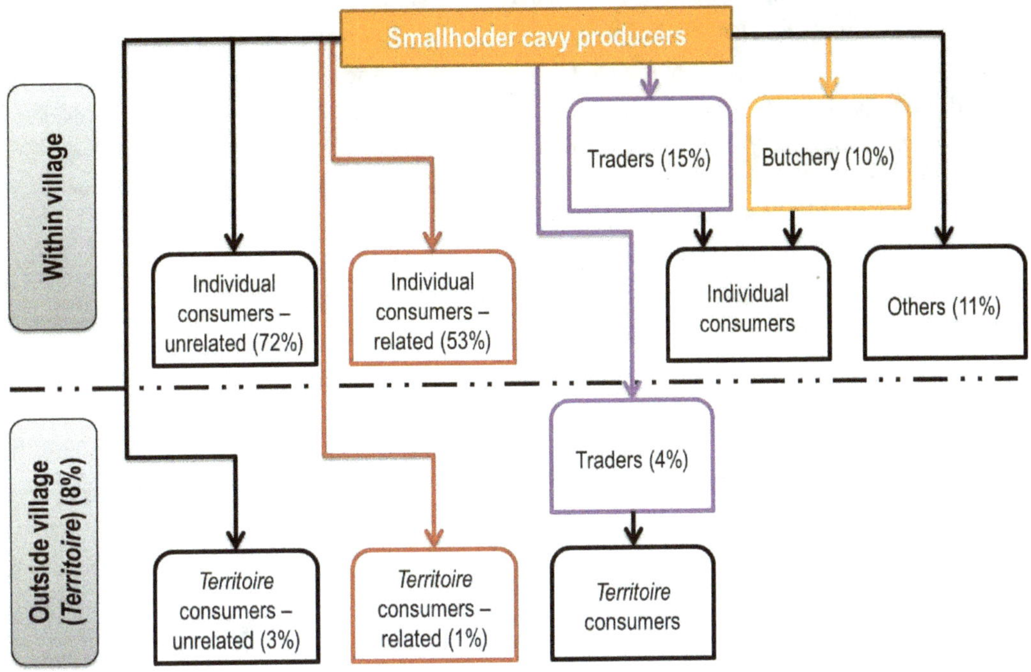


*Acknowledgements*

The unreserved collaboration by all survey respondents and dedicated survey execution by the enumerator team from DRC (especially D. Katunga-Musale and F. L. Muhimuzi) is gratefully acknowledged. We thank for financial support by the Australian Agency for International Development (AusAID) under the Africa Food Security Initiative and through the partnership between the Commonwealth Scientific and Industrial Research Organisation (CSIRO) and the Biosciences eastern and central Africa (BecA) Hub at the International Livestock Research Institute (ILRI), Nairobi, Kenya who were leading the project.

## References

Alene, A. D., Manyong, V. M., Omanya, G., Mignouna, H. D., Bokanga, M. & Odhiambo, G. (2008). Smallholder market participation under transactions costs: Maize supply and fertilizer demand in Kenya. *Food Policy*, 33 (4), 318–328.

Argote, F. E., Villada, H. S. & Argote, H. A. (2009). Investigación de mercado sobre el grado de aceptación de la carne de cuy (*Cavia porcellus*) en presentaciones de ahumado, croquetas y apanado en la ciudad de Pasto. *Biotecnología en el Sector Agropecuario y Agroindustrial*, 7 (1), 55–66.

Barrett, C. (2008). Smallholder market participation: Concepts and evidence from Eastern and Southern Africa. *Food Policy*, 33 (4), 299–317.

Bellemare, M. F. & Barrett, C. B. (2006). An ordered Tobit model of market participation: Evidence from Kenya and Ethiopia. *American Journal of Agricultural Economics*, 88 (2), 324–337.

Benfica, R., Tschirley, D. & Boughton, D. (2006). Interlinked transactions in cash cropping economies: The determinants of farmer participation and performance in the Zambezi River Valley of Mozambique. Paper presented at the 26th International Association of Agricultural Economists Conference, Gold Coast, Australia, August 12–18, 2006.

Chiuri, W., Birachi, E., Buruchara, R., Adekunle, W., Fatunbi, O., Pali, P. N., Wimba, B., Bizosa, A., Nyamurinda, B., Nyamwaro, S. O., Habumugisha, P., Tuyisenge, J., Bonabana-Wabbi, J., Karume, K., Kaenge, V. & Kamugisha, R. (2013). Market access for agro-enterprise diversity in the Lake Kivu Pilot Learning Site of the sub-Saharan Africa Challenge Programme. *African Journal of Agricultural and Resource Economics*, 8 (3), 120–134.

Cox, T. P. (2012). Farming the battle field: the meanings of war, cattle and soil in South Kivu, Democratic Republic of the Congo. *Disasters*, 36 (2), 233–248.

Cragg, J. G. (1971). Some statistical models for limited dependent variables with application to the demand for durable goods. *Econometrica*, 39 (5), 829–844.

Goetz, S. J. (1992). A selectivity model of household food marketing behavior in Sub-Saharan Africa. *American Journal of Agricultural Economics*, 74 (2), 444–452.

Holloway, G. & Ehui, S. (2002). Expanding market participation among smallholder livestock producers: A collection of studies employing Gibbs sampling and data from the Ethiopian highlands, 1998–2001. Socio-economics and Policy Research Working Paper 48. ILRI (International Livestock Research Institute), Nairobi, Kenya.

de Janvry, A., Fafchamps, M. & Sadoulet, E. (1991). Peasant household behavior with missing markets: Some paradoxes explained. *Economic Journal*, 101 (2), 1400–1417.

Key, N., Sadoulet, E. & de Janvry, A. (2000). Transactions costs and agricultural household supply response. *American Journal of Agricultural Economics*, 82 (2), 245–259.

Kouakou, N. D. V., Grongnet, J.-F., Assidjo, E. N., Thys, E., Marnet, P.-G., Catheline, D., Legrand, P. & Kouba, M. (2013). Effect of a supplementation of *Euphorbia heterophylla* on nutritional meat quality of Guinea pig (*Cavia porcellus* L.). *Meat Science*, 93 (4), 821–826. doi:10.1016/j.meatsci.2012.11.036.

Lammers, P. J., Carlson, S. L., Zdorkowski, G. A. & Honeyman, M. S. (2009). Reducing food insecurity in developing countries through meat production: the potential of the guinea pig (*Cavia porcellus*). *Renewable Agriculture and Food Systems*, 24 (2), 155–162.

Levinsohn, J. A. & McMillan, M. (2007). Does food aid harm the poor: household evidence from Ethiopia. *In:* Harrison, A. (ed.), *Globalization and Poverty*. University of Chicago Press, Chicago.

Maass, B. L., Chiuri, W. L., Zozo, R., Katunga-Musale, D., Metre, T. K. & Birachi, E. (2013). Using the 'livestock ladder' as a means for poor crop–livestock farmers to exit poverty in Sud-Kivu province, Eastern DR Congo. *In:* Vanlauwe, B., van Asten, P. & Blomme, G. (eds.), *Agro-Ecological Intensification of Agricultural Systems in the African Highlands*. Ch. 11, pp. 145–155, Earthscan, Routledge, London, UK.

Maass, B. L., Katunga-Musale, D., Chiuri, W. L., Gassner, A. & Peters, M. (2012). Challenges and opportunities for smallholder livestock production in post-conflict South Kivu, eastern DR Congo. *Tropical Animal Health and Production*, 44 (6), 1221–1232.

Maass, B. L., Metre, T. K., Tsongo, F., Mugisho, A. B., Kampemba, F. M., Ayagirwe, R. B. B., Azine, P. C., Bindelle, J. & Chiuri, W. L. (2014). From taboo to commodity: history and current situation of cavy culture in the Democratic Republic of the Congo. *Livestock Research for Rural Development*, 26 (8), Article 151. Available at: http://www.lrrd.org/lrrd26/8/maas26151.htm (accessed on: 2 February 2016).

Mètre, T. K. (2011). Small, healthy, high-yielding. *Rural21 – The International Journal for Rural Development*, 45 (1), 40–42. Available at: http://www.rural21.com/fileadmin/_migrated/content_uploads/Small_healthy_highyielding_01.pdf (accessed on: 2 February 2016).

Renkow, M., Hallstrom, D. & Karanja, D. (2004). Rural infrastructure, transactions costs and market participation in Kenya. *Journal of Development Economics*, 73 (1), 349–367.

Stephens, E. C. & Barrett, C. B. (2011). Incomplete credit markets and commodity marketing behavior. *Journal of Agricultural Economics*, 62 (1), 1–24. doi: 10.1111/j.1477-9552.2010.00274.x.

Ulimwengu, J., Funes, J., Headey, D. & You, L. (2009). Paving the way for development? The impact of transport infrastructure on agricultural production and poverty reduction in the Democratic Republic of Congo. IFPRI Discussion Paper 00944, International Food Policy Research Institute (IFPRI), Washington DC, USA. , 48 pp.

Yiva, C. H., Fon, D. E., Meutchieye, F., Niba, A. T., Manjeli, Y. & Djikeng, A. (2014). Cavies for income generation, manure for the farm and meat for the table. *Scholarly Journal of Agricultural Science*, 4 (5), 260–264. Available at: http://www.scholarly-journals.com/sjas/archive/2014/May/pdf/Herman%20et%20al.pdf

# Permissions

All chapters in this book were first published in JARDTS, by Kassel University Press; hereby published with permission under the Creative Commons Attribution License or equivalent. Every chapter published in this book has been scrutinized by our experts. Their significance has been extensively debated. The topics covered herein carry significant findings which will fuel the growth of the discipline. They may even be implemented as practical applications or may be referred to as a beginning point for another development.

The contributors of this book come from diverse backgrounds, making this book a truly international effort. This book will bring forth new frontiers with its revolutionizing research information and detailed analysis of the nascent developments around the world.

We would like to thank all the contributing authors for lending their expertise to make the book truly unique. They have played a crucial role in the development of this book. Without their invaluable contributions this book wouldn't have been possible. They have made vital efforts to compile up to date information on the varied aspects of this subject to make this book a valuable addition to the collection of many professionals and students.

This book was conceptualized with the vision of imparting up-to-date information and advanced data in this field. To ensure the same, a matchless editorial board was set up. Every individual on the board went through rigorous rounds of assessment to prove their worth. After which they invested a large part of their time researching and compiling the most relevant data for our readers.

The editorial board has been involved in producing this book since its inception. They have spent rigorous hours researching and exploring the diverse topics which have resulted in the successful publishing of this book. They have passed on their knowledge of decades through this book. To expedite this challenging task, the publisher supported the team at every step. A small team of assistant editors was also appointed to further simplify the editing procedure and attain best results for the readers.

Apart from the editorial board, the designing team has also invested a significant amount of their time in understanding the subject and creating the most relevant covers. They scrutinized every image to scout for the most suitable representation of the subject and create an appropriate cover for the book.

The publishing team has been an ardent support to the editorial, designing and production team. Their endless efforts to recruit the best for this project, has resulted in the accomplishment of this book. They are a veteran in the field of academics and their pool of knowledge is as vast as their experience in printing. Their expertise and guidance has proved useful at every step. Their uncompromising quality standards have made this book an exceptional effort. Their encouragement from time to time has been an inspiration for everyone.

The publisher and the editorial board hope that this book will prove to be a valuable piece of knowledge for researchers, students, practitioners and scholars across the globe.

# List of Contributors

**Cedric Mutyaba and Richard O. Ogwal**
National Agricultural Research Laboratories (NARL), Post-Harvest Research Programme, Kampala, Uganda

**Moses H. Lubinga**
Agricultural Research Council, Agricultural Economics and Capacity Development Division, Pretoria 0001, South Africa

**Steven Tumwesigye**
National Agricultural Research Laboratories (NARL), Food Bioscience Research Center, Kampala, Uganda

**Ahmad Maryudi, Emma Soraya, Teguh Yuwono, Bowo Dwi Siswoko, Budi Mulyana and Nunuk Supriyatnoa**
Faculty of Forestry, Universitas Gadjah Mada (UGM), Yogyakarta, Indonesia

**Ani A. Nawir**
Center for International Forestry Research (CIFOR), Bogor, Indonesia

**Purnomo Sumardamto**
District Forest Service of Gunungkidul, Indonesia

**Dewi Ayu Sekartaji**
Ministry of Environment and Forestry, Indonesia

**Tatek Woldu, André Markemann, Christoph Reiber, Philipp C. Mutha and Anne Valle Záratea**
University of Hohenheim, Institute of Agricultural Sciences in the Tropics (Hans-Ruthenberg-Institute), Department of Animal Breeding and Husbandry in the Tropics and Subtropics, Stuttgart, Germany

**Tatek Woldu**
Jimma University College of Agriculture and Veterinary Medicine, Jimma, Ethiopia

**Benjamin Tetteh Anang, Stefan Bäckman and Timo Sipiläinen**
Department of Economics and Management, University of Helsinki, FI-00014, Finland

**Benjamin Tetteh Anang**
Department of Agricultural and Resource Economics, Faculty of Agribusiness and Communication Sciences, University for Development Studies, Tamale, Ghana

**Giulio Castelli and Elena Bresci**
Department of Agricultural, Food and Forestry Systems, University of Florence, Italy

**Michael Ayamga, Richard W. N. Yeboah and Sylvester Nsobire Ayambila**
Faculty of Agribusiness and Communication Sciences, University for Development Studies, Tamale Ghana

**Lana A. Repar, Stephen Onakuse and Joe Bogue**
University College Cork, Ireland, College Road, Cork; Food Business and Development Department, Cork University Business School

**Ana Afonso**
Technical University of Madrid, Spain, s/n. 28040 Madrid; Technical College of Agriculture, Project Planning for Rural Development and Sustainable Management - GESPLAN research group

**Clemens M. Grünbühel**
School of Environment, Resources and Development, Asian Institute of Technology (AIT), Pathumthani, Thailand

**Liana J. Williams**
CSIRO Ecosystem Services, Commonwealth Scientific and Industrial Research Organisation, Brisbane, Australia

**Ivo Baur, Christian Gazzarin and Markus Lipsaa**
Agroscope; Institute for Sustainability Sciences, Tänikon, Switzerland

**Ivo Baur**
Ostrom Workshop, Indiana University Bloomington, USA

**McLoyd Banda**
Department of Agricultural Research Services, Mbawa Research Station, Embangweni, Mzimba, Malawi

**Florence Milan, Chu Thai Hoanh and Diana Suhardiman**
International Water Management Institute Southeast Asia Regional Office, Vientiane, Lao People's Democratic Republic

**Tran Thi Thu Huong and Manfred Zellerb**
Institute of Agricultural Sciences in the Tropics (Hans-Ruthenberg-Institute), University of Hohenheim, 70599 Stuttgart, Germany

**Tran Thi Thu Huong**
Vietnam National University of Agriculture, Trau Quy, Gia Lam, Ha Noi, Vietnam

**Nguyen Duy Phuong**
Soils and Fertilizers Research Institute, Dong Ngac, Tu Liem, Ha Noi, Vietnam

**Sagar Kisan Wadkar**
PMRDF Scheme, Tata Institute of Social Science, Mumbai, Maharashtra

**Khajan Singh and Asif Mohammad**
Division of Dairy Extension, ICAR - National Dairy Research Institute, Karnal, Haryana

**Ravinder Malhotra**
Division of Economics, Statistics and Management, ICAR - National Dairy Research Institute, Karnal, Haryana

**Rajiv Baliram Kaled**
Zonal Project Directorate, Zone VI, CAZRI Campus, Jodhpur, Rajasthan

**Nepomuscene Ntukamazina and Richard N. Onwonga**
Department of Land Resource Management and Agricultural Technology, University of Nairobi, Nairobi, Kenya

**Rolf Sommer**
International Center for Tropical Agriculture (CIAT), Nairobi, Kenya

**Jean Claude Rubyogo**
International Center for Tropical Agriculture (CIAT), Arusha, Tanzania

**Clare M. Mukankusi**
International Center for Tropical Agriculture (CIAT), Kampala, Uganda

**John Mburu**
Department of Agricultural and Applied Economics, University of Nairobi, Nairobi, Kenya

**Rahab Kariuki**
Agriculture and Climate Risk Enterprise, Nairobi, Kenya

**Joeri Kaal**
Institute for Heritage Science (Incipit), Spanish National Research Council (CSIC), Santiago de Compostela, Spain

**Joeri Kaal and Antonio Martínez Cortizas**
Departamento de Edafoloxía e Química Agrícola, Fac. Bioloxía, Universidade de Santiago de Compostela, Campus Sur, Santiago de Compostela, Spain

**Roberto Calvelo**
New Zealand Biochar Research Centre, Soil and Earth Sciences Group, Institute of Agriculture and Environment, Massey University, Palmerston North, New Zealand

**Art Donnelly**
SeaChar.Org, University Place, Seattle, USA

**Anna McBeath**
College of Science, Technology and Engineering, James Cook University, Cairns, Australia

**Hugh McLaughlin**
NextChar, LLC, Amherst, MA, USA

**Sanzidur Rahman and Brodrick O. Awerije**
School of Geography, Earth and Environmental Sciences, University of Plymouth, Drake Circus, Plymouth, United Kingdom

**Sintayehu Hailu Alemu, Luuk van Kempen and Ruerd Rubenac**
Department of Cultural Anthropology and Development Studies, Radboud University, Nijmegen, The Netherlands

**Sintayehu Hailu Alemua,**
Department of Economics, Hawassa University, SNNPRS, Hawassa, Ethiopia

**Ruerd Rubenac**
Agricultural Research Institute (LEI) - Wageningen University

**Silvia L. Saravia-Matus**
International Senior Consultant at European Commission - Joint Research Centre in Seville, Santiago, Chile

**Sergio Gomez y Paloma**
European Commission - Joint Research Centre in Seville, Spain

**Sebastien Mary**
Du Paul University, Chicago, U.S.A

**Sikhulumile Sinyolo**
Human Sciences Research Council (HSRC), Pretoria, 0001, South Africa

**Maxwell Mudhara and Edilegnaw Waleb**
Discipline of Agricultural Economics, University of KwaZulu-Natal, Scottsville, 3209, South Africa

**Franklin Simtowe**
International Maize and Wheat Improvement Center (CIMMYT), Nairobi, Kenya

**Birthe K. Paul b**
International Center for Tropical Agriculture (CIAT), Nairobi, Kenya

**Benjamin M. M. Wimba**
Institut National pour l'Etude et la Recherche Agronomiques (INERA)-Mulungu, Bukavu, Democratic Republic of Congo

**Samy B. Bacigale**
Université Evangélique en Afrique (UEA), Bukavu, Democratic Republic of Congo

**Wanjiku L. Chiuri**
International Center for Tropical Agriculture (CIAT), Kigali, Rwanda

# Index

www.ingramcontent.com/pod-product-compliance
Lightning Source LLC
LaVergne TN
LVHW081033210926
848485LV00004B/1012
*9781639890316*